Soil-Structure Interaction

Soil-Structure Interaction

Editors

Dominic E. L. Ong
Wen-Chieh Cheng
Hannah Zhou

MDPI • Basel • Beijing • Wuhan • Barcelona • Belgrade • Manchester • Tokyo • Cluj • Tianjin

Editors
Dominic E. L. Ong
Griffith University
Australia

Wen-Chieh Cheng
Xi'an University of Architecture and Technology
China

Hannah Zhou
University of Macau
China

Editorial Office
MDPI
St. Alban-Anlage 66
4052 Basel, Switzerland

This is a reprint of articles from the Special Issue published online in the open access journal *Geosciences* (ISSN 2076-3263) (available at: https://www.mdpi.com/journal/geosciences/special_issues/Soil_Structure_Interaction).

For citation purposes, cite each article independently as indicated on the article page online and as indicated below:

LastName, A.A.; LastName, B.B.; LastName, C.C. Article Title. *Journal Name* **Year**, *Volume Number*, Page Range.

ISBN 978-3-0365-6826-3 (Hbk)
ISBN 978-3-0365-6827-0 (PDF)

Contents

About the Editors

Dominic E. L. Ong

Dr. Dominic Ek Leong Ong obtained his Bachelor's of Civil Engineering degree from the University of Western Australia (UWA) in 1998 and his PhD from the National University of Singapore (NUS) in 2004. His research interests are in soil–structure interaction, deep excavation, tunneling, ground improvement, numerical and centrifuge modeling, and field observational methods. He is currently an Associate Professor at Griffith University, Australia, and the Immediate Past President of the Malaysian Geotechnical Society (MGS). He is a Fellow and Chartered Professional Engineer of both Engineers Australia (EA) and the Institution of Engineers Malaysia (IEM). He is also currently the Joint Editor-in-Chief of UK's Institution of Civil Engineers (ICE) journal, *Geotechnical Research*, and sits on the International Society for Soil Mechanics and Geotechnical Engineering (ISSMGE) Technical Committee TC104 Physical Modelling in Geomechanics and TC207 Soil-Structure Interaction and Retaining Walls. He won the American Society of Civil Engineers (ASCE) Outstanding Reviewer Award in 2018 and 2019.

Wen-Chieh Cheng

Dr. Wen-Chieh Cheng is a Professor at the Shanxi Key Laboratory of Geotechnical and Underground Space Engineering, Xi'an University of Architecture and Technology (XAUAT). He received his PhD in Geotechnical Engineering from the National Taipei University of Technology, Taiwan, in 2010. His current research focuses on underground space engineering and environmental geotechnics, specifically (i) performance and lubrication efficiency of pipe jacking within complex ground conditions, (ii) geohazard mitigation, (iii) contaminants migration and control measures, and (iv) environmental friendly materials. He is currently a Council Member of the Chinese Society for Rock Mechanics and Engineering (CSRME) and the Shaanxi Society for Rock-Soil Mechanics and Engineering (SXRSME). He also sits on the International Society for Soil Mechanics and Geotechnical Engineering (ISSMGE) Technical Committee TC213 Scour and Erosion and the Institute of Materials, Minerals and Mining (IOM3) Chartered Engineer Assessor Committee.

Hannah Zhou

Dr. Hannah Zhou is a Professor and the Head of the Department of Civil and Environmental Engineering, University of Macau. She received her PhD from Hong Kong Polytechnic University in 2008 and both her Master's and Bachelor's of Civil Engineering degrees from Zhejiang University, China, in 2005 and 2002, respectively. Prof. Zhou is a progressive and keen adopter of advanced techniques in the field of geotechnical engineering. She has international prominence in the fields of (i) machine learning in geotechnical engineering, (ii) experimental and model study of sensing systems of geotechnical structures, and (iii) intelligent predictions based on geological uncertainty. She won the ASCE Outstanding Reviewer Award in 2016 and 2017 and the Canadian Geotechnical Society's R.M. Quigley Award (2011 Honorable Mention).

Preface to "Soil-Structure Interaction"

The Editors are pleased to present this collection of 10 articles under the theme soil–structure interaction (SSI). The terminology SSI implies that an action from the surrounding ground would result in a reaction to the structure and vice versa, hence the interdependency of geotechnical engineering and structural engineering. This book immerses its readers into the specialized fields of

(i) Soil and rock characterization (two articles);
(ii) Ground improvement (two articles);
(iii) Pavement engineering (one article);
(iv) Shallow foundation (one article);
(v) Deep foundation (two articles);
(vi) Deep excavation (one article);
(vii) Earthquake engineering (one article).

After understanding these SSI concepts, readers are expected to be able to

- Appreciate the design of complex geo-structures interfacing naturally occurring geomaterials;
- Interpret nonlinear soil strength and surface roughness parameters for use in design;
- Articulate the use of sophisticated modeling tools and techniques to understand complex soil–structure interaction problems;
- Reflect on the prowess of modern-day IT technology in aiding the construction of complex geo-structures;
- Create strategies to decarbonize infrastructure in a net-zero economy.

At the time of writing, this SSI compilation has already gathered a collective view rate of close to **22,500 per annum** since its online debut in Dec 2021, making it one of the most sought-after Special Issues of Geosciences (MDPI) and the most popular SI of the Geomechanics Section! Happy reading!

Dominic E. L. Ong, Wen-Chieh Cheng, and Hannah Zhou
Editors

geosciences

Editorial

Editorial of Special Issue "Soil–Structure Interaction"

Dominic E. L. Ong [1,*], Wen-Chieh Cheng [2] and Hannah Zhou [3]

[1] School of Engineering and Built Environment—Civil and Environmental Engineering, Griffith University, Nathan, QLD 4111, Australia
[2] School of Civil Engineering, Xi'an University of Architecture and Technology, Xi'an 710055, China
[3] Department of Civil and Environment Engineering, Faculty of Science and Technology, University of Macau, Taipa, Macau, China
* Correspondence: d.ong@griffith.edu.au

In today's world, construction projects tend to be more complex due to one or a combination of the following reasons: climate change impacting design requirements; enhanced material properties; the scarcity of greenfield sites in highly built-up cities, leading to challenging underground construction; easily accessible scientific knowledge; the prowess of computational speed; the advent of new technologies driving innovation; and stringent health and safety requirements to safeguard the public. These factors have since become the new normal, thus increasing the expectations of scientists and engineers to deliver "unimaginable" products as envisioned by clients; however, the one aspect of any construction project that has a common link to these expectations is the ground or foundation that these structures sit on. Serving as a platform for buildings and infrastructure to be founded on, soils and rocks are important geomaterials that are challenging to characterize because they are naturally formed, and thus not subject to any quality control protocol. As such, it is timely that this Special Issue focuses on the complex topic of soil–structure interaction (SSI).

Out of the ten articles, three are state-of-the-art review articles documenting and describing how the concept of SSI is efficiently implemented. In their review of the effects of surface roughness and particle characteristics on soil–structure interactions, Wang et al. [1] found that the influential factors that affect the interface shear mechanism include (i) surface roughness, (ii) particle angularity, (iii) mean particle size, (iv) surface hardness, (v) particle breakage, and (vi) confinement condition. From various past studies and the authors' own research, it is found that interface shear strength is enhanced with an increment in surface roughness and gradually approaches the actual soil shear strength when the material surface is in a random or ribbed form. Similarly, the increasing peak friction angle progressively approaches the soil internal friction angle until the surface roughness reaches its critical roughness. These observations prove that surface roughness has a negligible influence on the residual friction angle.

In the second review article, Fatehi et al. [2] performed a comprehensive critique on (i) the environmental assessment of biopolymers as binders in soil improvement works, (ii) biopolymer-treated soil characteristics, and (iii) important factors affecting the strength behavior of biopolymer-treated soil. It was discovered that (i) a relatively low content of biopolymer is needed to achieve a comparable compressive strength when the same soil is treated with traditional binders, such as cement and lime, (ii) water content is the single most sensitive parameter that affects biopolymer-treated soil strength development, and that (iii) durability is affected by wetting–drying cycles, freeze–thaw cycles, microorganisms, and ultraviolet radiation.

Riaz et al. [3] reviewed the recent achievements in research on soil–structure interaction (SSI), with a main focus on numerical analyses. Their review defines the characteristics of a wide variety of treatments for SSI, from a basic model to a high-fidelity mode, and is based on the continuum mechanics theory as well as the utilization of high-performance

Citation: Ong, D.E.L.; Cheng, W.-C.; Zhou, H. Editorial of Special Issue "Soil–Structure Interaction". *Geosciences* **2023**, *13*, 54. https://doi.org/10.3390/geosciences13020054

Received: 28 January 2023
Accepted: 6 February 2023
Published: 10 February 2023

computing (HPC). According to continuum mechanics theory, all numerical analyses must resolve a mathematical conundrum that is consistent with the original Lagrangian problem of a soil–structure system. Based on the degree of accuracy and resolution of the needed answer, a suitable numerical analysis can then be selected.

In the three review articles mentioned earlier, it is clear that each article emphasizes the fundamental theories of soil and rock mechanics, supplemented by the use of advanced testing facilities and computing prowess. Detailed, well-documented explanations and references have been provided to allow for the easy comprehension and implementation of SSI techniques in our readers' day-to-day design or research endeavors.

The other seven articles involve original research findings related to common soil–structure interaction problems normally faced by engineers and scientists. In Aloisio et al. [4], non-destructive testing (NDT) using a seismic dilatometer was successfully carried out in a deep borehole to characterize the surrounding soils for the accurate estimation of shear wave velocity. Their paper proposes an enhancement of the standard procedure to address the importance of Lamb waves in soil characterization. The experimental results of the transverse propagating waves showed distinctive dispersion curves that are indicative of the type of Lamb waves. It is likely that the seismic dilatometer results do not refer to the pure shear waves, as is usually accepted in the field of geotechnics, but rather Lamb waves are a more appropriate classification. As a result, the conventional method that derives the shear wave velocity from the cross-correlation of the two signals obtained by two spaced geophones may be unreliable.

Liu et al. [5] explained that clay and sand particles behave in a way that is distinct from other particles, but that silt particles behave in a way that is intermediate between sand and clay. As a result, they investigated the following: (a) the impact of silt contents on the strength of soft soils; (b) the impact of silt contents on the strength of soft soils stabilized by cement; and (c) the microstructure of soft soil specimens with different particle size distributions. To accomplish these goals, a series of tests, including consolidated, isotropic undrained (CIU) triaxial tests as well as unconfined compressive strength (UCS) tests, and SEM images were carried out. In conclusion, there is a correlation between silt content and the behavior of soft soils in their critical states (both clay and silt particles). When the cement percentage is 10%, the unconfined compressive strength for the cement-stabilized specimens increases with the increase in silt content. When the cement percentage reaches 30%, however, the UCS falls as silt content rises. With a cement content ranging from 15 to 25 percent, the UCS rises initially as the silt content rises but falls after the silt content reaches a "saturation" point.

Since processed crushed rocks are frequently used as base or subbase layers, Sun et al. [6] studied the mechanical properties of naturally occurring unbound granular materials (UGMs) driven by their widespread applications. In many specifications and standards, the rutting and settlement caused by base and subbase layers are severely limiting. In order to study the behavior of UGMs under the fluctuation of the degree of saturation (DOS) (59–100%), a series of repeated load triaxial tests (RLTs) was conducted to test the dynamic behavior of the crushed rocks collected from Queensland in Australia, including the resilient modulus and the plastic strain. The OMC and 100% standard proctor maximum dry unit weight conditions were used in the laboratory to properly grade the RLT specimens. The RLT tests revealed that UGM specimens soaked at a higher DOS had a lower resilient modulus and a weaker resistance to high traffic volumes, as well as a significant accumulation of plastic strain.

Pantelidis and Gravanis [7] proposed an elastic settlement analysis method for rigid rectangular footings that was applicable to both clays and sands. The case of clay is distinguished from the case of sand by the use of different contact pressure distributions, whereas the modulus of elasticity for sands increases linearly with depth. Among the most intriguing findings is that sands produce "settlement x soil modulus/applied pressure" values that are approximately 10% higher than those produced by clays. Furthermore, for large Poisson's ratio (ν) values, the settlement of rigid footing is closer to the settlement of respective flexible footing. Finally, corrections for net applied pressure, footing rigidity, and soil nonelastic responses under loading are provided.

Wang et al. [8] reported that one of the most effective methods for accurately estimating pile bearing capacity is by performing a bidirectional static load test (BDSLT), in which the test pile is divided into two sections by activating a single-loading device welded along the pile shaft. In comparison to conventional static load tests, a BDSLT eliminates the safety concerns and space constraints imposed by the reaction system (kentledge). The conventional analytical solutions for test piles in preliminary designs, including the Alpha and Beta as well as semiempirical methods, are validated in this paper. In order to predict the load–settlement response in a soil stiffness reduction model, modified closed-form analytical solutions based on Randolph's analytical method have been successfully used.

In the paper of Al-Atroush et al. [9], field measurements of an axial-loaded large-diameter bored pile installed in OC stiff clay (Alzey Bridge Case Study, Germany) was load-tested to failure to evaluate the quality of two numerical models developed with the aim of simulating pile behavior in both drained and undrained conditions. It was found that the pile diameter influences the ultimate bearing stress below the large-diameter pile base; however, several codes and design standards proposed using constant bearing stress at a specific settlement value (i.e., 5% D) to estimate the ultimate capacity of the LDBP, regardless of pile geometry and without discrimination for any class of cohesive soils. This observation verifies the case study pile performance well.

The use of a 700 mm-diameter contiguous bored pile (CBP) wall for a main basement in a deep excavation project with a cut-and-cover tunnel is described by Chong and Ong [10]. The main basement was deemed "impermeable" due to the presence of cement grout columns between the successive piles behind the CBP wall; however, site observations revealed that the installation of ground anchors unintentionally punctured the wall water tightness, resulting in leakages through the CBP wall and possibly localized groundwater lowering, as evidenced by the relatively large settlements. It was observed that, in the absence of cement grout columns at the cut-and-cover tunnel section, the excavation rate resulted in immediate groundwater drawdown. Settlement caused by the excavation and groundwater drawdown was only slowed after the skinwall was cast to prevent groundwater from flowing through the wall. The unintentional groundwater leak caused minor wall deflection. The maximum settlement-to-maximum deflection ratio is unusual in comparison to that reported in the literature, hence the observed complexity of a soil–structure interaction problem. The analysis also revealed that the corner effect is significant, with smaller settlement recorded at the wall corners.

Conflicts of Interest: The authors declare no conflict of interest.

References

1. Wang, R.; Ong, D.E.L.; Peerun, M.; Jeng, D. Influence of Surface Roughness and Particle Characteristics on Soil—Structure Interactions: A State-of-the-Art Review. *Geosciences* **2022**, *12*, 145. [CrossRef]
2. Fatehi, H.; Ong, D.E.L.; Yu, J.; Chang, I. Biopolymers as Green Binders for Soil Improvement in Geotechnical Applications: A Review. *Geosciences* **2021**, *11*, 291. [CrossRef]
3. Riaz, M.; Motoyama, H.; Hori, M. Review of Soil-Structure Interaction Based on Continuum Mechanics Theory and Use of High Performance Computing. *Geosciences* **2021**, *11*, 72. [CrossRef]
4. Aloisio, A.; Totani, F.; Alaggio, R.; Totani, G. Dispersion Curves of Transverse Waves Propagating in Multi-Layered Soils from Experimental Tests in a 100 m Deep Borehole. *Geosciences* **2021**, *11*, 207. [CrossRef]
5. Liu, Y.; Liu, Z.; Oh, E.; Ong, D.E.L. Strength and Microstructural Assessment of Reconstituted and Stabilised Soft Soils with Varying Silt Contents. *Geosciences* **2021**, *11*, 302. [CrossRef]
6. Sun, J.; Oh, E.; Ong, D.E.L. Influence of Degree of Saturation (DOS) on Dynamic Behavior of Unbound Granular Materials. *Geosciences* **2021**, *11*, 89. [CrossRef]
7. Pantelidis, L.; Gravanis, E. Elastic Settlement Analysis of Rigid Rectangular Footings on Sands and Clays. *Geosciences* **2020**, *10*, 491. [CrossRef]
8. Wang, R.; Ong, D.E.L.; Zhou, J.; Liu, S.; Oh, E. Validation of Analytical Solutions for Predicting Drilled Pile Behaviour under Bi-Directional Static Load Tests. *Geosciences* **2022**, *12*, 284. [CrossRef]

9. Al-Atroush, M.; Hefny, A.; Zaghloul, Y.; Sorour, T. Behavior of a Large Diameter Bored Pile in Drained and Undrained Con-ditions: Comparative Analysis. *Geosciences* **2020**, *10*, 261. [CrossRef]
10. Chong, E.; Ong, D.E.L. Data-Driven Field Observational Method of a Contiguous Bored Pile Wall System Affected by Acci-dental Groundwater Drawdown. *Geosciences* **2020**, *10*, 268. [CrossRef]

geosciences

MDPI

Review

Influence of Surface Roughness and Particle Characteristics on Soil–Structure Interactions: A State-of-the-Art Review

Runshen Wang [1,2,*], Dominic E. L. Ong [1,2], Mohammud I. Peerun [1,2] and Dong-Sheng Jeng [3]

1 School of Engineering and Built Environment, Griffith University, Nathan 4111, Australia; d.ong@griffith.edu.au (D.E.L.O.); irfaan.peerun@griffithuni.edu.au (M.I.P.)
2 Cities Research Institute, Griffith University, Southport 4215, Australia
3 School of Engineering and Built Environment, Griffith University, Gold Coast Campus, Southport 4222, Australia; d.jeng@griffith.edu.au
* Correspondence: jason.wang2@griffithuni.edu.au

Abstract: The study of soil–structure interface behavior contributes to the fundamental understanding of engineering performance and foundation design optimization. Previous research studies the effect of soil characteristics and surface roughness property on the soil–material interface mechanism via interface shear test. The reviews utilizing past established laboratory studies and more recent tests based on state-of-the-art technologies reveal that surface roughness significantly affects interface shear performances in the studies of soil–structure interactions, especially in peak shear strength development. A preliminary but original investigative study by the authors was also carried out using a sophisticated portable surface roughness gauge to define the material surface roughness properties in order to study the interface behavior parametrically. Additionally, using the authors' own original research findings as a proof-of-concept innovation, particle image velocimetry (PIV) technology is applied using a digital single-lens reflex (DSLR) camera to capture sequential images of particle interactions in a custom-built transparent shear box, which validate the well-established four-stage soil shearing model. The authors also envisaged that machine learning, e.g., artificial neural network (ANN) and Bayesian inference method, amongst others, as well as numerical modeling, e.g., discrete element method (DEM), have the potential to also promote research advances on interface shear mechanisms, which will assist in developing a greater understanding in the complex study of soil–structure interactions.

Keywords: interface shear test; surface roughness; critical roughness; random or ribbed surface roughness; friction development; PIV technology

Citation: Wang, R.; Ong, D.E.L.; Peerun, M.I.; Jeng, D.-S. Influence of Surface Roughness and Particle Characteristics on Soil–Structure Interactions: A State-of-the-Art Review. *Geosciences* **2022**, *12*, 145. https://doi.org/10.3390/geosciences12040145

Academic Editors: Mohamed Shahin and Jesus Martinez-Frias

Received: 17 January 2022
Accepted: 17 March 2022
Published: 23 March 2022

1. Introduction

Soil–structure interfaces play key roles in transferring construction gravity and external loads from the substructures to the ground. The research on soil–structure interface mechanism and behavior is critical for geotechnical engineering applications. In recent years, advancements in technology have influenced how designs are conceptualized, considering sustainability aspects [1–4], challenges faced in laboratory soil tests [5,6], soil–structure simulations [7,8], and inferior in situ geomaterial conditions requiring biological [9–14] or chemical stabilization [15–17]. For instance, the study of the effects of the interface on soil–structure interactions have benefitted from this technological advancement to include the design of deep foundations [18–23], pavement–subgrade behavior [24,25], tunneling [26–33], mechanically stabilized geogrid retaining walls, and ground improvement in soft soils [34–36]. For example, in the design of deep foundations, static and dynamic pile load tests provide the necessary evidence-based field observational results to verify the accuracy and reliability of the initially adopted pile design parameters. Previous research conducted via the conventional static load tests evidenced that the designed pile shaft capacities in many projects had been grossly over-designed due to the lack of reliable pile–soil interface parameters [37,38]. As a result, it is not uncommon that the construction costs of deep foundations often account for 20–25% of the total costs of the project [37], which is economically unsustainable. Thus, understanding the soil–structure interface mechanism can help contribute to potentially greater savings in the design of pile foundations. In another example, it has been shown that installing underground sewers via the pipe jacking method would induce varying jacking forces according to the traversed geological layers during jacking operations [28,30,39]. Pipe jacking forces that are over-predicted will result in pipe wall fractures or excessive ground heave, whereas under-predicted jacking forces will prevent the micro-tunneling boring machine (MTBM) from breaking through the receiving shaft or cause the collapse of the tunnel face excavation, leading to ground loss [40]. As such, it is evident that the influence of surface roughness not only contributes to the economics of underground construction, but also seeks to enhance construction safety.

The research carried out on the influence of surface roughness is often closely associated with the understanding of particle characteristics as well. Normally, such studies are conducted via the conventional shear tests and interface shear tests. Based on conventional shear test results, Coulomb (1773) defined a linear failure envelope for soil limit states and found that fully developed shear resistance results in shear failure [41]. The fully developed shear resistance, also known as soil shear strength, has been determined using laboratory and field tests. Even though soil characteristics obtained from conventional shear tests may be used to broadly predict general soil–structure behavior of localized geostructures (e.g., retaining wall in the plane-strain direction), some engineering applications involving linear infrastructure, such as tunneling, pipe jacking, and pavement–subgrade interactions, would require greater accuracy due to the often varying geological profiles in the plane-stress (horizontal) direction. For example, Staheli (2006) conducted interface direct shear tests using granular materials and various pipe materials with varying surface roughness to analyze pipe jacking forces [42]. Based on the test results, the frictional jacking force model was established for granular materials, but it was only applicable to unlubricated jacking drives, hence the obvious limitation of the research to mimic real-life phenomena. Even though surface roughness [43,44] has significant influences on the interface shear mechanisms, other interface characteristics, such as surface hardness [45], particle angularity [46,47], mean particle size [48], particle breakage [49], and confinement condition [50,51], have also been found to affect the soil's shear behavior [28,29,52,53]. Peerun et al. (2020) successfully conducted direct shear tests on reconstituted rock spoil from various geological lithologies to develop a mathematical model to reliably determine the often-elusive effective jacking forces to propel MTBM forward in a typical tunneling project [30]. Similarly, in the research of [54], the strength parameters of the reconstituted rock spoil were determined and then integrated into the jacking force model using direct shear tests to provide an accurate jacking force prediction. From the discussions above, it is clear that the development of friction is complex, as it involves many other parameters that are difficult to quantify for consistent use in design.

Therefore, in this review paper, conventional shear tests, such as the (i) direct shear test, (ii) simple shear test, and (iii) ring shear test, are carefully reviewed in terms of their operating procedures, advantages, and disadvantages to study soil shear strength. This is then followed by the review of the critical factors, such as (i) surface roughness, (ii) particle angularity, (iii) mean particle size, (iv) surface hardness, (v) particle breakability, and (vi) confinement condition, that influence the complex behavior of soil–material interfaces in soil–structure interaction problems. Due to the historical challenges in observing and describing the microscopic-level interface interactions, a proof-of-concept innovative transparent Perspex shear box has been designed, fabricated, and commissioned by the authors to help with the visualization process, with preliminary results discussed. In tandem, the effectiveness of integrating the shear testing with technologically advanced visualization tools, such as (i) particle image velocimetry (PIV) and (ii) discrete element modeling (DEM), is reviewed and investigated. Finally, the potential use of machine learning techniques, such as (i) artificial neural network (ANN) and (ii) Bayesian updating process, are discussed with respect to practical engineering soil–material interface applications.

This review paper develops the understanding of soil–structure interactions due to surface roughness and particle characteristics. Understanding and results derived from past research, as well as the authors' own original results using the direct shear test (DST) method, are presented and discussed in detail. Particle image velocimetry (PIV) technology is the main analytical tool championed by the authors and has also been integrated into this review paper as a means for verification to increase the confidence levels of the results presented herein. Furthermore, other supplementary methods to enhance the understanding of soil–structure interactions are also briefly introduced, such as neural network, DEM, 3D printing, and Bayesian method.

2. Conventional Shear Tests

Conventional shear tests that will be discussed in this paper are (i) direct shear test, (ii) simple shear test, and (iii) ring shear test, which measure the soil shear strength.

2.1. Direct Shear Test

Direct shear tests (Figure 1) are simple to conduct and have typically been used for site investigation and foundation design. It has been widely used for shear strength characteristics of granular material. Additionally, the soil sample is prepared with a small thickness, which allows the pore water pressure to dissipate rapidly. The shear box is sheared along the predetermined failure plane, with constant tangential shear force or displacement while applying a normal force on the sample [55]. While the peak shear strength and residual strength are measured by S-type load cell, vertical and horizontal displacement are recorded by linear variable differential transformers (LVDTs) (see Figure 2). Based on a linear graph of shear stress against normal stress, the slope and the intercept value represent soil friction angle and cohesion stress. The equipment setup and test procedures of the direct shear test are fully detailed in ASTM D3080 and AS1289.

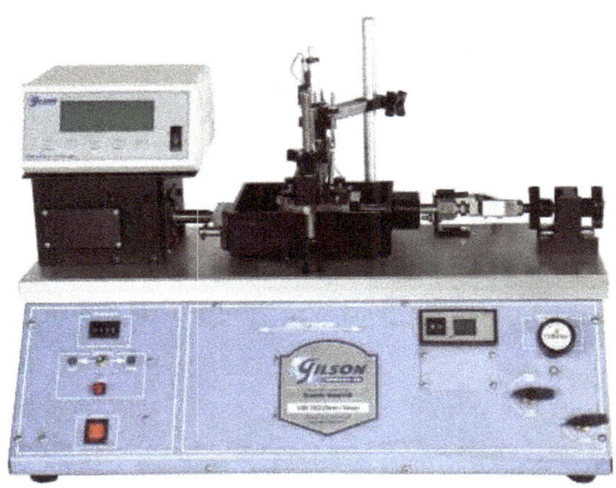

Figure 1. Direct shear test apparatus (HM-382F) produced by Gilson company. This original direct shear apparatus has since been modified by the authors for use as an interface shear apparatus.

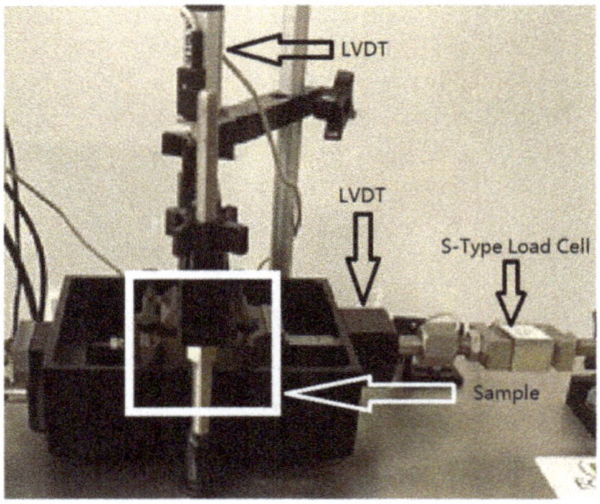

Figure 2. DST apparatus used in this research.

There are three kinds of DST methods: the slow shear test (or S-test), quick shear test (or Q-test, with the speed of 0.02 mm/min), and consolidated quick shear test. The Chinese standard [56,57] recommends the quick shear test for site investigation and requires that each soil type be tested at least four times. The specimen of DST is confined by a rectangular or circular metal shear box, usually circular, with 61.8-mm-diameter width and 20-mm height. Australian standards recommend slow shearing tests for peak shear strength and residual shear strength. The procedures of direct shear tests are shown below in Federal Highway Administration (FHWA) and AS 1289 6.22 [56]:

1. Place the top half of the shear box on the bottom one with two clamping screws fixed, as shown in Figure 3.
2. Fix the retaining plate and the perforated grid plate, and the shearing force is perpendicular to a serration orientation of the plate.
3. Fix the perforated grid plate and the loading pad on top, and then place the assembly into the water chamber; slot the end onto the pushrod.
4. Fix the horizontal and vertical LVDTs on the specific location.
5. Apply normal stress and wait for consolidation to be completed (cohesionless soil is consolidated instantaneously).
6. Shear the soil specimen with the constant rate (R_d), which is determined by the sample's hydraulic conductivity, and the formula is shown below [58].

$$R_d = \frac{d_f}{t_f} \quad (1)$$

where:

t_f—estimated elapsed time to specimen failure in shearing apparatus;
t_{50}—consolidation with 50 percent under the maximum normal stress;
t_{90}—consolidation with 90 percent under the maximum normal stress;
d_f—estimated relative lateral displacement at failure, $d_f = 10$ mm (for overconsolidated fine-grained soil), otherwise $d_f = 5$ mm.

Note: if the normal deformation versus logarithmic time is significant yield, $t_f = 50t_{50}$, otherwise, $t_f = 11.6t_{90}$. Due to the absence of soil consolidation data in the specific case, Section 9.10.3 of ASTM D3080 also provides a table for a minimum time to failure (Table 1).

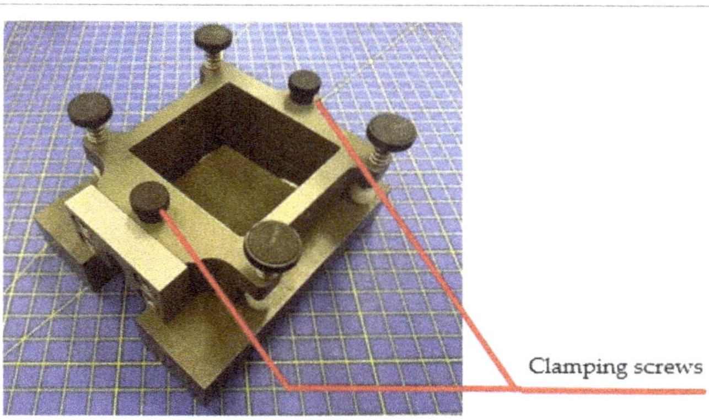

Figure 3. Typical shear box.

Table 1. Minimum time to failure [58].

USCS Classification (D2487)	Minimum Time to Failure, t_f
SW, SP (<5% fines)	10 min
SW-SM, SP_SM, SM (>5% fines)	60 min
SC, ML, CL, SP-SC	200 min
MH, CH	24 h

After several specimens are tested using the DST method, the peak shear strength and residual strength are obtained with different normal effective stresses. Then, according to Mohr–Coulomb constitutive relation, cohesion (c) and internal friction angle are determined by Equation (2).

$$S = c + \sigma \tan \varphi \quad (2)$$

where:

S—shearing strength of soil (kPa);
C—cohesion strength of soil (kPa);
σ—normal effective strength (kPa);
φ—internal friction angle (°).

The relative displacement of the top and bottom shear box along the predetermined failure plane reduces the shearing area. Thus, Yu et al. (2014) apply the modified method for direct shear test results, including the single-point area correction method and multipoint area correction method [59]. The single-point area correction method applies Newton's second law to determine the shear stress (see Equation (3)). For the strain-controlled shear test, the acceleration (a) is zero. Thus, Equation (3) is converted to Equation (4), which simply considers the effect of shear area reduction. On the other hand, the multipoint area correction method applies the principles of momentum and energy conservation (see Equations (5) and (6)). This method assumes the product of velocity variation in the initial state and mass is much lower than the time integral of horizontal load. Thus, Equations (5) and (6) are integrated into Equation (7).

Single-point area correction method:

$$F - \tau A = ma \quad (3)$$

$$\tau_f = \tau' = \frac{\tau A_0}{A_2} = \frac{CRA_0}{A_2} = \tau\beta \quad (4)$$

Multipoint area correction method:

$$\int_{t_2}^{t_1} [F(t)dt - \tau A(t)dt] = m\Delta v \quad (5)$$

$$\int_{s_2}^{s_1} [F(s)ds - \tau A(s)ds] = \Delta E \quad (6)$$

$$\tau = \frac{\int_{p_2}^{p_1} F(p)dp}{\int_{p_2}^{p_1} A(p)dp} \quad (7)$$

The results of a conventional direct shear test include the shear stress at peak and residual states, as well as volumetric deformation of the sample. Li and Aydin (2010) developed a four-stage shearing model based on direct shear test results and revealed that the process of direct shear tests could be divided into four stages: (1) end zone deformation: peak stress tends to concentrate at the shear zone's end, and the soil specimen is compressed until it reaches the lowest volumetric strain; (2) particle interlocking: this stage plays a major role in increasing volumetric strain until it reaches the peak shear stress; (3) shear zone formation: particles rearrange where smaller particles flow into the specimen void, and larger particles rotate; and (4) stable shear: there is no further compression or dilation at this stage, and the shear stress remains constant [60]. This four-stage model will subsequently be used to validate the preliminary tests carried out using the transparent shear box to verify its applicability to study the interaction between surface roughness and the in-contact material (soil) types.

2.2. Simple Shear Test

Bjerrum and Landva (1966) established the simple shear test, which is widely used to determine the constant volume strength and stress–strain relationship of cohesive soils [61,62]. Meanwhile, the understanding of the shear mechanism in simple shear tests contributes to solving the offshore gravity-based structure and slope stability problem [63]. Comparing soil specimen in the direct shear test, the soil specimen in the simple shear test is restricted by two parallel rigid platens, which maintain the cross-sectional area constantly. Thus, the failure mode of soil specimen in the simple shear is present not only in the horizontal predetermined failure plane, as well as direct shear tests, but exists along the vertical plane (see Figure 4). Due to the difference in shear mechanisms, direct shear tests provide a higher shearing strength and stiffness than simple shear tests [63].

According to ASTM D6528-17 requirements, the tested cohesive soil can be intact, laboratory reconstituted, and compacted soils [64]. The soil specimen is one-dimensionally consolidated until the excess pore water pressures are dissipated. Meanwhile, the pore water pressures remain constant during the simple shear test in the undrained condition. Thus, the variation in normal stress equals the variation in effective stress. However, reconstituted and compacted soil preparation is not detailed in the standard, thus limiting the study of soil–structure interactions in silty "sand-like" soils.

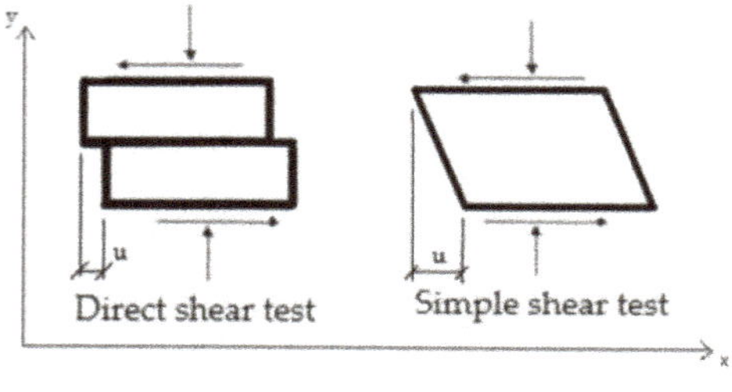

Figure 4. Schematic of direct shear test and simple shear test.

2.3. *Ring Shear Test*

The irreversible shearing direction of the ring direct shear test (see Figure 5) leads to large shear displacement, which contributes to the reorientation of the tested particles [65]. Thus, ring shear tests provide a more accurate residual shear strength than conventional direct shear tests and triaxial tests [66]. As there is no limitation on shearing displacement, the ring shear test has proven its reliability in the back analyses of landslides [67,68]. However, the low shearing rate of a ring shear test cannot be utilized to simulate "extremely rapid landslides" [66,69]. Meanwhile, the rubber edge of the ring shear apparatus causes an invisible area, which disrupts clear observations of the development of the shear band. The use of an opaque box makes particle image capture impossible and limits the application of state-of-the-art image processing techniques to develop an enhanced understanding of soil–structure interactions [66,70,71].

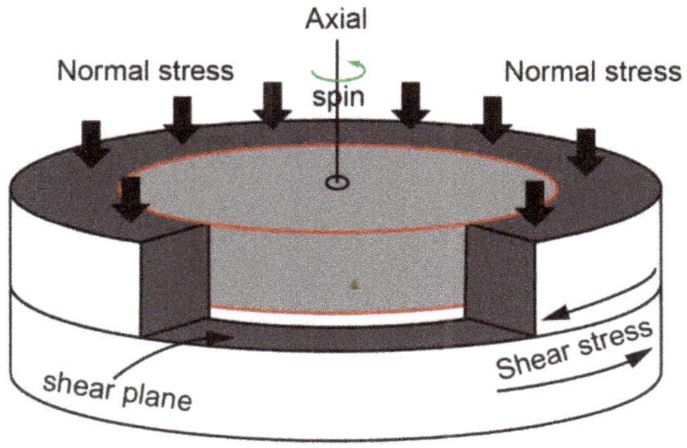

Figure 5. The principle of ring shear test.

3. Interface Shear Strength

The soil shear strength investigation contributes to the fundamental understanding of construction performance and deep foundation design. However, the research of soil–structure interface behavior cannot be simply limited to soil characteristics, but must also include surface roughness properties. Therefore, the interface shear test apparatus is manufactured based on conventional shear tests, including direct shear tests, simple shear tests, and ring shear tests. Potyondy (1961) first developed the research on soil–structure interface behavior and found that shear resistance is influenced by several characteristics, such as particle size distribution, soil moisture content, normal stress, structure material, and surface finish. As a result, the skin friction angle increases until the value of surface roughness reaches the soil internal friction angle. With advanced technology applied in interface shear tests, the current understanding reveals that several parameters affect the soil–structure interface behavior, including surface roughness [43,44], mean particle size [48], surface hardness [45], particle angularity [46,47], particle breakage [49], and confinement condition [50,51].

3.1. *Influence of Surface Roughness*

3.1.1. Background

Potyondy (1961) first investigated the effect of surface roughness on soil–structure interface shear behavior, where surface roughness is simply described as "rough" or "smooth". Based on this research approach, subsequent research determined the 2D cross-sectional profile of the material surface (see Figure 6) and defined the surface roughness with a series of roughness parameters (see Table 2) and applied the theory to engineering applications [72,73]. Reference values of roughness parameters include the maximum height of the profile (R_{zi} or R_z) and arithmetical mean height of the profile (or average roughness, R_a).

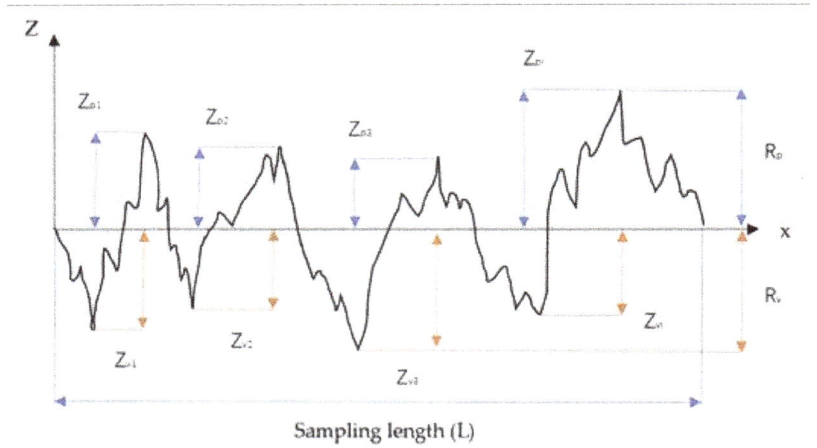

Figure 6. The maximum peak height and the maximum valley depth.

Table 2. Characteristics of surface roughness parameters.

Parameters	Characteristics	Formula
Z_{pi}	The ith peck height of the material roughness profile	-
Z_{vi}	The ith valley depth of the material roughness profile	-
R_{pi}	The maximum peak height of the material roughness profile within a sampling length (L or lr)	$R_{pi} = \max Z(x)$
R_{vi}	The maximum valley depth of the material roughness profile within a sampling length (L or lr)	$R_{vi} = \lvert \min Z(x) \rvert$
R_{zi}	The overall height of the roughness profile R_{zi} is the sum of R_{pi} and R_{vi} within a sampling length (L or lr)	$R_{zi} = R_{pi} + R_{vi}$
R_{max}	Maximum R_{zi} value over assessment length	-
R_z	Average R_{zi} value over assessment length	$R_z = \frac{\sum_{i=1}^{n} R_{zi}}{n}$
R_a	Roughness average R_a is the arithmetic mean of the absolute values of the roughness profile ordinates	$R_a = \frac{1}{l_r}\int_0^{l_r} \lvert z(x) \rvert dx$

Yoshimi and Kishida (1981) defined the surface roughness with the reference value, which is the maximum height of the profile (R_z) along the 2.5-mm sample length (L) [74]. Uesugi and Kishida (1986) also used the maximum height of the profile (R_z) as the reference value, but the sample length is set as 2 mm or average particle size (D_{50}). Current research applies the normalized surface roughness ($R_n = R_{max}(L = D_{50})/D_{50}$ or R_a/D_{50} or R_q/D_{50}) into the granular material and steel interface shear test [75–77].

Staheli et al. (2006) conducted interface shear tests using sand against various pipe materials, including wet cast concrete, vitrified clay, polycrete, and hobas, to determine the relationship between interface characteristics and construction behavior (see Table 3). The surface roughness (average roughness, R_a) of the pipe materials, as well as its corresponding friction coefficient (μ, the ratio of shear stress to normal stress), were measured using a Taylor Hobson Form Talysurf Series 2 stylus profilometer. The friction coefficient increases as R_a increases, as demonstrated by the test results.

Table 3. Effect of surface roughness on the friction coefficient with Ottawa sand at 40 kPa, 80 kPa and 120 kPa.

Parameters	Hobas	Polycrete	Wet Cast Concrete	Vitrified Clay	Packhead Concrete
Average Ra (μm)	6.5	16.9	18.7	24.8	55.1
Friction coefficient (peak) (40 kPa)	0.51	0.50	0.68	0.68	0.81
Friction coefficient (residual) (40 kPa)	0.43	0.42	0.49	0.49	0.54
Friction coefficient (peak) (80 kPa)	0.50	0.49	0.62	0.65	0.73
Friction coefficient (residual) (80 kPa)	0.44	0.43	0.44	0.48	0.53
Friction coefficient (peak) (120 kPa)	0.48	0.47	0.62	0.63	0.73
Friction coefficient (residual) (120 kPa)	0.42	0.43	0.47	0.45	0.52

Iscimen and Frost (2010) produced laboratory equipment to investigate the shear stresses at the pipe–soil interface for curved pipe surfaces [78]. It revealed that the shear strength of the soil–pipe interface was found to rise with increasing surface roughness up to a specific value, beyond which no effect on the shear strength was observed. Knappett and Craig (2012) concluded the previous research and demonstrated the relationship between various soil–material interfaces (Figure 7) [79]. These studies demonstrate that the increasing friction angle gradually approaches the soil internal friction angle until the surface roughness reaches a certain value, called critical roughness.

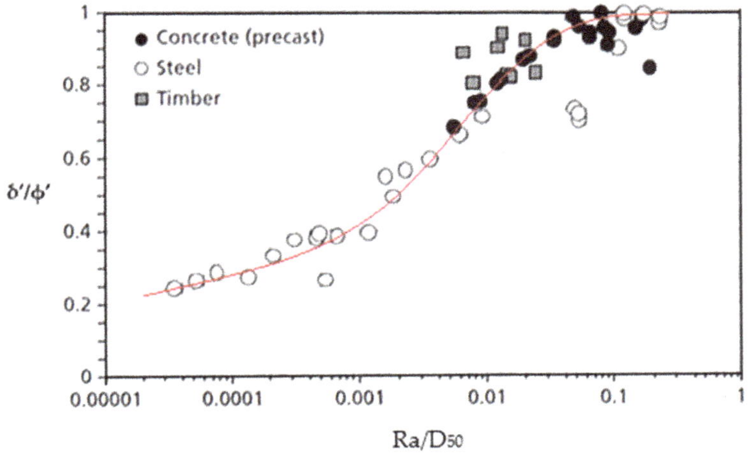

Figure 7. Interface friction angle for various construction materials.

3.1.2. Roughness Measurement

Before conducting interface shear tests, the microscopic surface roughness of a structural material must be determined to quantify its surface roughness with defined specific parameters that will affect soil–structure interface behavior. Roughness parameters are typically determined using a surface roughness tester and are quantified in terms of the average roughness (Ra, see Figure 8). Surface roughness testers consist of portable surface roughness gauges and the three-dimensional microscope profilometer. Compared to 3D optical profilometers, portable surface roughness gauges (see Figure 9) provide a better cost–performance ratio, including the surface profile gauge (with a broad range and poor resolution) and the surface roughness gauge (low range and high resolution). As a result, portable surface roughness gauges are usually applied in industry and laboratories.

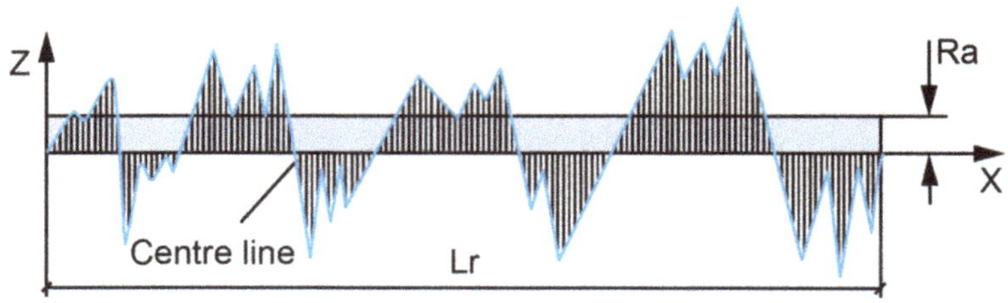

Figure 8. Schematic representation of average roughness, Ra.

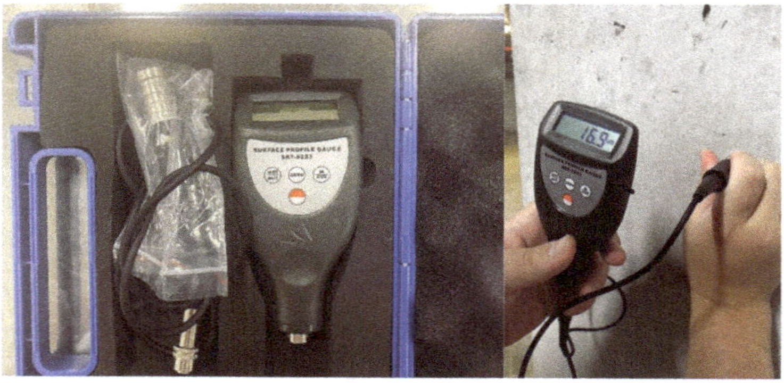

(**a**)

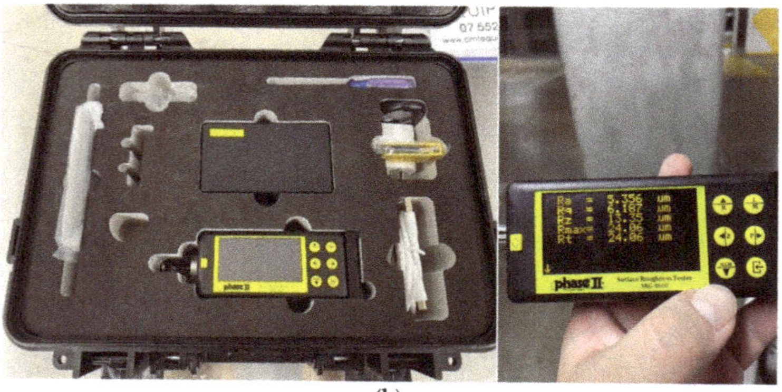

(**b**)

Figure 9. Portable surface roughness gauge: (**a**) surface profile gauge (large range and low resolution); (**b**) surface roughness gauge (low range and high resolution).

However, the probe contact and manipulative interferometry technique cause the inaccuracy of the surface roughness result, which may not satisfy the high requirement on the resolution [80]. Thus, a 3D optical profilometer (a chromatic confocal apparatus) is applied for high-resolution results with the chromatic confocal imaging method and nonintrusive technology [81]. The 3D optical profilometer consists of confocal scanning and high chromatic aberration of the objective lens, which can measure various materials' roughness with the latest white light interferometry (WLI). The white light, transmitted through the profilometer's objective lens, is used to scan the material, generating a wavelength on the single monochromatic point in terms of the refractive index (see Figure 10). The image can then be obtained as image results via the formation of single monochromatic points, which is much more accurate than merely using mean surface roughness. Additionally, the reflectivity and absorption of the scanned sample have no effect on results, even if the material is transparent or specular.

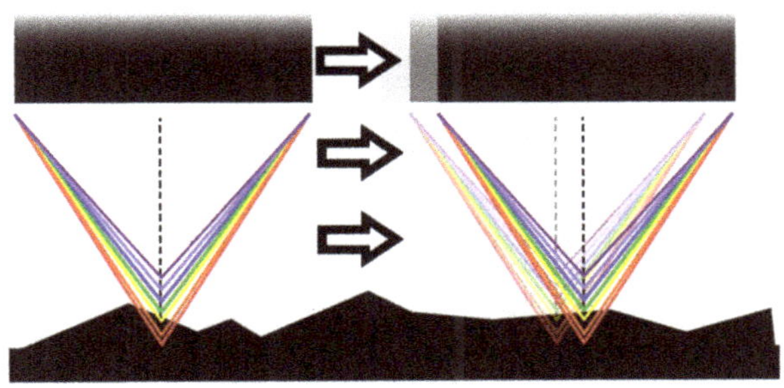

Figure 10. Various wavelengths on the single monochromatic point.

Sandpaper, a synthetic abrasive material, is manufactured with specified surface characteristics, which can be used to simulate a variety of materials, including vitrified clay pipe, precast concrete, polycrete concrete, Permalok steel casting pipe, and Hobas pipe [42]. However, previous research on sandpaper reveals that sandpapers of the same grit but manufactured by various companies have significantly variable surface roughness properties [81,82]. Thus, roughness characteristics should be determined using specialized instruments rather than simply defined by the company's marked grit level.

3.1.3. Definition of Surface Roughness Form

Current research indicates that the critical surface roughness cannot limit the shear strength development under certain conditions caused by the additional passive resistance induced by interface clogging. Thus, surface roughness is classified as random or ribbed surface roughness form (clogging) or structured surface roughness form (non-clogging) [83].

The random or ribbed roughness form (Figure 11a,b) is widely applied for research into the behavior between soil–material friction, especially for precast concrete surface and flat metal surface. Researchers found a critical state for interface behavior, called critical roughness. The critical roughness is determined by the global behavior of the interface shear test and the reference values for surface roughness. According to research on the interface shear behavior between normal consolidated Kawasaki clay and steel, Tsubakihara and Kishida (1993) found that surface roughness increment led to the developed shearing resistance and interface sliding until the reference value reached the critical steel roughness. Once the reference value is over the critical steel roughness, shear failure exists in the clay instead of the soil–material interface. Therefore, for random or ribbed surface roughness form, the maximum interface strength equals the soil strength, called "$\delta = \varnothing$ condition", and the critical roughness is about half of the mean particle size (D_{50}) [83,84].

On the other hand, some researchers believe that the interface friction angle, caused by a highly rough structured surface (Figure 11c), is larger than soil internal friction angle in some specific loading conditions, which is called "$\delta > \varnothing$ condition" (interface friction angle > soil internal friction angle), especially for cast-in-situ concrete surface and patterned metal surface [85–87]. Mitchell and Villet (1987) found that the shear strength of the interface shear test was not only developed by interfacial friction between the sand particle and structure surface as "δ = ∅ condition", but additional passive resistance generated [88]. Martinez and Frost (2017) defined two kinds of surface roughness forms, clogging-prone surface (random or ribbed roughness form) and non-clogging surface (structured roughness form), applied in sand–structure interface shear test [83]. Compared with structured roughness forms, random or ribbed roughness forms produce the interface clogging and restrict interface friction angle to soil internal friction angle.

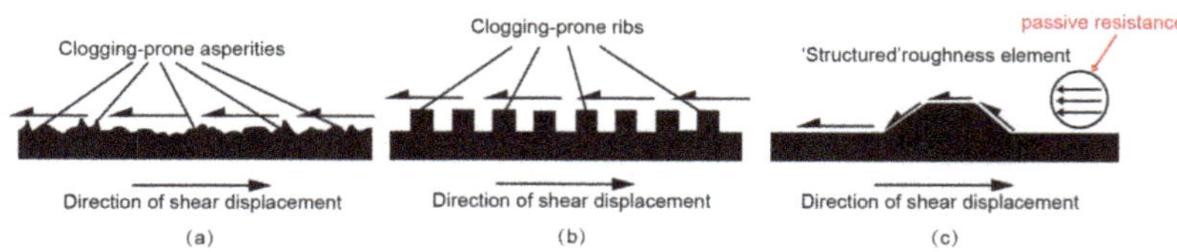

Figure 11. Load transfer mechanisms during shear against the surface of (**a**) random, (**b**) ribbed, and (**c**) structured roughness form.

The identification of clogging-prone and non-clogging surfaces helps researchers to determine the development of soil–material interface shear mechanism. Besides, it was also identified that the ratio of particle size to roughness height would affect the interface behavior, especially in terms of its induced shear forces. With the advanced technology recommended or applied in roughness measurement, surface form definition helps to quantify the surface roughness for more accurate and reliable use in soil–structure interactions.

3.2. *Influence of Other Factors on Interface Behavior*

3.2.1. Particle Angularity

The shapes of the particles have a significant impact on the shear behavior of granular materials [89–91]. Particle shapes are described by roughness, roundness, texture, and sphericity [92]. Uesugi and Kishida (1986) found that particle angularity greatly affected the pipe–soil interface shear strength, as angular particles interlock more strongly than rounded particles. Consequently, it is critical to describe the particle angularity during the analysis of frictional forces. Particles can be characterized according to their angularity as rounded, subrounded, angular, or subangular: (1) angular particles have relatively rough flat surfaces and sharp edges; (2) subangular particles have rounded edges; (3) rounded particles have smooth curved sides that lack edges; and (4) subrounded particles have well-rounded corners and edges, but almost flat sides [93]. Figure 12 illustrates the various particle shape classifications.

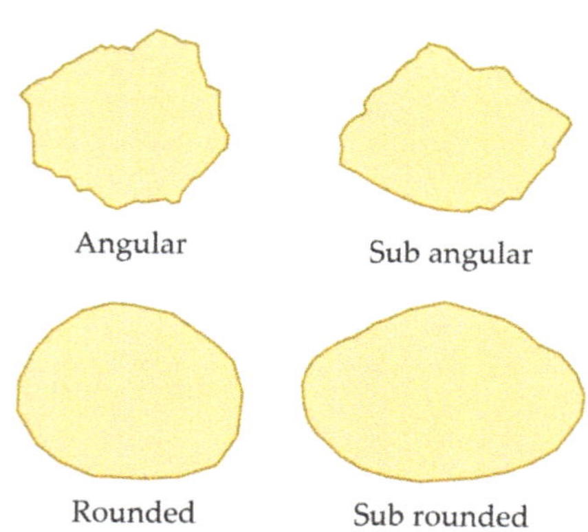

Figure 12. Classification of particle shape as angular, subangular, rounded and subrounded.

According to previous studies, increasing the angularity and decreasing the roundness can result in larger maximum (e_{max}) and minimum (e_{min}) void ratios, respectively [92,94]. Cho et al. (2007) described particle shape and surface roughness analysis as examining the relationship between particle shape, stiffness, packing density, and sand strength, whilst taking into account the particle's roundness, sphericity, and smoothness. When exposed to increasing particle irregularity, it was revealed that e_{max} and e_{min} increased, while the critical state friction angle and compressibility decreased.

Angular quarry materials demonstrated greater shear strength when compared to a combination of subrounded to subangular gravel and sand in triaxial tests [95]. Edil et al. (2006) conducted a series of research on interface shear tests with various particle angularities and particle sizes. They found that the increment in particle angularity leads to a more significant particle interlocking and higher peak shear strength. Additionally, the angular granular material requires a larger shearing displacement to reach peak strength. Li et al. (2013) studied the peak and constant-volume friction angles in a clay gravel combination by adjusting the gravel content. It was found that increasing the asymmetry and roughness of the gravel particles resulted in an increase in constant volume friction angle and a reduction in peak friction angle.

Peerun et al., (2019) investigated the shear behaviors of granular tunneling spoil by comparing rounded sand-based sandstone particles with angular shale particles with lower carbonaceous mineral contents (see Figure 13). Table 4 summarizes the data on vertical stress acting on the pipe. Due to the strong rounded particles of sandstone, higher dilation occurred, and friction angles of 40.5° and 28.6° were measured for pipe-jacking Drives C and D, respectively, in their study [30]. The two pipe-jacking drives had apparent cohesion values of 11.9 kN/m^2 and 17.3 kN/m^2, respectively. The vertical tension acting on the pipes was computed using the Pellet-Beaucour and Kastner (2002) jacking force model (Equation (8)). Pipe-jacking Drives C and D [30,96], in their study, encountered low positive vertical stresses, σ_{EV} of 15.3 kN/m^2 and 12.1 kN/m^2, respectively, indicating that less arching occurred due to the low σ_{EV} values.

(a) (b)

Figure 13. The difference in particle angularity between (**a**) shale (angular) and (**b**) sand (subrounded).

Note: Pellet-Beaucour and Kastner (2002) use Equation (8) to describe the vertical soil stress experience at the pipe or tunnel crown due to the arching effect.

$$\sigma_{EV} = \frac{b\left(\gamma - \frac{2C}{b}\right)}{2K\ \tan\Phi}\left(1 - e^{-2K\frac{h}{b}\ \tan\Phi}\right) \tag{8}$$

where:

C—soil cohesion;

Φ—soil internal friction angle;
γ—soil unit weight;
K—lateral earth pressure coefficient;
h—soil cover from the ground level to the pipe crown;
b—influencing soil width above the pipe.

Table 4. Assessment of vertical stresses acting on pipe crown.

Parameters	Symbol	Unit	Drive C	Drive D	Drive E	Drive F
Geology	-	-	Sandstone		Shale	
Soil friction angle	ϕ	degree	40.5	28.6	37.3	37.3
Soil cohesion	C	kPa	11.9	17.3	0.0	0.0
Arching	-	-	Low to moderately favorable		Least to not favorable	
Calculated vertical stress on pipe crown	σ	kN/m^2	15.3	12.1	34.8	42.9
Average volume of lubricant used	-	L/m	250	250	500	375
Average jacking speed	-	mm/min	29.2	22.4	10.4	18.9
Measured jacking force	JF	kN/m	10.1	18.9	29.0	28.5

On the other hand, angular particles of shale formed a high void ratio and were more vulnerable to compression [30]. As such, particle breakages were postulated, which may have contributed to the decrease in peak shear stresses and apparent cohesion [30]. According to Equation (8), soil cohesion and internal friction angle have significant effects on the vertical soil stress acting on the pipe, with vertical soil stresses, σ_{EV} = 34.8 kN/m^2 for Drive E (shale) and σ_{EV} = 42.9 kN/m^2 for Drive F (shale). In comparison to sandstone (pipe-jacking Drives C and D), pipe-jacking Drives E and F [30,96] in shale sustained much greater vertical stresses, implying a lesser arching state, and thus greater friction developed between pipe surface and shale. This would indicate that the jacked pipes came into contact with the surrounding geology, resulting in increased frictional resistance. This is supported by the substantially higher observed jacking forces (JF) for shale drives (JF = 29.0 kN/m and 28.5 kN/m) compared to sandstone drives (JF = 10.1 kN/m and 18.9 kN/m). Ong and Choo (2018) found similar observations as well [97]. These consistent observations indicate that the angularity of particles does affect the development of the load transfer mechanism and the mobilization of loads between soil–pipe (structure) surfaces.

3.2.2. Mean Particle Size

Subba Rao et al. (1998) found that particle size greatly affects soil internal friction angle, and simply classifying the structure surface with roughness parameters may not be reasonable. Thus, the assessment of interface shear test results requires a newly defined parameter, which is the normalized roughness, defined as the ratio of the average roughness and the mean particle size (D_{50}). Vangla and Gali (2016) demonstrated that the fitness between the mean particle size of granular material and material roughness contributes to the particle interlocking, and thus fully develops the interface shear strength [98]. Additionally, Punetha et al. (2017) found that the effect of mean particle size depends on the material texture [99]. The mean particle size produced a slight effect on soil–geomembrane interface behavior, but a significant effect on the soil–geotextile interface behavior. With mean particle size increment, the contact stress per particle is reduced for the soil–geotextile interface behavior. Sand–steel interface shear test reveals that the increment in mean particle size has a slight effect on friction development, and the friction angle from the interface shear test is much lower than soil internal friction angle [100]. These outcomes demonstrate that the mean particle size influences the interface shear mechanism corresponding to surface roughness and material texture.

3.2.3. Surface Hardness

Surface hardness is one of the essential factors dictating mechanical interface behavior in tribology [101]. Frost et al. (2002) demonstrated the effect of the material hardness on the soil–material interface behavior. According to the plot of peak secant friction coefficient (tan δ) versus normal stress (see Figure 14), the interface shear resistance dwindles at the initial state and rebounds after achieving critical normal stress. At the initial state, the contact area between each granular particle and material surface is increased with lower normal stress. Thus, the effective shear stress of the particle is decreased until the contact area reaches the specific value. Once the normal stress reaches the limit state, additional increments of normal stress cause particle movement and particle plowing, leading to shear stress increasing [51]. Thus, the design of the substructure with low overburden pressure (or normal stress) requires a further investigation on surface hardness, which affects plowing behavior.

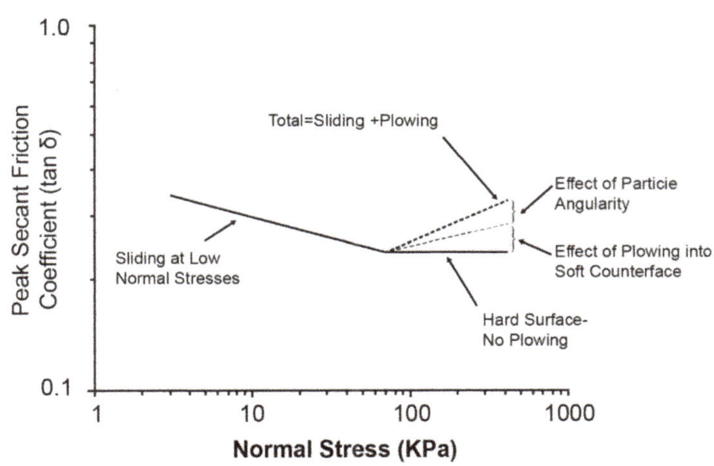

Figure 14. Normal stress versus friction coefficient for soil–continuum material interfaces.

3.2.4. Particle Breakage

For the large-displacement shear test, the plowing and sliding in the interface shear test leads to significant particle breakage, which converts the soil mechanical property, changes the frictional behavior, and contributes to shear zone development [49]. A soil sample with a high void ratio and angular particles is more likely to break and experience a decrease in the internal friction angle of the specimen [29,30,102]. Peerun (2016) researched metagraywacke and shale specimens using direct shear tests. The specimens have low dry densities and high void ratios due to their rough-surfaced angular particles. The higher void ratios and poorer mineralogy of the particles resulted in anticipated particle breakages, which drastically decreased the beneficial particle interlocking phenomenon. Thus, using the flaky metagraywacke and shale particles as examples, low interlocking activity has resulted in reduced apparent cohesion, and thus less opportunity for strength development. Particle breakage remains an extreme challenge to measure, especially when tests are performed in the small shear box; thus, the use of discrete element modeling (DEM) may help solve this problem. DEM shall be discussed later in the paper.

3.2.5. Confinement Condition

According to various research requirements, interface shear apparatus is manufactured based on conventional shear techniques, which include direct shear apparatus, simple shear apparatus, and ring shear apparatus. Thus, confinement conditions vary in interface shear tests, owing to the different limitations of the shear apparatus. Other than the restrictions of the apparatus, the shear mechanism is also affected by the normal stress applied to the soil specimen. At higher normal stresses, a soil sample will experience increased compression and particle interlocking, resulting in a greater specimen shear strength [102,103]. Thus, it is critical to match the interface shear apparatus and normal stress range to the actual site conditions.

4. Potential Application of Particle Image Velocimetry (PIV) Technology for Research on Interface Friction

4.1. Background

Particle image velocimetry (PIV) technology was developed by Adrian (1991) and based on laser speckle velocimetry, which measures the velocity of solid or fluid via an illumination system and two consecutive particle images [104,105]. White and Take (2002) introduced the principle of PIV technology (see Figure 15) and provided a Matlab module (GeoPIV) to analyze geotechnical displacement [106].

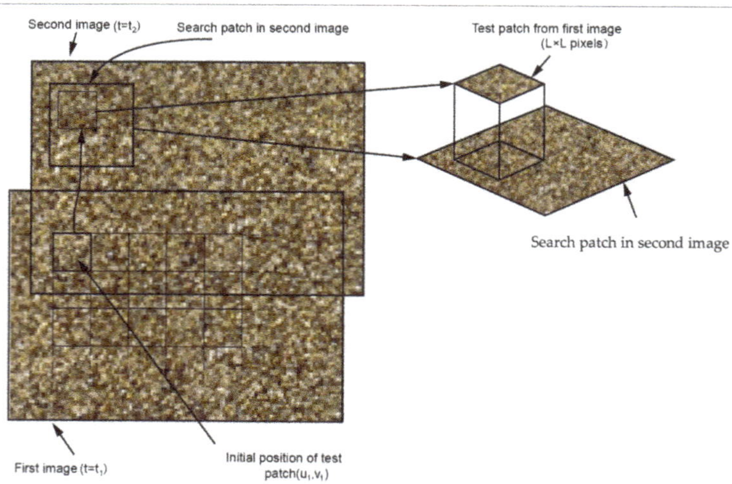

Figure 15. The principles of PIV technology.

Numerous researchers have investigated particle behavior along with shear bands. Laboratory studies and numerical modeling are often used to better understand particle behavior during shearing [29,30,107–109]. Peerun et al., (2019) revealed that such investigations have several limitations, including the following: (1) numerical simulations reveal the nonuniformity of stress generated during direct shear testing [110,111]; (2) restricted views of shear zones caused by equipment fittings [112]; (3) sporadic observations of shear band formation and particle activity [113,114]; and (4) studies on granular material are constrained to plane-strain assumptions [115]. These known constraints impede the physical testing requirements for experimental validation of predicted particle behaviors and for providing continuous and unobstructed views of granular materials during shearing when advanced image processing techniques are used to develop an in-depth understanding of surface roughness.

Peerun et al., (2019) addressed the difficulties associated with obtaining clear evidence of test specimens during shearing in order to examine particle movements along the shear zone. This was achieved by continually capturing high-resolution photographs during direct shear testing on actual granular materials using a fully transparent shear box. Although particle image velocimetry (PIV) is not a technological evolution, the supplemental use of PIV in validating a conventional shear box test and then utilizing interpreted parameters and techniques developed for proper interpretation of actual soil–structure interactions, i.e., pipe jacking operations [29,96,97,116,117], is novel and significant in terms of developing basic knowledge in the field of study of the development of interface friction.

4.2. Introduction

The success of the above-mentioned tests and observations on the reliable use of PIV technology have led the authors to explore the possibility of further modifying the interface direct shear test to examine, quantify, and verify the specific relationship between soil–structure interaction and structural surface roughness. In order to investigate the sand–material interface shear mechanism under high normal stress, interface direct shear apparatus is conducted based on small-scale direct shear apparatus. In addition, this conceptual idea can be in the form of replacing the bottom shear box with the solid block. Furthermore, PIV technology is applied to obtain particle movement by replacing the opaque metal shear box with a custom-made, transparent Perspex (resin) shear box. The random or ribbed surface roughness of any structural surfaces can then be assessed using a portable surface roughness gauge, which includes a surface profile gauge (with a wide range and low resolution) or a surface roughness gauge (with a low range and high resolution).

The interface direct shear test consists of a series of parametric studies using sand and sandpaper interface for calibration (Figure 16a), and then performing a validation study using the same sand but with the sandpaper being replaced by a structural material interface (Figure 16b), each of which utilizes a specific bottom shear box. Parametric research was conducted using the results of interface direct shear tests to determine the relationship between the defined roughness parameters (sandpaper grit is readily available or can be easily confirmed with a profilometer) and the soil–sandpaper interface shear mechanism. The interface behavior between soil and structural material is then validated with the four-stage model, corresponding to parametric research.

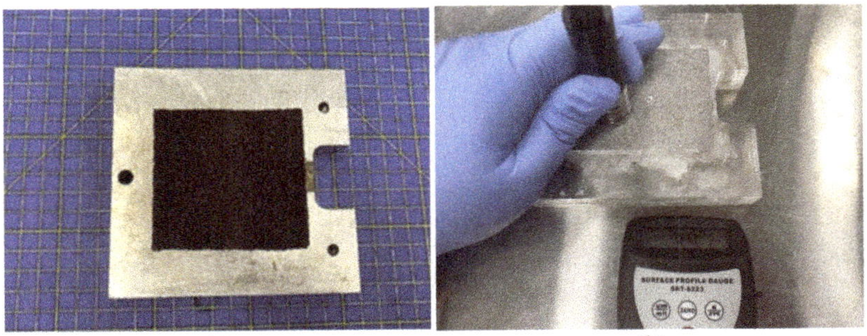

(a) (b)

Figure 16. (**a**) Sandpaper glued on the bottom shear box; (**b**) material block cast by concrete.

4.3. Tested Material

The granular material (coarse sand) collected from the Logan River system of Queensland is categorized as poorly graded sand (UCSC) and subrounded particles (ASTM D2488, see Figure 17), which is used for the interface direct shear test with Australia standard. Based on soil compaction and density test standard (AS1289.5.5.1), the maximum and minimum void ratios are 0.62 and 0.46, respectively. The sieve test reveals that the mean particle size of granular material is 0.833 mm. The coefficient of uniformity (Cu) and coefficient of gradation (Cc) are 1.44 and 1.01, respectively. Previous research on interface shear tests shows that the relationship between particle size and surface roughness is roughly estimated [46,78]. Thus, the critical surface roughness determined in this research may range from 13 μm to 415 μm. Additionally, fundamental research requires a portable surface roughness gauge with a large measurement range and high resolution. Thus, surface roughness gauge (SRG) (see Figure 9b and Table 5) with a low range and high resolution is applied for determining smooth surface roughness (sandpaper with grit 180 and grit 360) in this research. Meanwhile, due to the limited measurement range of surface roughness gauge, the surface profile gauge (SPG) (see Figure 9a and Table 6) with a wider range but poorer resolution is applied for measuring the rough surface (concrete block, sandpaper with grit 40 and grit 80). The sandpaper and the structural material roughness parameters are provided in Table 7.

Figure 17. Particle shape extraction of tested material using a microscope.

Table 5. General specifications of the surface roughness gauge SRG-4600.

Measures grooves and recessions:	wider than 0.16 in (4 mm)
Roughness parameters:	Ra, Rz, Rq (Rms), Rt, Rs, Rsm, Rmax, Rpc, Rmr.
Roughness standards:	ISO/DIN/JIS/ANSI
Display resolutions:	0.01 μm/±20 μm 0.02 μm/±40 μm 0.04 μm/±80 μm
Measuring accuracy:	≤±10%
Repeatability:	Less than or equal to 6%
Measuring range:	Ra, Rq: 0.01–40 μm Rz, Rt, Rm: 0.02–160 μm
Maximum drive range:	0.7 inch (17.5 mm)
Cut-off length:	0.009/0.03/0.09 inch (0.25/0.8/2.5 mm)
Tracing speeds:	speed 1 mm/s (sampling length 2.5 mm) speed 0.5 mm/s (sampling length 0.8 mm) speed 0.135 mm/s (sampling length 0.25 mm)
Sensor:	Inductance type
Sensor-stylus:	Diamond, radius 5 μm
Pick up Force:	<4 mN
Filter:	RC, PC-RC, GAUSSIAN, and D-P
Data output:	USB
Operating temperature:	41–104 °F (5–40 °C)
Weight:	0.97 lb (440 g)
Dimensions:	(119 × 47 × 65 mm)
Power:	Li-Ion rechargeable battery

Table 6. General specifications of the surface profile gauge SRT-6223.

Display:	Four digits, LCD
Range:	0 µm to 800 µm (0 mils to 30 mils)
Accuracy:	±5% or ±5 µm, (whichever is the greater)
Resolution:	1 µm (0.1 mils)
Measurement speed:	>30 readings per minute
Weight:	280 g
Dimensions:	162 × 65 × 28 mm (6.4 × 2.6 × 1.1 inch)
Operating temperature:	0 °C to 50 °C, <80%RH
Batteries:	4 × 1.5v (AAA) battery

Table 7. Material roughness parameters.

	Ra	Rq	Rz	Rmax	D_{50} (CS)	Rn
Structural material	N/A	N/A	422	N/A	833	0.507
Grit 40 Sandpaper	N/A	N/A	722	N/A	833	0.867
Grit 80 Sandpaper	N/A	N/A	460	N/A	833	0.552
Grit 160 Sandpaper	20.07	26.63	81.66	123.3	833	0.148
Grit 320 Sandpaper	10.96	13.37	29.39	58.0	833	0.070

Note: "N/A" means "Not within measuring range".

4.4. Equipment Setup

4.4.1. Transparent Shear Box

A 60-mm square shape shear box (see Figure 18a) produced by Gilson Company consists of a top and bottom shear box, connected with two location screws. Four lifting screws of the shear box adjust the gap between the top and bottom shear box. As the conventional shear box is made of steel, the behavior of particles along the predetermined shear band thickness is, thus, invisible. To detect how geomaterial particles shear and flow around the interface, a transparent shear box (see Figure 18b) made of high-strength plastic resin is manufactured for this conceptual study. Compared with the initial shear box, three location screws are applied on the transparent shear box to analyze the interface behavior. The modification includes extending the maximum shear displacement from 8 mm to 15 mm, which ensures that effective shearing would encompass the development of residual strength [70].

(a)

(b)

Figure 18. (**a**) A 60-mm (2.36 in) shear box (HMA-734MS) and (**b**) custom-made 60-mm (2.36 in) transparent shear box by the authors for preliminary interface shear study.

4.4.2. Camera and Illumination System

A Canon EOS 90D Digital Single Lens Reflex (DSLR) camera is installed on the aluminum clamp with a bottom screw and applied to capture the interface behavior between the top and bottom shear box, which provides high-resolution photos with a 32.5-megapixel APS-C size sensor. However, the camera installation is limited by several factors, such as the shear box displacement, the testing space, the loading frame capacity, the dimensions, and focal length of the camera. After several trial tests, the window dimension of the modified loading frame and water chamber is produced at 110 mm × 40 mm (Figure 19a) and 140 mm × 40 mm (Figure 19b), respectively. Meanwhile, the displacement between the modified loading frame and the camera body is 260 mm.

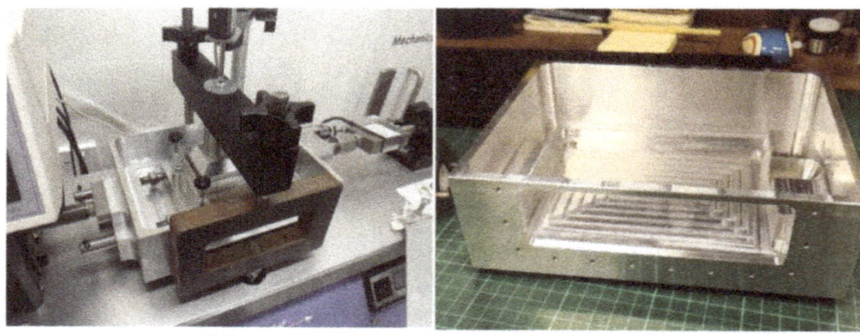

(**a**) (**b**)

Figure 19. Custom-made (**a**) modified loading frame and (**b**) modified water chamber by the authors for preliminary interface shear study.

The film sensitivity of Canon EOS 90D leads to a large expandable ISO speed, which represents a high "light gathering" ability. However, the film's sensitivity is not significant enough to capture high-resolution images in darkness. Meanwhile, the original laboratory environment light causes the shadow to form over part of the critical image. Thus, a new illumination system is set up using a 152-mm ring light and an Elgato multi-mount kit (see Figure 20). Furthermore, to avoid the disturbances caused by the manual operation on the camera, Canon support software (EOS Utility) is used to remotely control the automatic and sequential capture of the images.

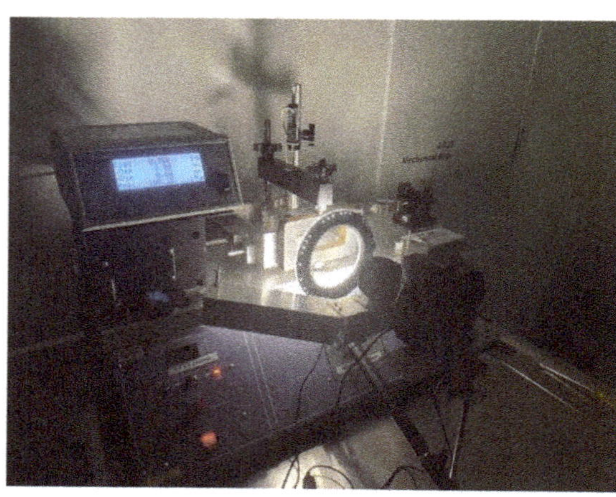

Figure 20. Custom-installed illumination system by the authors for preliminary interface shear study.

4.4.3. Modified Loading Frame

Previous research found that the horizontal force of the direct shear test causes the nonuniform stress on the inner side of the shear box and interrupts the accuracy of the measurements [29,52,58,70]. Preparatory experiments demonstrated that the conventional direct shear test with a free top plate would result in an uneven loading plate when the large normal stress is applied on the soil sample (see Figure 21a). Jewell (1989) found that a free loading plate leads to a higher peak shear strength than a fixed loading plate due to the free particle movement of the interlocking stage (see Figure 22) [118]. Thus, a fixed connection (see Figure 21b), instead of a ball bearing, is customized and manufactured between the loading frame and shear box, providing the symmetric normal stress and limiting the top plate rotation. A similar improvement of the loading frame has been adopted in the study by Peerun et al. (2019).

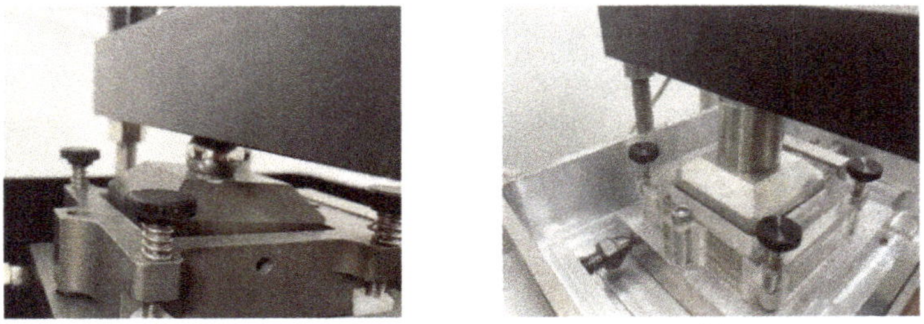

(a) (b)

Figure 21. (**a**) Conventional free loading plate and (**b**) custom-made fixed loading plate by the authors for preliminary interface shear study.

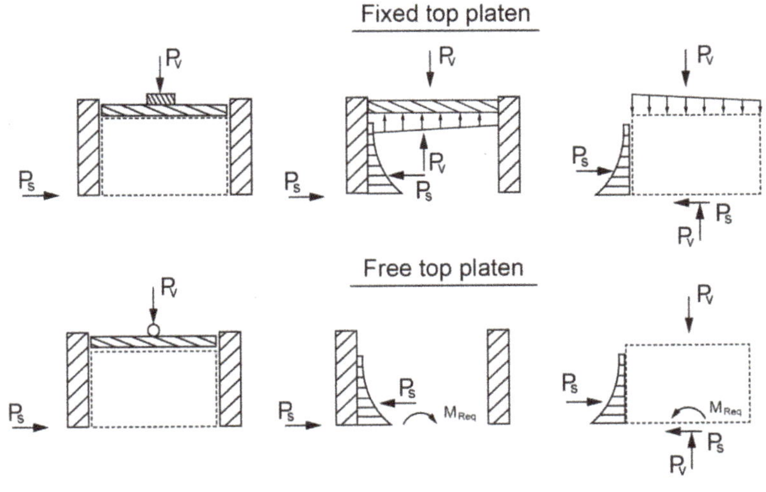

Figure 22. Stress distribution for free loading plate and fixed loading plate.

4.4.4. Procedure of Interface Direct Shear Test

The soil sample is prepared with the dry pluviation method or compaction method for different relative densities. The shearing rate, the maximum horizontal displacement, and the shear box gap are preliminarily set as the aforementioned direct shear test. However, since the residual shear strength cannot be determined with the limited horizontal displacement, the maximum horizontal displacement is expanded in the interface shear test. The procedures of the interface shear test are shown below:

1. For abrasive material, clean the modified bottom shear box with acetone in the fume cupboard (Figure 23), and stick the sandpaper on the modified bottom shear box with proper location.
2. For reinforcing material, put the sample block into the modified bottom shear box.
3. Adjust the lifting screw and clamping screws to suit the 0.6 mm steel plate between the top and bottom shear box, and remove the steel plates.
4. Fix the base plate and the perforated grid plate, and the shearing force is perpendicular to a serration orientation.
5. Fill the soil sample with 14.5-mm thickness with dry pluviation method or compaction method, and slightly put a porous stone on the sample.
6. Clean the bottom shear box and water chamber.
7. Measure the total weight of the shear box, and then place the assembly into the water chamber, slotting the end onto the pushrod.
8. Fix the horizontal and vertical LVDTs on the specific location and return records to zero.
9. Apply normal stress and wait until consolidation is completed (cohesionless soil is consolidated instantaneously).
10. Remove clamping screws.
11. Shear the soil specimen with the constant rate (R_d).

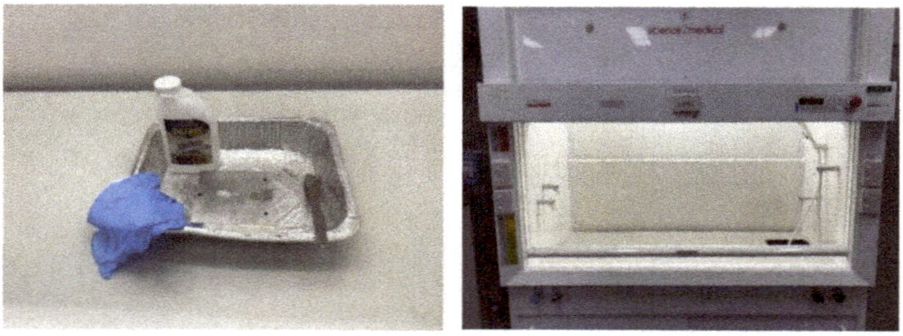

(a) (b)

Figure 23. (**a**) Acetone for cleaning the metal block and (**b**) premium fiberglass fume cupboard used during the author's preliminary interface shear study.

4.5. Application of PIV in Interface Shear Tests

Test results reveal that the interface shear strength is enhanced with the increment in surface roughness and gradually approaches soil shear strength (see Figure 24 and Table 8). In these cases, the coarse sand particles are observed to be much bigger and coarser than all the sandpapers used; hence, the results are intuitively correct. Therefore, it is evident that the increment in surface roughness leads to greater improvement on peak friction angle than residual friction angle (see Table 8).

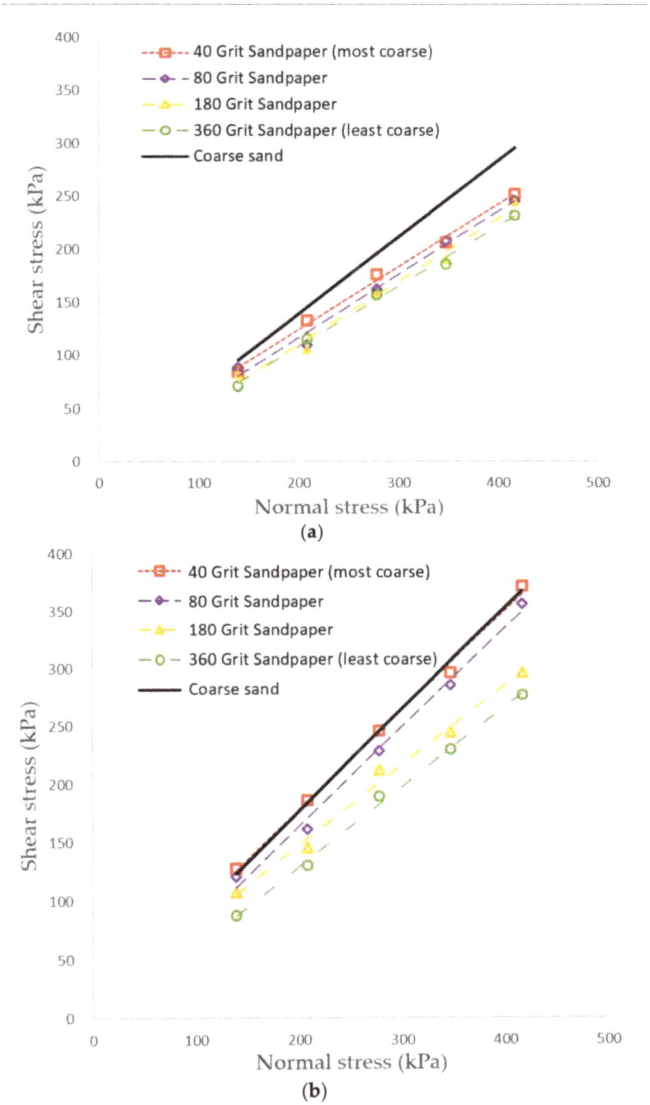

Figure 24. Measured soil internal friction angles and soil–material interface friction angles at (**a**) peak state and (**b**) residual state in the authors' preliminary interface shear study.

Table 8. Results of the soil internal friction angles (sand only) and soil–material interface friction angles (sand–sandpaper).

Soil Type	Material Type	D_{50} or R_{max} (Micron)	Peak Friction Angle (deg)	Residual Friction Angle (deg)
Coarse sand	Coarse sand only	833 (D_{50})	41.3 (sand only)	35.8 (sand only)
Coarse sand	Grit 40 sandpaper	722 (R_{max})	40.8 (sand–sandpaper)	30.5 (sand–sandpaper)
Coarse sand	Grit 80 sandpaper	460 (R_{max})	40.6 (sand–sandpaper)	30.7 (sand–sandpaper)
Coarse sand	Grit 180 sandpaper	81.66 (R_{max})	34.6 (sand–sandpaper)	30.5 (sand–sandpaper)
Coarse sand	Grit 360 sandpaper	29.39 (R_{max})	34.5 (sand–sandpaper)	29.5 (sand–sandpaper)

Particle image velocimetry (PIV) technology is applied to capture the particle microstructure and study the shear mechanism on different soil–reinforcement interactions. The interface shear responses (see Figure 25a), such as contraction and dilation of the soil sample, are validated by analyzing the particle interlocking, rotation, and possibly breakage behaviors. The four-stage model explains particle movement (see Figure 25b) corresponding to the stress–strain relationship of the soil sample. During stage 1 (end zone deformation), particles rearrange and move to the left, with slight vertical movement. During stage 2 (particle interlocking), the vector plot shows that localized particle interlocking at the left end of the shear box leads to 0.19 mm dilation through the soil sample. The particle interlocking prevents particle displacement and further develops the shear resistance to reach the peak state. During stage 3 (shear zone formation), small particles filled into the soil void and rotated large particles readjust the soil sample structure, which reduces the shear resistance and increases volumetric strain. Finally, during stage 4 (steady shear), the shear strain plot reveals that the soil sample generates a stable shear zone along the predetermined failure plane, causing nearly constant shear stress.

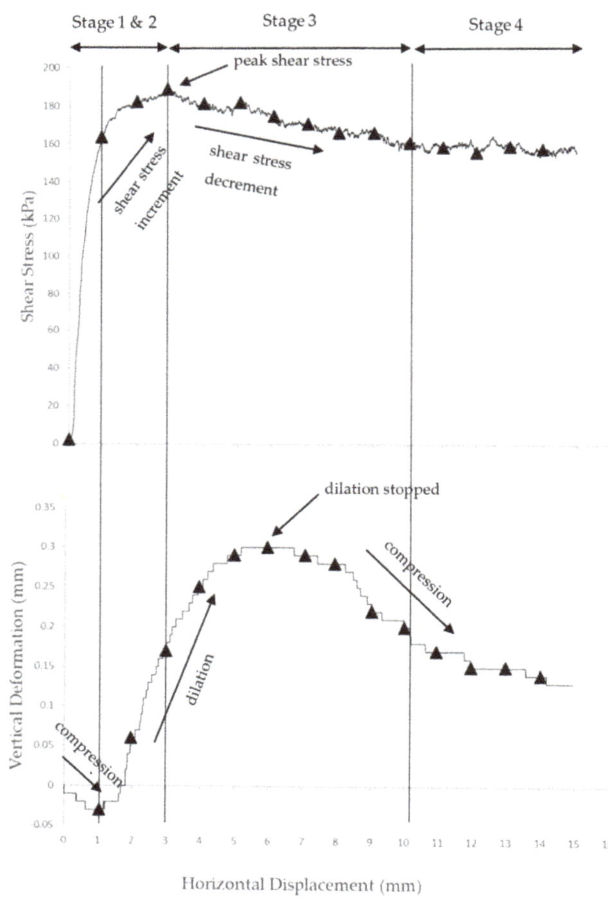

(a)

Figure 25. *Cont.*

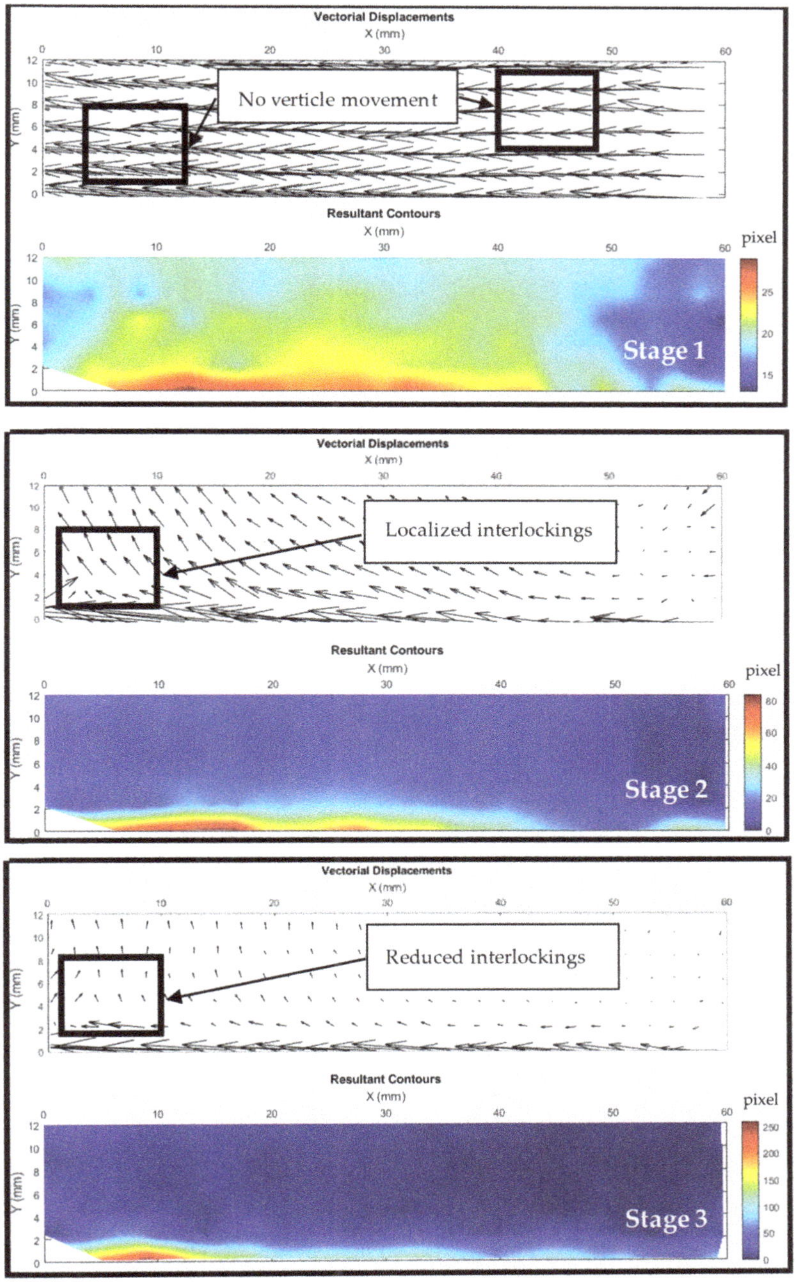

Figure 25. *Cont.*

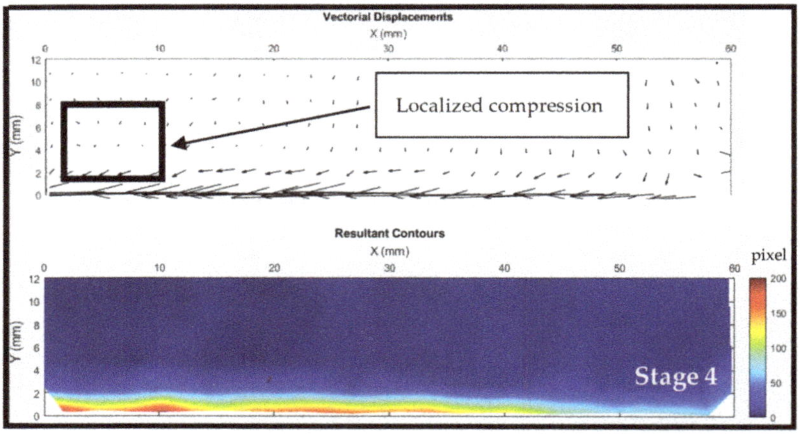

(b)

Figure 25. (a) Interface shear test results and (b) corresponding GeoPIV-RG results for Grit 360 sandpaper at 278 kPa normal stress in the authors' preliminary interface shear study (one pixel is equal to 15.79 micron).

The interface shear responses (see Figure 25a) validate the previous research on the effects of surface roughness. To further explore the coupling effect of particle size and surface hardness, granular material with different particle size distribution and recent advances can be applied in future research for greater accuracy. PIV analysis (see Figure 25b) reveals that the particle behavior of the interface shear test is similar to that of the direct shear test. Shear zone development in interface shear tests contributes to the understanding of soil–material interface behavior.

5. Recent Advances in Shear Strength Characterization

Interface shear test is a fundamental testing method to investigate soil–structure interactions. The use of the reliable PIV technique in interface shear tests is well-established and serves as the main verification strategy in this review paper to improve confidence levels and the justification of particle movements and interlocking behavior that help explain friction development in the shear box. Recent advances in shear strength characterization are not limited within the understanding of PIV application only, but also in other up-and-coming methods, such as artificial neural network (ANN), DEM, 3D printing, and Bayesian machine-learning method. These advanced methods also have great potential for applications for shear strength characterization and, thus, shall be reviewed in this paper as well.

5.1. Neural Network

The aforementioned research found that the mutual coupling effect of surface roughness and other factors on the interface shear mechanism requires a greater understanding instead of the effect caused by the single factor. Meanwhile, the influence factors on interface shear mechanism are determined from laboratory tests and geotechnical investigations, and then applied for preliminary design. Even the standards normally provide a conservative design to ensure construction safety. Geotechnical engineers still require a method to accurately predict substructure performance to save construction costs [119].

Artificial neural network (ANN), a computational method based on empirical data, is widely applied as an alternative and as stochastic algorithms in the engineering field, such as material modeling, damage assessment, structural analysis and design, seismic liquefaction assessment, and ocean monitoring [120]. It is thought that machine learning methods such as ANN can be developed to consider the various influencing factors so that the model can be used to predict the often elusive interface shear forces during soil–structure interactions as a result of understanding interface shear behavior. ANN imitates the nervous system (Figure 26a) to acquire knowledge and make conclusions that automatically deal with complex and challenging work [121]. ANN produces a nonlinear relationship by partially or fully connecting input variables (influence factors) and output variables (construction performance, e.g., interface shear) with one or more hidden layers (Figure 26b), corresponding mathematical functions, and Monte Carlo simulation [122].

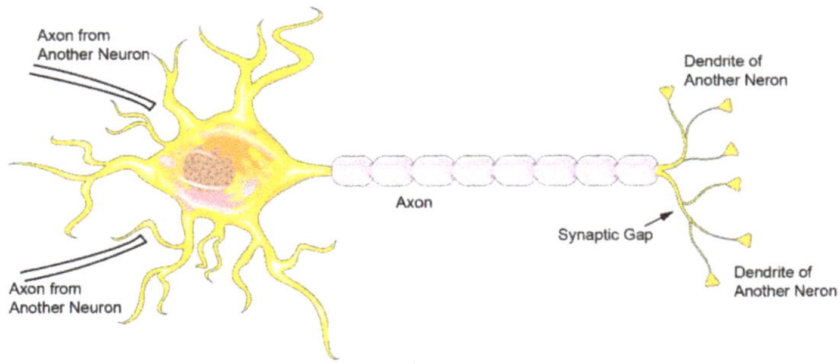

(a)

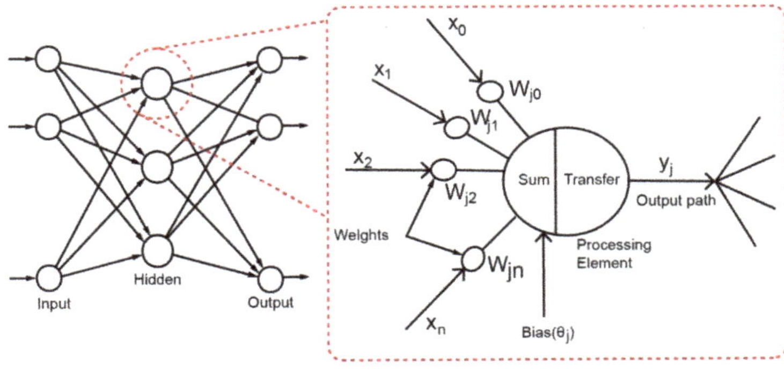

(b)

Figure 26. (**a**) Typical structure of biological neuron; (**b**) typical structure and operation of ANN.

5.2. DEM and 3D Printing

Due to the challenges in maintaining identical test consistencies while performing parametric analyses involving interface shear tests, the discrete element method (DEM) can be used to perform virtual parametric studies based on the influences of different surface hardness on tested construction materials so as to develop a better understanding of complex interface shear mechanisms, such as the cone-penetration testing and pipe-jacking [107,123]. Additionally, instead of the spherical model often used in DEM, a micro-CT-scanned 3D model of the actual particle shapes can improve the simulation quality. Furthermore, calibration of interface shear tests using 3D printing of synthetic particles on different material surfaces can nowadays be readily executed in order to validate the DEM results [31,106,108,109]. For example, Figure 27 shows (a) a sandstone particle that was scanned using micro-CT to obtain its (b) particle shape in a triangular mesh format, and the latter was used to (c) 3D print a synthetic sandstone particle using polyamide material.

Figure 27. Particle shape extraction using micro-CT and printed synthetic particles for calibration of DEM: (**a**) a sandstone particle, (**b**) similar particle in a triangular mesh format, and (**c**) 3D-printed synthetic sandstone particle using polyamide printing material.

DEM is a numerical technique that was developed specifically for the purpose of resolving issues involving discontinuous materials. A jointed rock consists of discrete rock blocks and joint surfaces that enable the rock blocks to move, rotate, and deform while the joint surfaces may be squeezed, detached, or slid. The rock mass is considered a discontinuous discrete media, and its interior may be separated by large displacement, rotation, slide, and breakage, allowing it to more accurately simulate the nonlinear large deformation features of jointed rock. The discrete element method provides a series of physical parameters, such as speed, acceleration, angular velocity, line displacement, and rotation angle.

However, the majority of DEM models rely on limited experimental results for calibration and validation at the granular scale [124,125]. Numerous combinations of microstructures and constitutive laws are widely applied to match constitutive responses during the calibration process, which makes use of just a few parameters, owing to data scarcity. Friedman (1997) refers to these arbitraries as the "curse of dimensionality", in which calibrated DEM models are unable to provide accurate prediction [126–128]. Hanaor et al. (2013) developed specialized particle geometries for 3D printing by combining three different shape descriptors: fractal surface overlay, directed polyhedral aggregation, and contour rotation interpolation [129]. Thus, the discrete element method is implemented utilizing 3D printing technology. The method of contour rotation interpolation was determined to be the most appropriate, and synthetic particles were manufactured in bulk and submitted to triaxial testing. Three-dimensional geometries were employed to simulate triaxial testing using DEMs in order to compare them to experimental tests. The three-dimensional printed particles exhibited shear behavior and the effect of particle form on the bulk material responses. During the macro-mechanical research of particle behavior, the introduction of synthetic particles enabled the separation of particle shape and material characteristics. Currently, the printing resolution and commercially accessible printing materials limit the use of 3D printing for geotechnical testing. The bulk-printed granular materials are composed of irregular angular forms. Gupta et al. (2018) extracted particle geometries from micro-CT scans and utilized 3D printing to reproduce sand particles. The DEM simulations were performed using the particle morphologies acquired from micro-CT scans and compared to the use of shape descriptors in numerical modeling. Odometric experiments were conducted using the 3D-printed particles, and the findings were utilized to calibrate and validate the DEM models. Figure 28 shows typical preliminary DEM results. The direct shear test was simulated using sand particles in DEM and compared to the laboratory direct shear test result. The normal stress applied on the specimen was 100 kPa, and sphere particles were used to represent the sand particles in the simulation. From Figure 28, it can be seen that DEM is able to simulate the direct shear test, with an initial increase in shear stress until peak state is achieved and then a reduction to the residual state. However, there is a noticeable difference between the DEM simulation and the laboratory test results. This difference is most likely due to the use of spheres instead of the actual sand particle shapes.

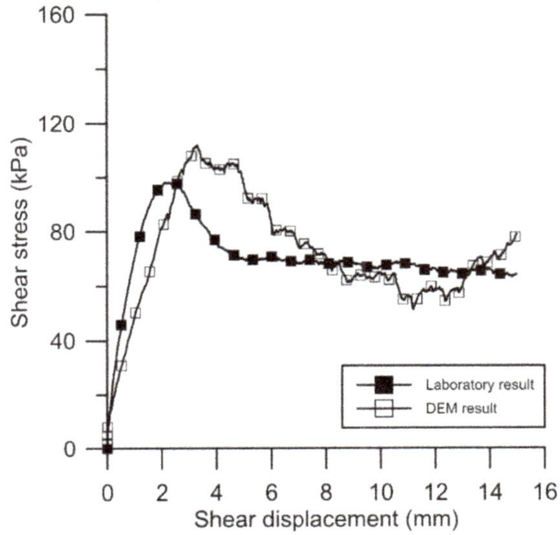

Figure 28. Typical DEM shear stress results compared against laboratory results for spheres or sphere clumps being used in DEM modeling.

Figure 29 shows state-of-the-art technologies being developed by the authors to improve the shear stress prediction of particles interacting with different material surfaces. The realistic particles obtained from micro-CT scanning can now be exported to DEM without the use of conventional spheres or sphere clumps. This technique is considered revolutionary at this stage of research. In this case, particle breakages can now be modeled in DEM to reflect the complex, true interparticle behavior. Hence, the proposed methodology of scanned particle shapes using micro-CT and printed particles for DEM calibration has much potential to contribute to more refined and accurate DEM results, which the authors will eventually publish. When sample extraction cannot be achieved due to site restrictions or unavailability of laboratory tests, the DEM technique will one day provide an extensive advantage to accurately model soil–structure interactions.

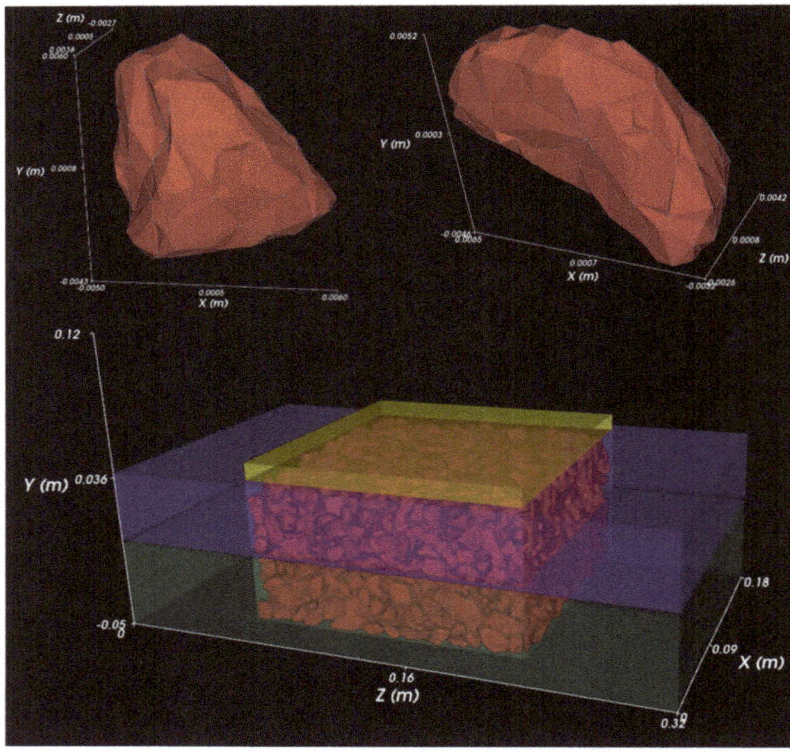

Figure 29. Realistic particle shapes are being used in DEM modeling as a revolutionary technique currently being developed by the authors.

5.3. Bayesian Method

Bayesian method, a machine learning technique based on Bayes' theorem [130], is widely used due to its ability to quantify parameter uncertainty using probability. It computes the evolution of probability in the parameters, before observing the data, and after incorporating the data in the analysis [131]. It is found to be superior in updating, for example, soil parameters [132,133] or, in this case, surface roughness from prior studies, because it considers these parameters as random variables instead of fixed constants. Thus, the Bayesian updating process provides a robust approach in back analyzing or predicting soil parameters based on the field observational method compared to deterministic techniques that produce outputs with fixed values [134]. The probabilistic estimates obtained from a Bayesian model are used to quantify the parameter uncertainties to inform the confidence levels associated with the model predictions, thus making it a potentially suitable method to be applied to the study of interface surface roughness as a function of different soil types for the design of soil–structure interactions.

6. Conclusions and Recommendations

Interface shear mechanisms have been investigated through various shear tests and applied in geotechnical engineering applications. Previous research found that several influential factors affect the interface shear mechanism, such as (i) surface roughness, (ii) particle angularity, (iii) mean particle size, (iv) surface hardness, (v) particle breakage, and (vi) confinement condition. The main contributions of this review paper are in highlighting the following important observations:

- To parametrically study the interface shear mechanism, portable surface roughness gauges are required to evaluate surface roughness. For random or ribbed surface form, the maximum interface strength equals the soil strength, called "$\delta = \varnothing$ condition". For structured surface form, the shear strength of the interface shear test is not only developed by interfacial friction between the sand particle and structure surface as "$\delta = \varnothing$ condition", but additional passive resistance generated. Thus, the maximum interface strength is greater than the soil strength, called "$\delta > \varnothing$ condition".
- Particle angularity influences the friction development and mobilization of loads between soil–material interfaces. The increment in particle angularity leads to a more significant particle interlocking and higher peak shear strength.
- Mean particle size influences the interface shear mechanism corresponding to surface roughness and material texture.
- Surface hardness affects plowing behavior on the soil–material interface. The effective shear stress of the particle decreases until the contact area reaches the specific value. Once the normal stress reaches the limit state, an additional increment in normal stress causes particle movement and particle plowing, leading to shear stress increasing.
- The higher void ratios and poorer mineralogy of the particles resulted in anticipated particle breakages, significantly reducing the beneficial particle interlocking phenomena. Consequently, using flaky metagraywacke and shale particles as examples, a lack of interlocking activity results in decreased apparent cohesion, and thus a lower opportunity for strength development.
- The shearing boundaries (or limits) for soil specimens placed in the shear box vary with different shear apparatus set up. Moreover, the higher the normal stress, the more significant the particle interlocking and shear strength values would be. Thus, the selection of shear apparatus and normal stress range should match the real-life site conditions.

PIV technology is the main verification tool adopted in this paper to track and justify particle movements and interlocking during shear testing. According to the shear test results, interface shear strength is enhanced with the increment in surface roughness and gradually approaches the actual soil shear strength, when the material surface is in a random or ribbed form. Similarly, the increasing peak friction angle progressively approaches the soil internal friction angle until the surface roughness reaches its critical roughness. These observations prove that surface roughness has a negligible influence on residual friction angle. In parallel, the application of PIV technology demonstrates that the four-stage model is highly applicable and reliable to describe the development of soil shear strength in the direct shear tests and the interface shear tests. With the development of other advanced testing and verification methods (e.g., ANN, DEM, 3D printing, and Bayesian machine learning), the research of interface shear mechanisms is becoming more important and relevant in the study of soil–structure interactions. The conclusions and recommendations are presented below:

- Artificial neural network has the potential to establish a nonlinear relationship to investigate the mutual coupling effect of surface roughness and other factors on the interface shear mechanism.
- Rather than the spherical model that is frequently used in DEM, a micro-CT-scanned 3D model of the actual particle morphologies can be employed to enhance the simulation quality.
- In the calibration of interface shear tests, 3D printing of synthetic particles is recommended to validate the DEM results.
- Bayesian method considers soil and interface parameters as random variables and provides probabilistic estimates to reasonably quantify the parameter uncertainties and may be used in interface shear studies to develop a greater understanding in the study of soil–structure interactions.

Author Contributions: Conceptualization, R.W., D.E.L.O. and M.I.P.; methodology, R.W., D.E.L.O. and M.I.P.; software, R.W.; validation, R.W., D.E.L.O. and M.I.P.; formal analysis, R.W. and M.I.P.; investigation, R.W. and M.I.P.; resources, R.W., D.E.L.O., M.I.P. and D.-S.J.; data curation, R.W.; writing—original draft preparation, R.W. and D.E.L.O.; writing—review and editing, R.W., D.E.L.O., M.I.P. and D.-S.J.; visualization, R.W. and; supervision, D.E.L.O. and D.-S.J.; project administration, D.E.L.O. and D.-S.J.; funding acquisition, D.E.L.O. All authors have read and agreed to the published version of the manuscript.

Funding: This research received no external funding.

Institutional Review Board Statement: Not applicable.

Informed Consent Statement: Not applicable.

Data Availability Statement: The original experimental data are available on request from the corresponding author.

Conflicts of Interest: The authors declare no conflict of interest.

References

1. Leong, H.Y.; Ong, D.E.L.; Sanjayan, J.G.; Nazari, A. A genetic programming predictive model for parametric study of factors affecting strength of geopolymers. *RSC Adv.* **2015**, *5*, 85630–85639. [CrossRef]
2. Leong, H.Y.; Ong, D.E.L.; Sanjayan, J.; Nazari, A. The effect of different Na_2O and K_2O ratios of alkali activator on compressive strength of fly ash based-geopolymer. *Constr. Build. Mater.* **2016**, *106*, 500–511. [CrossRef]
3. Leong, H.Y.; Ong, D.E.L.; Sanjayan, J.; Nazari, A. Suitability of Sarawak and Gladstone fly ash to produce geopolymers: A physical, chemical, mechanical, mineralogical and microstructural analysis. *Ceram. Int.* **2016**, *42*, 9613–9620. [CrossRef]
4. Ngu, L.; Song, J.W.; Hashim, S.S.; Ong, D.E. Lab-scale atmospheric CO_2 absorption for calcium carbonate precipitation in sand. *Greenh. Gases Sci. Technol.* **2019**, *9*, 519–528. [CrossRef]
5. Mehdizadeh, A.; Disfani, M.M.; Evans, R.; Arulrajah, A.; Ong, D.E.L. Discussion of "Development of an Internal Camera-Based Volume Determination System for Triaxial Testing" by S.E. Salazar, A. Barnes and R.A. Coffman. *Geotech. Test. J. 2015, 38; reprinted in Geotech. Test. J.* **2016**, *39*, 165–168. [CrossRef]
6. Mehdizadeh, A.; Disfani, M.M.; Evans, R.; Arulrajah, A.; Ong, D.E.L. Mechanical Consequences of Suffusion on Undrained Behaviour of a Gap-Graded Cohesionless Soil—An Experimental Approach. *Geotech. Test. J.* **2017**, *40*, 1026–1042. [CrossRef]
7. Ong, D.E.L. Benchmarking of FEM technique involving deep excavation, pile-soil interaction and embankment construction. In Proceedings of the 12th International Conference of International Association for Computer Methods and Advanced in Geomechanics (IACMAG), Goa, India, 1–6 October 2008.
8. Ong, D.E.L.; Yang, D.; Phang, S. Comparison of finite element modelling of a deep excavation using SAGECRISP and PLAXIS. In Proceedings of the 2006 International Conference on Deep Excavations, Singapore, 28–30 June 2006; pp. 50–63.
9. Omoregie, A.; Khoshdelnezamiha, G.; Senian, N.; Ong, D.E.L.; Nissom, P.M. Experimental optimisation of various cultural conditions on urease activity for isolated Sporosarcina pasteurii strains and evaluation of their biocement potentials. *Ecol. Eng.* **2017**, *109*, 65–75. [CrossRef]
10. Omoregie, A.I.; Ngu, L.H.; Ong, D.E.L.; Nissom, P.M. Low-cost cultivation of Sporosarcina pasteurii strain in food-grade yeast extract medium for microbially induced carbonate precipitation (MICP) application. *Biocatal. Agric. Biotechnol.* **2019**, *17*, 247–255. [CrossRef]
11. Omoregie, A.I.; Ong, D.E.L.; Nissom, P.M. Assessing ureolytic bacteria with calcifying abilities isolated from limestone caves for biocalcification. *Lett. Appl. Microbiol.* **2019**, *68*, 173–181. [CrossRef] [PubMed]
12. Omoregie, A.I.; Palombo, E.A.; Ong, D.E.; Nissom, P.M. Biocementation of sand by Sporosarcina pasteurii strain and technical-grade cementation reagents through surface percolation treatment method. *Constr. Build. Mater.* **2019**, *228*, 116828. [CrossRef]
13. Omoregie, A.I.; Palombo, E.A.; Ong, D.E.; Nissom, P.M. A feasible scale-up production of Sporosarcina pasteurii using custom-built stirred tank reactor for in-situ soil biocementation. *Biocatal. Agric. Biotechnol.* **2020**, *24*, 101544. [CrossRef]
14. Omoregie, A.I.; Senian, N.; Li, P.Y.; Hei, N.L.; Leong, D.O.E.; Ginjom, I.R.H.; Nissom, P.M. Ureolytic bacteria isolated from Sarawak limestone caves show high urease enzyme activity comparable to that of Sporosarcina pasteurii (DSM 33). *Malays. J. Microbiol.* **2016**, *12*, 463–470.
15. Leong, H.Y.; Ong, D.E.L.; Sanjayan, J.G.; Nazari, A. Strength Development of Soil–Fly Ash Geopolymer: Assessment of Soil, Fly Ash, Alkali Activators, and Water. *J. Mater. Civ. Eng.* **2018**, *30*, 04018171. [CrossRef]
16. Leong, H.Y.; Ong, D.E.L.; Sanjayan, J.G.; Nazari, A.; Kueh, S.M. Effects of Significant Variables on Compressive Strength of Soil-Fly Ash Geopolymer: Variable Analytical Approach Based on Neural Networks and Genetic Programming. *J. Mater. Civ. Eng.* **2018**, *30*, 04018129. [CrossRef]
17. Liu, Y.; Liu, Z.; Oh, E.; Ong, D. Strength and Microstructural Assessment of Reconstituted and Stabilised Soft Soils with Varying Silt Contents. *Geosciences* **2021**, *11*, 302. [CrossRef]

18. Chong, E.; Ong, D. Data-Driven Field Observational Method of a Contiguous Bored Pile Wall System Affected by Accidental Groundwater Drawdown. *Geosciences* **2020**, *10*, 268. [CrossRef]
19. Ong, D.E.L.; Choo, C. 2011, Sustainable construction of a bored pile foundation system in erratic phyllite. In Proceedings of the 2011 ASEAN-Australian Engineering Congress, Kuching, Malaysia, 25–27 July 2011; pp. 50–65.
20. Ong, D.E.L.; Leung, C.; Chow, Y. Piles subject to excavation-induced soil movement in clay. In Proceedings of the 13th European Conference on Soil Mechanics and Geotechnical Engineering, Prague, Czech Republic, 25–28 August 2003; pp. 777–782.
21. Ong, D.E.L.; Leung, C.; Chow, Y. Time-dependent Pile Behavior due to Excavation-Induced Soil Movement in Clay. In Proceedings of the 12th Pan-American Conference on Soil Mechanics and Geotechnical Engineering, Cambridge, MA, USA, 22–26 June 2003; Massachusetts Institute of Technology: Boston, MA, USA, 2003; Volume 2, pp. 2035–2040.
22. Ong, D.E.L.; Leung, C.F.; Chow, Y.K.; Ng, T.G. Severe Damage of a Pile Group due to Slope Failure. *J. Geotech. Geoenviron. Eng.* **2015**, *141*, 04015014. [CrossRef]
23. Ong, D.E.L.; Sim, Y.S.; Leung, C.F. Performance of Field and Numerical Back-Analysis of Floating Stone Columns in Soft Clay Considering the Influence of Dilatancy. *Int. J. Géoméch.* **2018**, *18*, 04018135. [CrossRef]
24. Liu, Z.; Ong, D.E.L.; Liu, Y.; Bolton, M.W.; Oh, E. Effect of Cement and Bentonite Mixture on the Consolidation Behavior of Soft Estuarine Soils. *Int. J. GEOMATE* **2020**, *18*, 49–54. [CrossRef]
25. Sun, J.; Oh, E.; Ong, D. Influence of Degree of Saturation (DOS) on Dynamic Behavior of Unbound Granular Materials. *Geosciences* **2021**, *11*, 89. [CrossRef]
26. Cheng, W.-C.; Li, G.; Ong, D.E. Lubrication characteristics of pipejacking in soft alluvial deposits. In *Geotechnical Aspects of Underground Construction in Soft Ground*; CRC Press: Boca Raton, FL, USA, 2021; pp. 12–18. [CrossRef]
27. Cheng, W.-C.; Li, G.; Ong, D.E.; Chen, S.-L.; Ni, J.C. Modelling liner forces response to very close-proximity tunnelling in soft alluvial deposits. *Tunn. Undergr. Space Technol.* **2020**, *103*, 103455. [CrossRef]
28. Peerun, M.I.; Ong, D.E.L.; Choo, C.S. Calibration and parametric studies using geopiv technology to track particle movements in a transparent shear box. In Proceedings of the Young Geotechnical Engineers Conference 2016: Contributions of Young Geotechnical Engineers to Nation Building, Petaling Jaya, Selangor, Malaysia, 30 May 2016. [CrossRef]
29. Peerun, M.I.; Ong, D.E.L.; Choo, C.S. Interpretation of Geomaterial Behavior during Shearing Aided by PIV Technology. *J. Mater. Civ. Eng.* **2019**, *31*, 04019195. [CrossRef]
30. Peerun, M.; Ong, D.; Choo, C.; Cheng, W. Effect of interparticle behavior on the development of soil arching in soil-structure interaction. *Tunn. Undergr. Space Technol.* **2020**, *106*, 103610. [CrossRef]
31. Peerun, M.I.; Ong, D.E.L.; Desha, C.; Oh, E.; Choo, C. Influences of geological characteristics on the construction of tunnels. In Proceedings of the WEC2019: World Engineers Convention 2019, Melbourne, VIC, Australia, 20–22 November 2019.
32. Ong, D.E.L. Pile Behaviour Subject to Excavation-Induced Soil Movement in Clay. Ph.D. Thesis, National University of Singapore, Singapore, 2005.
33. Peerun, M.I.; Ong, D.E.L.; Choo, C.S.; Phangkawira, F. *Novel Methods in Estimating Pipe-Jacking Forces in Highly Fractured Rocks*; Indian Society for Trenchless Technology, NoDIG: Delhi, India, 2017.
34. Kang, G.; Tsuchida, T.; Athapaththu, A. Strength mobilization of cement-treated dredged clay during the early stages of curing. *Soils Found.* **2015**, *55*, 375–392. [CrossRef]
35. Liu, Y.; Jiang, Y.J.; Xiao, H.; Lee, F.H. Determination of representative strength of deep cement-mixed clay from core strength data. *Géotechnique* **2017**, *67*, 350–364. [CrossRef]
36. Yao, K.; Pan, Y.; Jia, L.; Yi, J.T.; Hu, J.; Wu, C. Strength evaluation of marine clay stabilized by cementitious binder. *Mar. Georesour. Geotechnol.* **2020**, *38*, 730–743. [CrossRef]
37. Zhou, J. Performance of Full Scale Tests of Piles in Different Soil Conditions. Ph.D. Thesis, Griffith University, Brisbane, Australia, 2018.
38. Zhou, J.; Zhang, X.; Zhang, L.; Dong, F.; Oh, E. Static load tests of driven concrete piles under CFRP confinement. *Géoméch. Geoengin.* **2020**, *15*, 159–171. [CrossRef]
39. Cheng, W.-C.; Ni, J.C.; Huang, H.-W.; Shen, J.S. The use of tunnelling parameters and spoil characteristics to assess soil types: A case study from alluvial deposits at a pipejacking project site. *Bull. Eng. Geol. Environ.* **2019**, *78*, 2933–2942. [CrossRef]
40. Ji, X.; Zhao, W.; Ni, P.; Barla, M.; Han, J.; Jia, P.; Chen, Y.; Zhang, C. A method to estimate the jacking force for pipe jacking in sandy soils. *Tunn. Undergr. Space Technol.* **2019**, *90*, 119–130. [CrossRef]
41. Coulomb, C. Sur une Application des Regles de Mximus et Mnimus a Qelques Poblemes de Satique Rlatits a L'architecture, Academie Royale des Sciences. In *Memoires de Mathermatiques et de Physique par Divers Sowans*; Academy of Sciences: Paris, France, 1773; Volume 7, pp. 343–382.
42. Staheli, K. Jacking Force Prediction: An Interface Friction Approach Based on Pipe Surface Roughness. Ph.D. Thesis, Georgia Institute of Technology, Atlanta, GA, USA, 2006.
43. Paikowsky, S.; Player, C.; Connors, P. A Dual Interface Apparatus for Testing Unrestricted Friction of Soil Along Solid Surfaces. *Geotech. Test. J.* **1995**, *18*, 168–193. [CrossRef]
44. Uesugi, M.; Kishida, H. Frictional Resistance at Yield between Dry Sand and Mild Steel. *Soils Found.* **1986**, *26*, 139–149. [CrossRef]
45. Frost, J.; DeJong, J.; Recalde, M. Shear failure behavior of granular–continuum interfaces. *Eng. Fract. Mech.* **2002**, *69*, 2029–2048. [CrossRef]

46. Edil, T.B.; Bosscher, P.J.; Sundberg, A.J. Soil-Structure Interface Shear Transfer Behavior. In *Geomechanics II: Testing, Modeling, and Simulation*; Springer: Berlin/Heidelberg, Germany, 2006; pp. 528–543. [CrossRef]
47. Uesugi, M.; Kishida, H. Influential Factors of Friction between Steel and Dry Sands. *Soils Found.* **1986**, *26*, 33–46. [CrossRef]
48. Jardine, R.J.; Lehane, B.M.; Everton, S.J. Friction Coefficients for Piles in Sands and Silts. In *Offshore Site Investigation and Foundation Behaviour. Advances in Underwater Technology, Ocean Science and Offshore Engineering*; Ardus, D.A., Clare, D., Hill, A., Hobbs, R., Jardine, R.J., Squire, J.M., Eds.; Springer: Dordrecht, The Netherlands, 1993; Volume 28, pp. 661–677. [CrossRef]
49. Ho, T.; Jardine, R.; Anh-Minh, N. Large-displacement interface shear between steel and granular media. *Géotechnique* **2011**, *61*, 221–234. [CrossRef]
50. Boulon, M. Physical and numerical simulation of lateral shaft friction along offshore piles in sand. In Proceedings of the 3rd Internation Conference on Numerical Methods in Offshore Piling, Paris, France; 1986; pp. 127–147. [CrossRef]
51. Dove, J.E.; Frost, J.D. Peak Friction Behavior of Smooth Geomembrane-Particle Interfaces. *J. Geotech. Geoenviron. Eng.* **1999**, *125*, 544–555. [CrossRef]
52. Peerun, I.; Ong, D.; Choo, C. Behaviour of reconstituted sand-sized particles in direct shear tests using PIV technology. *Jpn. Geotech. Soc. Spéc. Publ.* **2016**, *2*, 354–359. [CrossRef]
53. Peerun, M.I.; Ong, D.E.L.; Choo, C.S. Effect of Particle Shapes on Shear Strength during Direct Shear Testing using GeoPIV Technology. In Proceedings of the 19th Southeast Asian Geotechnical Conference & 2nd Agssea Conference, Deep Excavation and Ground Improvement Dorsett Grand Subang, Subang Jaya, Malaysia, 31 May–3 June 2016.
54. Pellet-Beaucour, A.-L.; Kastner, R. Experimental and analytical study of friction forces during microtunneling operations. *Tunn. Undergr. Space Technol.* **2002**, *17*, 83–97. [CrossRef]
55. Samtani, N.C.; Nowatzki, E.A. *Soils and Foundations: Reference Manual–Volume I, Report, United States*; Federal Highway Administration: Washington, DC, USA, 2006.
56. *AS1289*; Methods of Testing Soils for Engineering Purposes. Standards Association of Australia: Sydney, NSW, Australia, 1998.
57. *GB/T50123*; Geotechnical Test Method Standard. China Planning Press: Beijing, China, 1999. (In Chinese)
58. *ASTM, D3080*; Standard Test Method for Direct Shear Test of Soils Under Consolidated Drained Conditions. ASTM International: West Conshohocken, PA, USA, 2004.
59. Yu, K.; Yao, X.; Zhang, Y.; Li, C.; Ou, L. Analysis of direct shear test data based on area and stress correction. *Chin. J. Rock Mech. Eng.* **2014**, *33*, 118–124. (In Chinese)
60. Li, Y.; Aydin, A. Behavior of rounded granular materials in direct shear: Mechanisms and quantification of fluctuations. *Eng. Geol.* **2010**, *115*, 96–104. [CrossRef]
61. Bjerrum, L.; Landva, A. Direct Simple-Shear Tests on a Norwegian Quick Clay. *Géotechnique.* **1966**, *16*, 1–20. [CrossRef]
62. Shimamura, A.; Miyamoto, Y.; Kobayashi, T. Experimental Study on Lateral Resistance of Steel Pipe Pile with Wings Using the New Composite Geo-material with Magnesium Acrylate. In *Dynamic Soil-Structure Interaction for Sustainable Infrastructures. GeoMEast 2018. Sustainable Civil Infrastructures*; Choudhury, D., El-Zahaby, K., Idriss, I., Eds.; Springer: Cham, Switzerland, 2018; pp. 62–74. [CrossRef]
63. Hanzawa, H.; Nutt, N.; Lunne, T.; Tang, Y.; Long, M. A Comparative Study Between the NGI Direct Simple Shear Apparatus and the Mikasa Direct Shear Apparatus. *Soils Found.* **2007**, *47*, 47–58. [CrossRef]
64. *ASTM, D6528*; Standard Test Method for Consolidated Undrained Direct Simple Shear Testing of Fine Grain Soils. ASTM International: West Conshohocken, PA, USA, 2017.
65. Bishop, A.W.; Green, G.E.; Garga, V.K.; Andresen, A.; Brown, J.D. A New Ring Shear Apparatus and Its Application to the Measurement of Residual Strength. *Géotechnique* **1971**, *21*, 273–328. [CrossRef]
66. Bromhead, E.N.; Dixon, N. The field residual strength of London Clay and its correlation with loboratory measurements, especially ring shear tests. *Géotechnique* **1986**, *36*, 449–452. [CrossRef]
67. Stark, T.D.; Eid, H.T. Modified Bromhead ring shear apparatus. *Geotech. Test. J.* **1993**, *16*, 100–107. [CrossRef]
68. Varnes, D.J. Slope movement types and processes. In *Transportation Research Board, Special Report*; Transportation Research Board: Washington, DC, USA, 1978; Volume 176, pp. 11–33.
69. Peerun, M.I. Behaviour of Reconstituted Sand-Sized Tunnelling Rock Spoils during Shearing Using GeoPIV Technology for the Assessment of Soil Arching Effect during Pipe-Jacking Works. Master's Thesis, Faculty of Engineering, Computing and Science, Swinburne University of Technology, Hawthorn, VIC, Australia, 2016.
70. Keramatikerman, M.; Chegenizadeh, A.; Nikraz, H.; Yilmaz, Y. Mechanical Behaviour of Atrazine-Contaminated Clay. *Appl. Sci.* **2020**, *10*, 2457. [CrossRef]
71. Potyondy, J.G. Skin Friction between Various Soils and Construction Materials. *Géotechnique* **1961**, *11*, 339–353. [CrossRef]
72. *ISO 4278:1977*; Geometrical Product Specifications (GPS)—Surface Texture: Profile Method—Terms, Definitions and Surface Texture Parameters. International Organization for Standardization: Geneva, Switzerland, 1977.
73. Rudawska, A.; Danczak, I.; Müller, M.; Valasek, P. The effect of sandblasting on surface properties for adhesion. *Int. J. Adhes. Adhes.* **2016**, *70*, 176–190. [CrossRef]
74. Yoshimi, Y.; Kishida, T. A Ring Torsion Apparatus for Evaluating Friction Between Soil and Metal Surfaces. *Geotech. Test. J.* **1981**, *4*, 145–152. [CrossRef]
75. Kishida, H.; Uesugi, M. Tests of the interface between sand and steel in the simple shear apparatus. *Géotechnique* **1987**, *37*, 45–52. [CrossRef]

76. Subba Rao, K.; Allam, M.; Robinson, R.; Paikowsky, S.; Player, C. Discussion on a dual interface apparatus for testing unrestricted friction of soil along solid surfaces. Authors' reply. *ASTM Geotech. Test. J.* **1996**, *19*, 446–451. [CrossRef]
77. Zhou, X.; Gong, W.; Ding, H.; Rao, X. Large-scale simple shear test on mechanical properties of interface between concrete face and gravel underlayer. *Chin. J. Geotech. Eng.* **2005**, *27*, 876–880.
78. Iscimen, M.; Frost, J. Shearing behaviour of curved interfaces. In *Characterization and Behavior of Interfaces: Proceedings of the Research Symposium on Characterization and Behavior of Interfaces, Atlanta, GA, USA, 21 September 2008*; IOS Press: Amsterdam, The Netherlands, 2010; pp. 35–43.
79. Knappett, J.; Craig, R.F. *Craig's Soil Mechanics*, 9th ed.; CRC Press: London, UK, 2019; Volume 8. [CrossRef]
80. Stępień, K. Testing the Accuracy of Surface Roughness Measurements Carried out with a Portable Profilometer. *Key Eng. Mater.* **2015**, *637*, 69–73. [CrossRef]
81. Mell, B. *Sandpaper Roughness Measurement Using 3D Profilometry*; Nanovea Technical Report; Nanovea: Irvine, CA, USA, 2010.
82. Liu, F. *Roughness and Particle Diameter of Sandpaper*; Nanovea Technical Report; Nanovea: Irvine, CA, USA, 2018.
83. Martinez, A.; Frost, J.D. The influence of surface roughness form on the strength of sand-structure interfaces. *Géotechnique Lett.* **2017**, *7*, 104–111. [CrossRef]
84. Tsubakihara, Y.; Kishida, H. Frictional Behaviour between Normally Consolidated Clay and Steel by Two Direct Shear Type Apparatuses. *Soils Found.* **1993**, *33*, 1–13. [CrossRef]
85. Chen, X.; Zhang, J.; Xiao, Y.; Li, J. Effect of roughness on shear behavior of red clay—Concrete interface in large-scale direct shear tests. *Can. Geotech. J.* **2015**, *52*, 1122–1135. [CrossRef]
86. Hryciw, R.D.; Irsyam, M. Behavior of Sand Particles Around Rigid Ribbed Inclusions During Shear. *Soils Found.* **1993**, *33*, 1–13. [CrossRef]
87. Koval, G.; Chevoir, F.; Roux, J.-N.; Sulem, J.; Corfdir, A. Interface roughness effect on slow cyclic annular shear of granular materials. *Granul. Matter* **2011**, *13*, 525–540. [CrossRef]
88. Mitchell, J.K.; Villet, W.C. *Reinforcement of Earth Slopes and Embankments*; NCHRP Report; Transportation Research Board: Washington, DC, USA, 1987.
89. Dyskin, A.V.; Estrin, Y.; Kanel-Belov, A.; Pasternak, E. Toughening by Fragmentation—How Topology Helps. *Adv. Eng. Mater.* **2001**, *3*, 885–888. [CrossRef]
90. Miura, K.; Maeda, K.; Furukawa, M.; Toki, S. Mechanical Characteristics of Sands with Different Primary Properties. *Soils Found.* **1998**, *38*, 159–172. [CrossRef]
91. Shimobe, S.; Moroto, N. A new classification chart for sand liquefaction. *Earthq. Geotech. Eng.* **1995**, 315–320.
92. Cho, G.-C.; Dodds, J.; Santamarina, J.C. Closure to "Particle Shape Effects on Packing Density, Stiffness, and Strength: Natural and Crushed Sands" by Gye-Chun Cho, Jake Dodds, and J. Carlos Santamarina. *J. Geotech. Geoenvironmental Eng.* **2007**, *133*, 1474. [CrossRef]
93. *ASTM, D2488-17*; Standard Practice for Description and Identification of Soils (Visual-Manual Procedure). ASTM International: West Conshohocken, PA, USA, 2017. [CrossRef]
94. De Graft-Johnson, J.; Bhatia, H.; Gidigasu, D. The strength characteristics of residual micaceous soils and their application to stability problems. In Proceedings of the 7th International Conference on Soil Mechanics and Foundation Engineering, Mexico City, Maxico; 1969; Volume 2.
95. Holtz, W.; Willard, M. *Triaxial Shear Characteristics of Clayey Gravel Soils*; US Bureau of Reclamation: Washington, DC, USA, 1961.
96. Choo, C.S.; Ong, D.E.L. Evaluation of Pipe-Jacking Forces Based on Direct Shear Testing of Reconstituted Tunneling Rock Spoils. *J. Geotech. Geoenviron. Eng.* **2015**, *141*, 04015044. [CrossRef]
97. Ong, D.; Choo, C. Assessment of non-linear rock strength parameters for the estimation of pipe-jacking forces. Part 1. Direct shear testing and backanalysis. *Eng. Geol.* **2018**, *244*, 159–172. [CrossRef]
98. Vangla, P.; Gali, M.L. Effect of particle size of sand and surface asperities of reinforcement on their interface shear behaviour. *Geotext. Geomembr.* **2016**, *44*, 254–268. [CrossRef]
99. Punetha, P.; Mohanty, P.; Samanta, M. Microstructural investigation on mechanical behavior of soil-geosynthetic interface in direct shear test. *Geotext. Geomembr.* **2017**, *45*, 197–210. [CrossRef]
100. Su, L.-J.; Zhou, W.-H.; Chen, W.-B.; Jie, X. Effects of relative roughness and mean particle size on the shear strength of sand-steel interface. *Measurement* **2018**, *122*, 339–346. [CrossRef]
101. Williams, J.A.; Kennedy, F.E. Engineering Tribology. *ASME J. Tribol.* **1998**, *120*, 644. [CrossRef]
102. Xiao, Y.; Liu, H.; Chen, Y.; Jiang, J. Strength and Deformation of Rockfill Material Based on Large-Scale Triaxial Compression Tests. I: Influences of Density and Pressure. *J. Geotech. Geoenviron. Eng.* **2014**, *140*, 04014070. [CrossRef]
103. Terzaghi, K.; Peck, R.B.; Mesri, G. *Soil Mechanics in Engineering Practice*; John Wiley & Sons: Hoboken, NJ, USA, 1996.
104. Adrian, R.J. Particle-Imaging Techniques for Experimental Fluid Mechanics. *Annu. Rev. Fluid Mech.* **1991**, *23*, 261–304. [CrossRef]
105. Grant, I. Particle image velocimetry: A review. *Proc. Inst. Mech. Eng. Part C J. Mech. Eng. Sci.* **1997**, *211*, 55–76. [CrossRef]
106. White, D.; Take, W. *GeoPIV: Particle Image Velocimetry (PIV) Software for Use in Geotechnical Testing*; Cambridge University: Cambridge, UK, 2002.
107. Ng, N.Y.J.; Peerun, M.I.; Ong, D.E.L.; Choo, C.S. *Assessment of Tunnelling Rock Spoils from Highly Weathered Lithology Using Rolling Friction Contact Models in Discrete Element Modelling*; The Institution of Engineers Malaysia: Kuala Lumpur, Malaysia, 2020.

108. Peerun, M.I.; Ong, D.E.L.; Desha, C.; Oh, E.; Choo, C.S. Advances in the Study of Micromechanical Behaviour for Granular Materials Using Micro-CT Scanner and 3D Printing. In *Challenges and Innovations in Geomechanics;* IACMAG 2021. Lecture Notes in Civil Engineering; Barla, M., Di Donna, A., Sterpi, D., Eds.; Springer: Cham, Switzerland, 2021; Volume 125, pp. 911–918. [CrossRef]
109. Peerun, M.I.; Ong, D.E.L.; Desha, C.; Oh, E.; Choo, C.S. Recent Advancements in Fundamental Studies of Particulate Interaction and Mechanical Behaviour using 3-D Printed Synthetic Particles. In Proceedings of the 1st Malaysian Geotechincal Society (MGS) and Geotechnical Society of Singapore (GeoSS) Conference 2019, Petaling Jaya, Malaysia, 24–26 June 2019.
110. Indraratna, B.; Ngo, T.; Rujikiatkamjorn, C.; Vinod, J.S. Behavior of Fresh and Fouled Railway Ballast Subjected to Direct Shear Testing: Discrete Element Simulation. *Int. J. Géoméch.* **2014**, *14*, 34–44. [CrossRef]
111. Potts, D.M.; Dounias, G.T.; Vaughan, P.R. Finite element analysis of the direct shear box test. *Géotechnique* **1987**, *37*, 11–23. [CrossRef]
112. Fukuoka, H.; Sassa, K.; Wang, G.; Sasaki, R. Observation of shear zone development in ring-shear apparatus with a transparent shear box. *Landslides* **2006**, *3*, 239. [CrossRef]
113. Agung, M.W.; Sassa, K.; Fukuoka, H.; Wang, G. Evolution of Shear-Zone Structure in Undrained Ring-Shear Tests. *Landslides* **2004**, *1*, 101–112. [CrossRef]
114. Wang, G.; Sassa, K. Post-failure mobility of saturated sands in undrained load-controlled ring shear tests. *Can. Geotech. J.* **2002**, *39*, 821–837. [CrossRef]
115. Yuan, Q.; Wang, Y.H.; Tam, P.O.; Li, X.; Gao, Y. Experimental Characterizations of Contact Movement in Two-Dimensional Rod Assembly Subjected to Direct Shearing. *Int. J. Géoméch.* **2017**, *17*, 04016032. [CrossRef]
116. Choo, C.S.; Ong, D.E.L. Impact of highly weathered geology on pipe-jacking forces. *Geotech. Res.* **2017**, *4*, 94–106. [CrossRef]
117. Ong, D.E.L.; Choo, C.S. Back-analysis and finite element modeling of jacking forces in weathered rocks. *Tunn. Undergr. Space Technol.* **2016**, *51*, 1–10. [CrossRef]
118. Jewell, R.A. Direct shear tests on sand. *Géotechnique* **1989**, *39*, 309–322. [CrossRef]
119. Shahin, M.; Maier, H.; Jaksa, M. Predicting Settlement of Shallow Foundations using Neural Networks. *J. Geotech. Geoenvironmental Eng.* **2002**, *128*, 785–793. [CrossRef]
120. Fatehnia, M.; Amirinia, G. A review of Genetic Programming and Artificial Neural Network applications in pile foundations. *Int. J. Geo-Eng.* **2018**, *9*, 2. [CrossRef]
121. Fausett, L. *Fundamentals of Neural Networks: Architectures, Algorithms, and Applications;* Prentice-Hall, Inc.: Hoboken, NJ, USA, 1994.
122. Maier, H.; Dandy, G. The effect of internal parameters and geometry on the performance of back-propagation neural networks: An empirical study. *Environ. Model. Softw.* **1998**, *13*, 193–209. [CrossRef]
123. Ecemis, N.; Bakunowicz, P. Feasible packing of granular materials in discrete-element modelling of cone-penetration testing. *Géoméch. Geoeng.* **2018**, *13*, 198–216. [CrossRef]
124. Liu, Y.; Sun, W.; Fish, J. Determining Material Parameters for Critical State Plasticity Models Based on Multilevel Extended Digital Database. *J. Appl. Mech.* **2016**, *83*, 011003. [CrossRef]
125. Ng, T.-T. Input Parameters of Discrete Element Methods. *J. Eng. Mech.* **2006**, *132*, 723–729. [CrossRef]
126. Friedman, J.H. On bias, variance, 0/1—Loss, and the curse-of-dimensionality. *Data Min. Knowl. Discov.* **1997**, *1*, 55–77. [CrossRef]
127. Gupta, R.; Salager, S.; Wang, K.; Sun, W. Open-source support toward validating and falsifying discrete mechanics models using synthetic granular materials—Part I: Experimental tests with particles manufactured by a 3D printer. *Acta Geotech.* **2019**, *14*, 923–937. [CrossRef]
128. Wang, K.; Sun, W.; Salager, S.; Na, S.; Khaddour, G. Identifying Material Parameters for a Micro-Polar Plasticity Model via X-ray Micro-Computed Tomographic (ct) Images: Lessons Learned from the Curve-Fitting Exercises. *Int. J. Multiscale Comput. Eng.* **2016**, *14*, 389–413. [CrossRef]
129. Hanaor, D.A.H.; Gan, Y.; Einav, I. Effects of surface structure deformation on static friction at fractal interfaces. *Géotechnique Lett.* **2013**, *3*, 52–58. [CrossRef]
130. Bayes, T. LII. An essay towards solving a problem in the doctrine of chances. By the late Rev. Mr. Bayes, F.R.S. communicated by Mr. Price, in a letter to John Canton, A.M.F.R.S. *Philos. Trans. R. Soc. Lond.* **1763**, *53*, 370–418. [CrossRef]
131. Jin, Y.; Biscontin, G.; Gardoni, P. A Bayesian definition of 'most probable' parameters. *Geotech. Res.* **2018**, *5*, 130–142. [CrossRef]
132. Jong, S.; Ong, D.; Oh, E. State-of-the-art review of geotechnical-driven artificial intelligence techniques in underground soil-structure interaction. *Tunn. Undergr. Space Technol.* **2021**, *113*, 103946. [CrossRef]
133. Jong, S.C.; Ong, D.E.L.; Oh, E.; Choo, C.S. Prediction of Frictional Jacking Forces Using Bayesian Inference. In *Challenges and Innovations in Geomechanics;* Barla, M., Di Donna, A., Sterpi, D., Eds.; IACMAG 2021. Lecture Notes in Civil Engineering; Springer: Cham, Switzerland, 2021; Volume 125, pp. 878–885. [CrossRef]
134. Qi, X.-H.; Zhou, W.-H. An efficient probabilistic back-analysis method for braced excavations using wall deflection data at multiple points. *Comput. Geotech.* **2017**, *85*, 186–198. [CrossRef]

geosciences

Article

Dispersion Curves of Transverse Waves Propagating in Multi-Layered Soils from Experimental Tests in a 100 m Deep Borehole

Angelo Aloisio *, Ferdinando Totani, Rocco Alaggio and Gianfranco Totani

Department of Civil, Construction-Architectural and Environmental Engineering, Università degli Studi dell'Aquila, Via G. Gronchi, 18, 67100 L'Aquila, Italy; ferdinando.totani@gmail.com (F.T.); rocco.alaggio@univaq.it (R.A.); gianfranco.totani@univaq.it (G.T.)

* Correspondence: angelo.aloisio1@univaq.it

Abstract: The estimate of the velocity of shear waves (V_s) is essential in seismic engineering to characterize the dynamic response of soils. There are various direct methods to estimate the V_s. The authors report the results of site characterization in Macerata (Italy), where they measured the V_s using the seismic dilatometer in a 100 m deep borehole. The standard V_s estimation originates from the cross-correlation between the signals acquired by two geophones at increasing depths. This paper focuses on the estimate of the dependence of V_s on the wavenumber. The dispersion curves reveal an unexpected hyperbolic dispersion curve typical of Lamb waves. Interestingly, the contribution of Lamb waves may be notable up to 100 m depth. The amplitude of surface waves decrease rapidly with depth; still, their influence may be essential up to depths considered unusual for standard geotechnical investigations, where their effect is generally neglected. Accordingly, these waves may bias the outcomes of the standard V_s estimations, which ignore frequency-dependent phenomena. The paper proposes an enhancement of the accepted procedure to estimate V_s and addresses the importance of Lamb waves in soil characterization.

Keywords: dispersion curve; shear wave; seismic dilatometer; soil mechanics; in situ test

Citation: Aloisio, A.; Totani, F.; Alaggio, R.; Totani, G. Dispersion Curves of Transverse Waves Propagating in Multi-Layered Soils from Experimental Tests in a 100 m Deep Borehole. *Geosciences* **2021**, *11*, 207. https://doi.org/10.3390/geosciences11050207

Academic Editors: Dominic E. L. Ong, Wen-Chieh Cheng, Hannah Zhou and Jesus Martinez-Frias

Received: 22 February 2021
Accepted: 3 May 2021
Published: 8 May 2021

1. Introduction

The Navier's equilibrium equations of an infinite isotropic elastic medium admit two types of waves: volumetric waves, involving no rotation and rotational waves, involving no volume changes. Soil mechanics and seismology name the longitudinal waves P (Pressure) waves and the transverse waves S (Shear) waves [1]. In the considered infinite idealized medium, the transverse and longitudinal motion of the particles is uncoupled: P and S waves are nondispersive and propagate with their constant velocities, independent on the wavenumber. In an isotropic elastic plane, a third class of solutions may satisfy the equilibrium equations: the Lamb waves [2]. After the pioneering works of Lamb and Rayleigh, the study of surface waves fed a vast amount of theoretical, experimental and technical investigations. A medium with a finite dimension admits the propagation of dispersive waves, which exhibit the coupling between the longitudinal and transverse motion. Therefore, Lamb waves do not originate from a modification of the constitutive properties, but descend from the boundaries modifications.

Several scholars [3–9] attributed the dispersive nature of soils to their constitutive properties. As explained in the following sections, the current research follows the classical formulations in soil mechanics, which considers the soil as an isotropic homogeneous elastic medium [10]. This assumption may be very restrictive, and several studies transcended the limits of the classical theories by formulating ad hoc differential equations for granular media. The interpretation of the experimental data in the light of the granular micromechanics would entail dedicated experimental tests. Therefore this research stands on the groundwork of classical theory of Lamb waves.

In the last decades, there has been a revived interest in methods used to measure the shear wave velocity (V_s) due to the inclusion of the value up to 30 m depth ($V_{s,30}$) in several building codes [11–15]. Accordingly, several direct and indirect methods were developed to evaluate $V_{s,30}$. Direct methods stem from direct measurements with the depth, while indirect methods estimate the soil properties from measurements on the free surface [16–18]. Direct methods, like the seismic dilatometer [19], determine the time lag between two signals acquired by two spaced sensors, triggered by an impulsive force, using frequency or time-domain methods.

On the contrary, indirect methods are based on measuring the dispersion features of surface waves leading to an estimate of the shear wave profile from inversion algorithms [20,21]. Modern surface wave testing relies on advanced signal processing and inversion algorithms to extract information about the shear wave velocity profile from observations of Rayleigh wave propagation by the free surface [22]. Frequently, the inversion derives from the comparison between a theoretical dispersion curve and the experimentally one. The V_s profile is iteratively updated until the agreement between the theoretical and experimental dispersion curves is satisfactory [23].

Rayleigh waves are nondispersive in a homogeneous isotropic elastic medium [1]. They become dispersive if the mechanical properties of the medium change with depth (layered medium). Accordingly, the dispersion curve is an indicator of the mechanical properties of the soil [24,25]. Rayleigh waves travel in a zone of about one wavelength from the free surface, and hence they are informative about this zone [26,27].

Still, the experimental tests carried out by the authors revealed that transverse waves generated on the free surface could manifest a dispersive nature up to significant depths. Specifically, the amplitude of these waves does not decrease with depth according to their wavelength. Accurately, waves with a wavelength less than 1m are detectable up to 100 m depth. Reasonably, these waves are not Rayleigh ones, which should extinguish rapidly with depth according to their wavenumber. They possibly belong to the broader class of Lamb waves propagating within each soil layer [28]. Lamb waves are always dispersive due to the nature of the boundaries. They propagate in solid plates, or spheres [29–31]. Lamb's wave theory, developed by Horace Lamb in 1916 [30], describes the characteristics of waves propagating in thin, flat or curved plates, having a thickness of the order of magnitude of the guided wavelength. These waves, remaining constrained within the thickness, have the advantage of propagating in long distances with reduced attenuation. Lamb waves arise from the superposition of longitudinal and transverse waves. They result in the excitation of symmetric and antisymmetric modes. The symmetrical modes are labelled longitudinal modes, since the average displacement across the plate thickness is parallel to the propagating direction. The antisymmetric modes show an average displacement in the transverse direction, orthogonal to the propagating direction.

Lamb waves have multiple applications. Specifically, in the last decades, they have been widely used for damage detection purposes [32–34] (e.g., identifying cracks in thin materials and tubular products). Extensive developments in Lamb wave applications provide a basis for controlling many industrial products in the aerospace and transport sectors [35,36].

Lamb waves can also propagate in multi-layered soils [28]. Their use in geotechnics is still limited and confined to research activities [37,38]. However, to the authors' knowledge, except for the recent paper by [39], no scholar endeavoured to estimate the propagation of Lamb waves in the multi-layered soils from their direct experimental measure within each layer [40]. Specifically, the authors measured the transverse response of a layered soil up to 100 m depth using the Seismic Flat Dilatometer (SDMT).

The soil, excited by a shear, directional impulse on the free surface, like the one generated by a pendulum hammer hitting an anvil pressed to the ground, maybe approximately viewed as a problem in the x-z plane. The y direction orthogonal to that of the excitation may be neglected in the mathematical formulation, and the problem befalls in the case of propagating Lamb waves. A certain amount of energy disperses in the y direction, thus causing a consistent decay of the amplitude with depth. Still, the directional nature of the shear impulse may endorse a mathematical formulation on the x-z plane.

The authors noticed that the waves measured in a 100 m deep borehole exhibited a nondispersive nature. Therefore, they will attempt to grasp the presence of Lamb waves by estimating the dependence of V_s with the wavenumber. This paper analyses the nature of waves propagating in the vertical direction. Unfortunately, the discussion is limited by the available experimental data, which refer to velocity measurements in the horizontal direction with increasing depth.

This paper has the following structure: the first section presents the mathematical background of wave propagation in an isotropic elastic space and isotropic elastic plane. The plane case leads to propagating Lamb waves characterized by hyperbolic dispersion curves. The third section introduces the experimental details of the in situ tests. The fourth section presents the results acquired from the 100 m deep borehole and addresses several issues arisen from the discussion of the dispersion curves.

2. Mathematical Background

This section presents the general theory of wave propagation in isotropic elastic media. The equations derived in this part support the interpretation of the experimental data. Additionally, the authors describe the method followed to calculate the dispersion curves from the acquired time series.

2.1. Waves in Infinite Isotropic Elastic Media

The Navier's equilibrium equations of an isotropic elastic medium in vector notation are [1]:

$$(\lambda + \mu)\nabla\nabla \cdot \boldsymbol{u} + \mu\nabla^2\boldsymbol{u} + \rho\boldsymbol{f} = \rho\ddot{\boldsymbol{u}}, \tag{1}$$

where λ and μ are the Lamè constants, ∇ is the gradient operator, $\nabla\cdot$ is the divergence operator, $\boldsymbol{u} = \{u, v, w\}$ collects the particle displacements in the x,y,z directions, ρ is the mass density, $\boldsymbol{f}$ is the vector of external forces, $\ddot{\boldsymbol{u}}$ is the double derivative of $\boldsymbol{u}$ with respect to time.

If the body forces $\boldsymbol{f}$ are null, the divergence and curl of Equation (1), yield the dilatation (Equation (2)) and distorsional (Equation (3)) wave equations [1]:

$$\nabla^2\Delta = \frac{1}{c_p^2}\frac{\partial^2\Delta}{\partial t^2} \tag{2}$$

$$\nabla^2\omega = \frac{1}{c_s^2}\frac{\partial^2\omega}{\partial t^2}, \tag{3}$$

where $\Delta = \frac{\partial u}{\partial x} + \frac{\partial v}{\partial y} + \frac{\partial w}{\partial z}$, $\boldsymbol{u} = \nabla \times \boldsymbol{u}$ and c_1, c_2 are respectively [1]:

$$c_p = \left(\frac{\lambda + 2\mu}{\rho}\right)^{0.5} \tag{4}$$

$$c_s = \left(\frac{\mu}{\rho}\right)^{0.5}. \tag{5}$$

Volumetric waves, involving no rotation, and rotational waves, involving no volume changes, propagate in an isotropic elastic medium with two distinct velocities, c_p and c_s, respectively. Poisson, Kirchhoff, Love and Rayleigh contributed significantly to the solution of Equation (1), by solving the initial value and boundary value problems for given body forces.

2.2. Waves in an Isotropic Elastic Plane

Dilatational and distortional waves do propagate in an infinite elastic medium. A third solution of Equation (1) arises in an elastic space. This section restraints Equation (1) to the

x-z plane, where the x axis is parallel to the free plane surface and z explores the depth of the plane. The third solution relates to the coupling between longitudinal and transverse waves. Therefore, Equation (1) restricted to the x-z plane yields the two following equations in terms of the Φ and Ψ potential functions:

$$\nabla^2\Phi = \left(\frac{\lambda + 2\mu}{\rho}\right)\frac{\partial^2\Phi}{\partial t^2} \tag{6}$$

$$\nabla^2\Psi = \left(\frac{\mu}{\rho}\right)\frac{\partial^2\Psi}{\partial t^2}, \tag{7}$$

where

$$u = \frac{\partial\Phi}{\partial x} + \frac{\partial\Psi}{\partial z} \tag{8}$$

$$w = \frac{\partial\Phi}{\partial z} - \frac{\partial\Psi}{\partial x}. \tag{9}$$

It is assumed that two waves with pulsation ω and wavenumber ν propagates in the z direction, with amplitude F and G dependent on the distance x:

$$\Phi = F(x)\exp[i(\nu z - \omega t)] \tag{10}$$

$$\Psi = G(x)\exp[i(\nu z - \omega t).] \tag{11}$$

The two waves, which express the coupling between the longitudinal and transverse motion are a class of the Lamb waves. The substitution of Equations (10) and (11) in Equations (6) and (7) returns two ordinary differential equations in terms of the two coefficients F and G:

$$F''(x) - qF(x) = 0 \tag{12}$$

$$G''(x) - sG(x) = 0 \tag{13}$$

where q and s are:

$$q = \left(\nu^2 + \frac{\omega^2}{c_p^2}\right) \tag{14}$$

$$s = \left(\nu^2 + \frac{\omega^2}{c_s^2}\right). \tag{15}$$

The solutions of Equations (11) and (12) are

$$F(x) = A_1\exp(-qx) + A_2\exp(qx) \tag{16}$$

$$G(x) = B_1\exp(-sx) + B_2\exp(sx). \tag{17}$$

Since the solution cannot reach infinity when the depth tends to zero, the authors consider the sole first terms:

$$F(x) = A_1\exp(-qx) \tag{18}$$

$$G(x) = B_1\exp(-sx). \tag{19}$$

A_1 and B_1 originate from the solution of the boundary value problem. The current investigation focuses on the wave propagation along the z axis corresponding to the position of the excitation source, when $x = 0$. The dispersion law of the coupled u-w wave originates by subtracting Equation (14) to Equation (15):

$$c^2 = \left(\frac{c_p^2 c_s^2}{c_s^2 - c_p^2}\right)\frac{q - s}{\nu^2}. \tag{20}$$

Thus, c reads

$$c = \left(c_p c_s \sqrt{\frac{q - s}{c_s^2 - c_p^2} \frac{1}{\nu}} \right) = \frac{A}{\nu}. \tag{21}$$

2.3. Empirical Estimate of the Dispersion Law from Acquired Signals

A conventional procedure to estimate the velocity of the shear waves using the SDMT originates from the cross-correlation between two time-series recorded by the two spaced sensors.

The authors estimate the occurrence of dispersion phenomena using a frequency-domain method based on the repeated filtering of the acquired signals. The use of filter-banks is an acknowledged procedure to estimate the frequency properties of structural systems. There are numerous applications of these techniques on structural systems different to soils [41]. The main idea is that the scholar isolates the harmonic contributions to the experimental response by using a sequence of narrow-banded filters, see [42,43]. The authors estimate the dependence of the velocity on the wavenumber: a rectangular band-pass filter with a certain width δF spans a given frequency domain. The lower bound of the frequency domain arises from the natural frequency of the geophones: the lower frequency corresponds to the frequency characterized by acceptable linearity of the characteristic frequency response function of the geophone. The upper bound originates from the inspection of the frequency spectrum of the acquired signals: the authors arbitrarily chose a frequency value associated with a significant frequency content. The dispersion curve stems from the relation between the phase velocity and the central natural frequency of the rectangular filtering window. Accurately, the dispersion curve is the relation between the phase velocity and the wavenumber. The wavenumber derives from the frequency using the following relation:

$$c = c(\nu) \quad \nu = \frac{2\pi f}{c}, \tag{22}$$

where c is the phase velocity, ν is the wavenumber, and f is the central frequency of the rectangular-like band filter.

3. Experimental Tests

In March–April 2019, the authors carried out an experimental investigation in Macerata (Italy). The details of the experimental campaigns are detailed in [39]. This paper focuses on the data acquired in a specific borehole 100 m deep using the seismic dilatometer in Figure 1. The SDMT is the combination of the mechanical flat dilatometer (DMT), introduced by Marchetti [44–48], with an add-on seismic module used to measure the V_s, first introduced by Hepton [49]. The DMT procedure follows the recommendations contained in the documents ASTM [50], Eurocode 7 [51] and ISO [51]. The SDMT is a seismic module for recording seismic waves in the soil to evaluate shear wave velocity V_S. The device is equipped with two geophone receivers with a 0.50 m vertical offset. The instrument may be combined with a DMT blade, a dummy cone or a CPT probe. The S wave is usually generated with a hammer striking in the horizontal direction. There are no international standards for the execution of SDMT tests. Still, there are several research papers that show a possible procedure for the reliable estimation of the shear wave velocities [52–59]. Specifically, according to [52], the ratio between the difference of the distances source-receivers ($S_2 - S_1$) and the delay between the two-time histories (Δt) is an estimate of the shear wave velocity [49]:

$$V_s = \frac{S_2 - S_1}{\Delta t}. \tag{23}$$

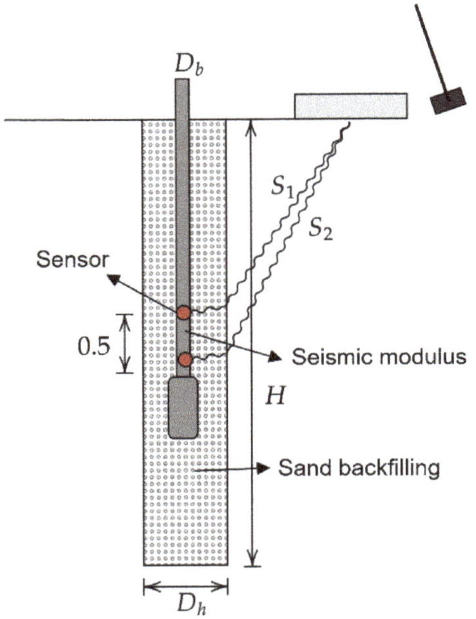

(**a**)

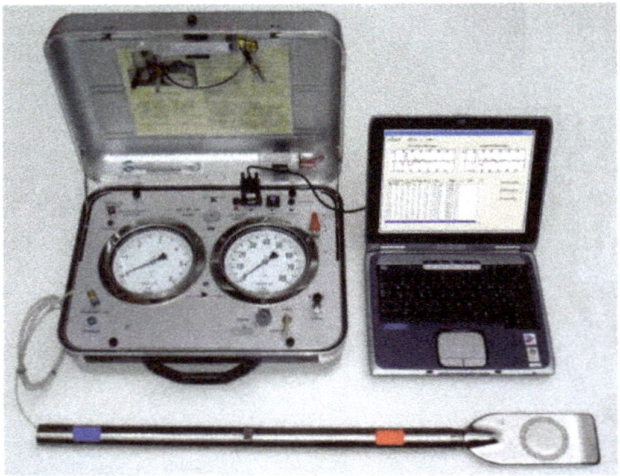

(**b**)

Figure 1. (**a**) Description of the SDMT procedure in non-penetrable soils, (**b**) SDMT equipment.

The delay time could be estimated from the cross-correlation between the time-series acquired by two spaced geophones. Since SDMT barely penetrates very hard soils, Totani et al. [19] developed an alternative procedure leading to the V_s estimates in non-penetrable soils based on the following process: drilling of a borehole to the required depth; insertion of the SDMT to the maximum depth; backfilling of the borehole with sand; backward penetration of the SDMT along the backfilled borehole and measurements of the shear waves in a 0.50 m step pattern. in this investigation, the experimenters followed the above procedure for non-penetrable soils.

4. Results

This section elaborates the signals acquired by the geophones at increasing depths. Figure 2a depicts sample signals pre-processing. After an initial high-frequency part, lower frequency waves appear from the extremity of the time-series. The two signals' direct examination proves that the impulsive input propagates with a visible distortion: the higher frequency waves propagate with a velocity different from, the lower frequency ones. The dispersion curve may display lowering velocities corresponding to increasing wavenumbers. The single side Fourier spectrum of the two signals in Figure 2a declares that the frequency content decreases as the frequency rises. It suggests that high-frequency waves carry higher energy content and energy decay in a shorter distance than low-frequency ones. The Fourier spectrum of each series is peculiar, like the sampled one, which exhibits an unexpected frequency content in the range 500–800 Hz. In general, the amplitude of the harmonic waves with frequency higher than 500/1000 Hz could be considered negligible, as confirmed by the phase spectrum of the same data and Figure 2d, which shows the difference in phase between the two signals.

Consequently, the authors decided to estimate the dispersion curve up to 100 Hz: at a 100 m depth, in the considered range, the authors are confident in the consistency of the results.

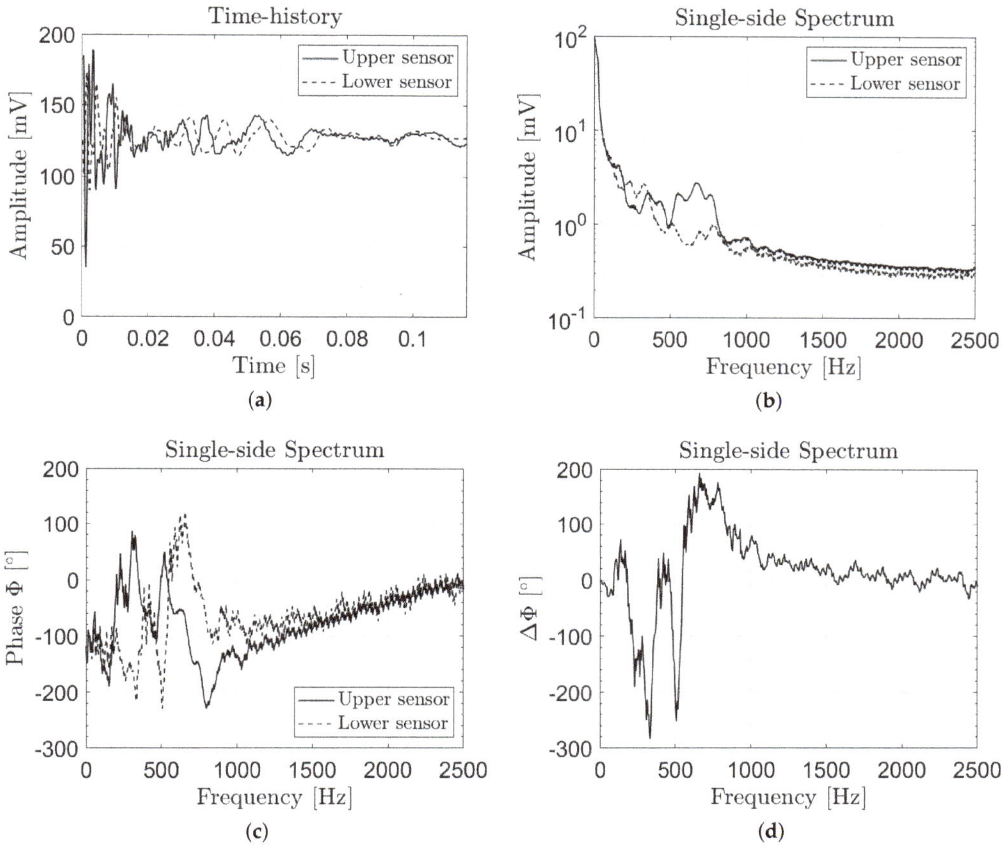

Figure 2. (**a**) Time history, (**b**) Amplitude, phase (**c**,**d**) difference of phase of two recorded signal corresponding to a 10 m depth.

The V_s descends from the re-phasing of the two signals. The re-phasing corresponds to a mutual shift equal to the time lag associated with the maximum value of the cross-correlation, as represented by Figure 3. The cross-correlation can be very noisy (Figure 3), and many competing peaks may stand alongside the highest one. This evidence further proves that the waves do propagate with distortion, thus causing a possible ambiguity in selecting a unique time lag, valid for all wavenumbers.

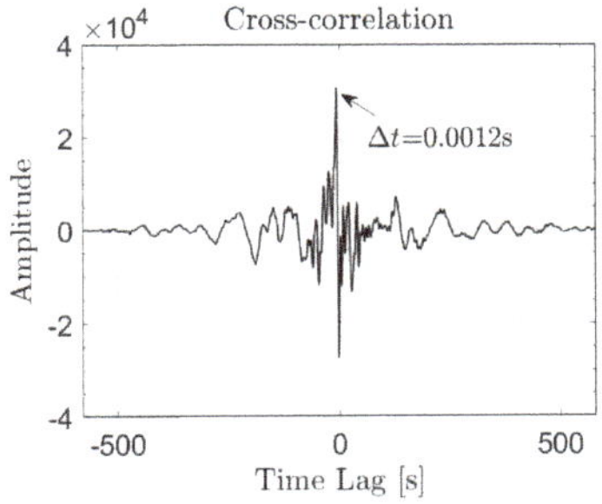

Figure 3. Cross-correlation between the two recorded signal corresponding to a 10 m depth.

The filtering of the two signals in Figure 4 with a bandpass ranging between 20 and 22 Hz shows an improvement of the results: the cross-correlation exhibits a single peak in a time lag shorter than the period of the lowest harmonic wave of the bandpass.

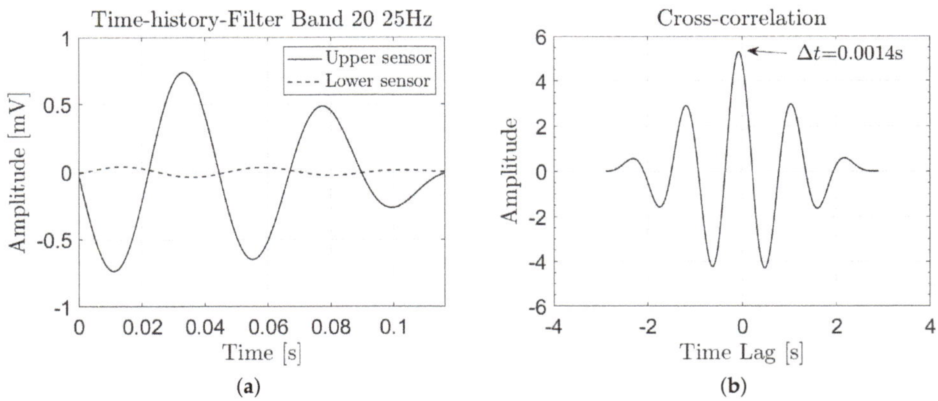

Figure 4. (**a**) Time history of the filtered signals, (**b**) Cross-correlation between the two filtered signals.

Figure 5 shows the final results: the vertical profile of the shear waves velocities determined according to the standard method described in the above paragraphs. The red dots correspond to the values, which yield the highest correlation. The black dots correspond to the data characterized by a lower correlation. This picture further enlightens the practical difficulties faced during the experimental tests: the authors repeated the tests as many times as needed to obtain a satisfactory quality of the cross-correlation.

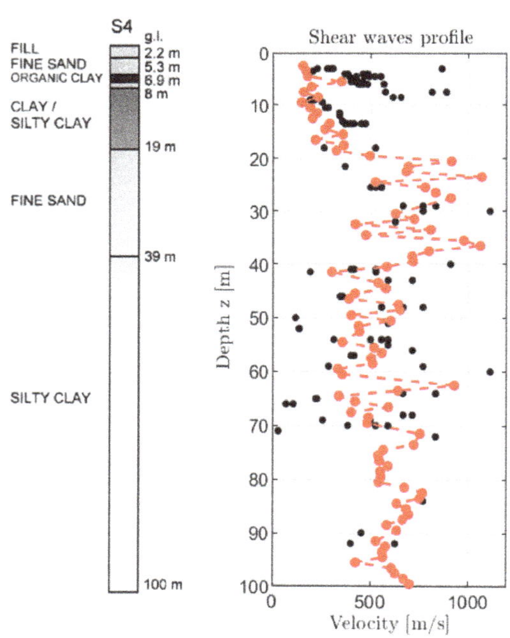

Figure 5. Profile of the shear waves: the red dots correspond to the time series which yield the maximum correlation, the black dots derive from cross-correlations with lower values.

Additionally, Figure 5 displays the schematic soil stratigraphy obtained from the S4 borehole to a 100 m depth.

Likely, the V_s generally grows with depth, showing higher values in correspondence of fine sand or sandstone layers, where values of $V_s > 800$ m/s may be encountered. The V_s obtained by SDMT is about 200–300 m/s in the upper ≈20 m, increase to about 700–800 m/s between ≈20 m and 40 m. Then, it reduces to about 400 m/s and below 40 m increases almost linearly with depth to about 600–700 m/s up to 100 m.

The Effect of Dispersion: Results and Discussion

Figure 6 illustrates the results obtained from the procedure described in Section 2. The sequence of band-pass filters used for the analyses has the following characteristics: bandwidth 5 Hz with 1 Hz overlapping. There is a marked dependency of the velocity of the shear waves on the wavenumber. As the wavenumber lowers, the velocity blows up; Conversely, as the wavenumber increases the velocity lowers and likely tends to zero or a horizontal asymptote. The results may grow in inaccuracy as the wavenumber increases. Still, in the considered range, the results can be considered reliable: the amplitude of the harmonics up to 100 Hz is considerable, as proved by the spectra in Figure 2. The four different colours identify the sequence of the four sections of the investigated vertical: 0–20 m, 20–40 m, 40–60 m, 60–100 m. Four hyperbolae, described by the following equation, manage to fit the experimental points adequately:

$$c = \frac{A}{\nu}, \tag{24}$$

where A is a constant.

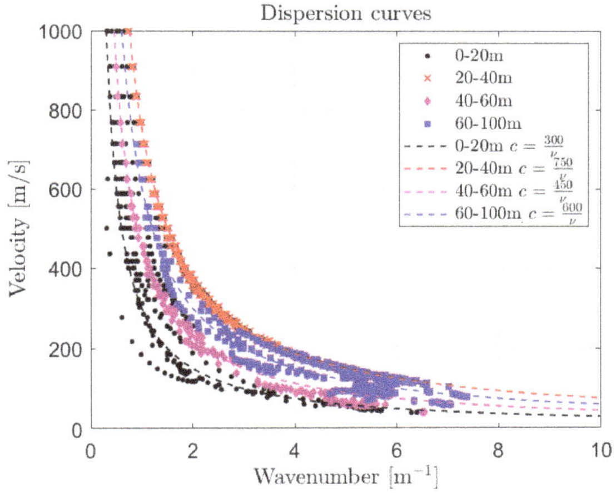

Figure 6. Dispersion curves in the considered four sections of the investigated vertical.

In contrast with the significant scatter of the results obtained from the standard procedure in Figure 5, the dispersion curves present a minimum scatter. The experimental dots gather very close to the fitting hyperbola. The results in Figure 6 are free of ambiguities: the acquired pulses are characterized by a notable dispersion which follows Equation (24). Pure shear waves do not manifest such behavior: they are nondispersive waves. Therefore two possible interpretations may follow.

- A first possibility is that the acquired pulses do not correspond to pure shear waves, but they exhibit coupling between the longitudinal and transverse motion mostly explicated in the x-z plane. The directional nature of the input may force the soil grans to move in the x-z plane, while they do not manifest a significant displacement in the y direction. These waves belong to the class of Lamb waves, characterized by a dispersion curve derived in the second section in Equation (21). Therefore, the leading cause of dispersion may stand in the excitation, which forces the particles to move in a plane, as occurs in Lamb waves which can propagate in plates and spheres.

- It is also plausible that the experimenters measure almost pure shear waves and that the primary source of dispersion stands in the granular nature of the medium. Granular media may exhibit hyperbolic-like dispersive curves [60,61]. Therefore, the constitutive nature of the soil, rather than the boundary conditions may determine the observed behavior. There are copious theoretical and numerical researches on the dispersion of continuum models representative of random granular assemblies [3–9]. These studies mostly deal with high-order deformation gradients, the constitutive relations descend from the Cosserat theory and the grains interact through Hertz-Mindlin contacts [60,61].

However, can a non-penetrable soil, like the one investigated in this research, be considered a granular medium? From a theoretical viewpoint, soils are micro granular media. However, the effect of granularity is scale-dependent. The coarser grains present in the boreholes lay in the rage of sands, between 2 and 0.063 mm. The wavenumber reached in the analyses do not exceed 8, that is, the considered wavelength does not exceed 0.125 m. The ratio between the lowest wavelength and the largest grain size would higher than 60. Still, these sand samples are characterized by tiny grains, approaching the dimensions of the silts. Therefore, the ratio between the considered lowest wavelength and the dimensions of the grains exceeds the thousands. In this perspective, the standard approximation, which assumes soils as isotropic elastic media may be consistent. Besides, the tested soils are classified as non-penetrable, due to the difficulty in penetrating them with an instrumented device at a controlled rate (e.g., Cone Penetration Test). The grains in fine non-penetrable soils are characterized by a certain amount of cementation, which makes them more similar to rocks than soils. Therefore, in this circumstance, the authors believe that the first hypothesis may be the most likely: the directional hammer may generate waves similar to Lamb waves. Besides, the dispersion curve is in excellent accordance with the model in Section 2.

Theoretically, damping phenomena may produce dispersion effects on the propagating shear waves. However, the damping causes dispersion phenomena different from those observed in this investigation, as illustrated in the recent work by [62].

Interestingly, A, the coefficient of the fitting hyperbola, approximately corresponds to the velocity of the shear waves estimated from the standard practice, without band-pass filtering on the entire series. The legend of Figure 6 details the four expressions of the fitting hyperbola. The experimental data collected from the S4 borehole comprise measurements in the transverse direction. Acquisitions in the longitudinal one might have supported the unequivocal assessment of the nature of the measured signals. It is not comfortable with the available information to study the correspondence between the terms appearing in Equation (21) and the A coefficient. This analysis would have supported the understanding of the physical nature of A. The authors will attempt to carry out experimental tests, where responses in the x,y and z directions would be possibly collected, as partially achieved by [63].

Figure 7 illustrates the maximum amplitude of the response as a function of the wavenumber and the depth. Accurately, Figure 7 presents the tridimensional plot, where the gradient colours express the varying wavenumbers. The comprehension of the tridimensional figure is challenging: Figure 7b show the projection of the dots on two planes.

Figure 7b reveals that the soil behaves like a filter which does not admit wavenumbers higher than five. There is almost a discontinuity, between the number of dots in the range 0–5 and those exceeding 5. Besides, the dots exceeding 5 refer to the first meters (see Figure 7a). It is likely that wavenumbers exceeding five do not propagate as a consequence of the microstructure of the soil.

Figure 7c bestows a piece of different information: except for the first 40 m where there are harmonics with high amplitude, there is not an evident curve which shows the decaying of the amplitude with depth (like the exponential decay in surface waves). This aspect may support the fact that the measured signals may be representatives of Lamb waves. The solution of the governing equations yielded an exponential decay along the x-axis. Interestingly, contemporary measures of the waves on the free surface would have granted a better comprehension.

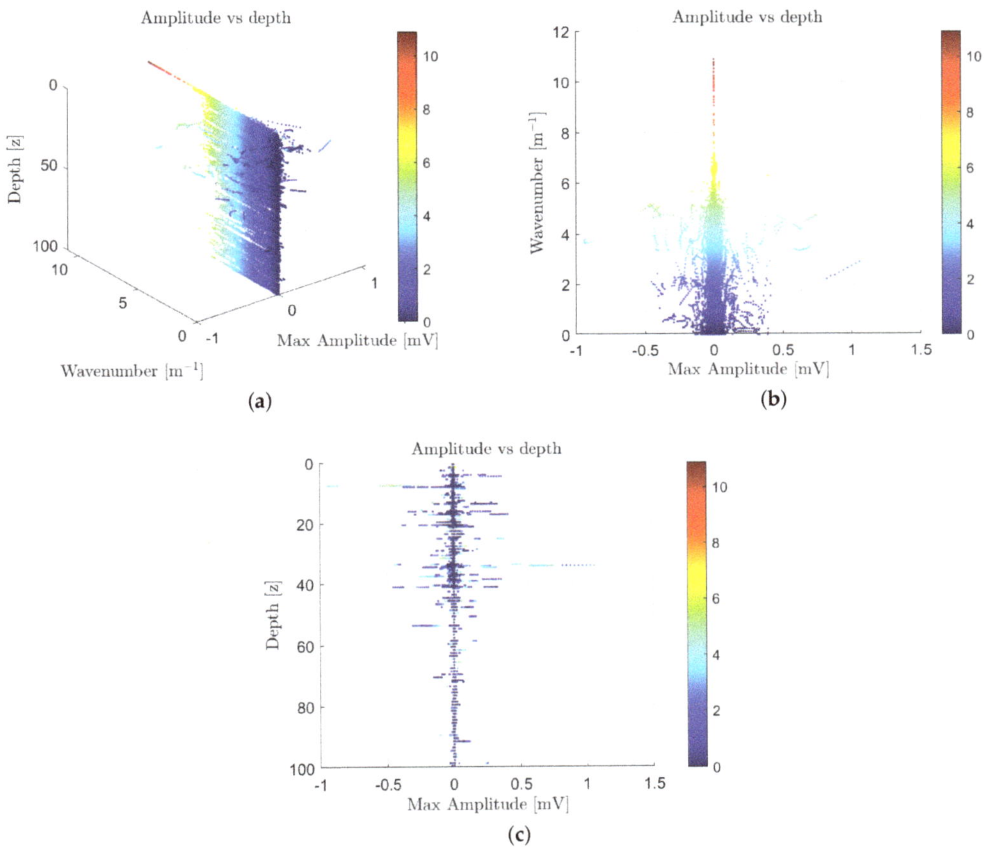

Figure 7. 3-D dispersion curves where the phase velocity is function of the signal amplitude in [mV], the wavenumber [m^{-1}] and the depth [m]. The three images (**a**–**c**) are different views of the same scatter plot.

5. Conclusions

The authors carried out the geotechnical characterization of a site in Macerata (Italy). They investigated a 100 m deep borehole within an extended experimental campaign using the Seismic Flat Dilatometer (SDMT). The conventional estimation of the shear wave velocities using SDMT originates from the cross-correlation between the recordings of two spaced geophones at increasing depth triggered by a hammer's strike on the surface. The authors attempted to estimate the dependence of the shear wave velocity on the wavenumber of the propagating wave by highlighting the dispersive soil features. They proposed a procedure based on the use of narrow-banded filter banks applied to the acquired signal. The dispersion curve stems from the relation between the phase velocity and the central natural frequency of the rectangular filtering window. The experimental data representative of the transverse propagating waves revealed characteristic dispersion curves typical of a class of Lamb waves. Likely, the seismic dilatometer measurements do not refer, as generally acknowledged in the field of geotechnics, to pure shear waves, but they may be more properly considered as Lamb waves. Therefore, the standard practice based on the estimation of the shear wave velocity from the cross-correlation of the two signals acquired by two spaced geophones may be inaccurate. Further analyses in the frequency domain by processing the data with moving filters may enrich the conventional practice in SDMT by estimating the dispersion curves of the propagating waves. Interestingly, the shear waves' conventional estimates correspond to the phase velocities with a wavelength close to 1 m. Except for the first 40 m where there are harmonics with high amplitude, there is no evident curve that shows the decaying of the amplitude with depth (like the exponential decay in surface waves). This aspect may support the fact that the measured signals may be representatives of Lamb waves. Future research efforts will focus on estimating the dispersion curves of penetrable soils characterized by more uniform mechanical properties. The analysis of the simultaneous recordings on the free surface and inside the borehole will possibly shed light on the nature of the waves generated by an impulsive shear excitation.

Author Contributions: A.A.: Conceptualization, Methodology, Software, Data curation, Experimental tests, Writing and Editing. F.T.: Experimental tests, Supervision. R.A.: Conceptualization, Supervision. G.T.: Experimental tests, Supervision. All authors have read and agreed to the published version of the manuscript.

Funding: This research received no external funding.

Institutional Review Board Statement: Not applicable.

Informed Consent Statement: Not applicable.

Data Availability Statement: Experimental data are available from the corresponding author upon reasonable request.

Acknowledgments: The authors acknowledge the significant support given by Paola Monaco and Giovanni Bosco, the contribution of Diego Marchetti, who provided the experimenters with the SDMT modulus and data acquisition system, and Massimo Bistocchi who executed the experimental tests with the help of his assistants.

Conflicts of Interest: The authors declare no conflict of interest.

References

1. Graff, K.F. *Wave Motion in Elastic Solids*; Courier Corporation: Chelmsford, MA, USA, 2012.
2. Worden, K. Rayleigh and Lamb Waves-Basic Principles. *Strain* **2001**, *37*, 167–172. [CrossRef]
3. Courant, R.; Hilbert, D. *Methoden der Mathematischen Physik*; Springer: Berlin/Heidelberg, Germany, 2013.
4. Digby, P. The effective elastic moduli of porous granular rocks. *J. Appl. Mech.* **1981**, *48*, 803–808 [CrossRef]
5. Muhlhaus, H.B.; De Borst, R.; Aifantis, E. Constitutive models and numerical analyses for ineslastic materials with microstructure. In Proceedings of the 7th International Conference on Computer Methods and Advances in Geomechanics, Cairns, Australia, 6–10 May 1991; pp. 377–385.
6. Vardoulakis, I.; Aifantis, E. Gradient dependent dilatancy and its implications in shear banding and liquefaction. *Ing. Arch.* **1989**, *59*, 197–208. [CrossRef]
7. Walton, K. The effective elastic moduli of a random packing of spheres. *J. Mech. Phys. Solids* **1987**, *35*, 213–226. [CrossRef]
8. Jenkins, J.T. Volume change in small strain axisymmetric deformations of a granular material. In *Studies in Applied Mechanics*; Elsevier: Amsterdam, The Netherlands, 1988; Volume 20, pp. 245–252.
9. Jenkins, J.T. Anisotropic elasticity for random arrays of identical spheres. In *Modern Theory of Anisotropic Elasticity and Applications*; SIAM: Philadelphia, PA, USA, 1991; pp. 368–377.
10. Lambe, T.W.; Whitman, R.V. *Soil Mechanics*; John Wiley & Sons: Hoboken, NJ, USA, 1991; Volume 10.
11. Borcherdt, R.D. Estimates of site-dependent response spectra for design (methodology and justification). *Earthq. Spectra* **1994**, *10*, 617–653. [CrossRef]
12. Castellaro, S.; Mulargia, F.; Rossi, P.L. VS30: Proxy for seismic amplification? *Seismol. Res. Lett.* **2008**, *79*, 540–543. [CrossRef]
13. Campbell, K.W. Empirical near-source attenuation relationships for horizontal and vertical components of peak ground acceleration, peak ground velocity, and pseudo-absolute acceleration response spectra. *Seismol. Res. Lett.* **1997**, *68*, 154–179. [CrossRef]
14. Wills, C.; Petersen, M.; Bryant, W.; Reichle, M.; Saucedo, G.; Tan, S.; Taylor, G.; Treiman, J. A site-conditions map for California based on geology and shear-wave velocity. *Bull. Seismol. Soc. Am.* **2000**, *90*, S187–S208. [CrossRef]

15. Allen, T.I.; Wald, D.J. On the use of high-resolution topographic data as a proxy for seismic site conditions (VS 30). *Bull. Seismol. Soc. Am.* **2009**, *99*, 935–943. [CrossRef]
16. Tokeshi, K.; Harutoonian, P.; Leo, C.J.; Liyanapathirana, S. Use of surface waves for geotechnical engineering applications in Western Sydney. *Adv. Geosci.* **2013**, *35*, 37. [CrossRef]
17. Karl, L.; Fechner, T.; Schevenels, M.; François, S.; Degrande, G. Geotechnical characterization of a river dyke by surface waves. *Near Surf. Geophys.* **2011**, *9*, 515–527. [CrossRef]
18. Long, M.; Trafford, A.; McGrath, T.; O'Connor, P. Multichannel analysis of surface waves (MASW) for offshore geotechnical investigations. *Eng. Geol.* **2020**, *272*, 105649. [CrossRef]
19. Totani, G.; Monaco, P.; Marchetti, S.; Marchetti, D. VS measurements by seismic dilatometer (SDMT) in non-penetrable soils. In *Proceedings of the 17th International Conference on Soil Mechanics and Geotechnical Engineering*; IOS Press: Amsterdam, The Netherlands, 2009; Volume 2, pp. 977–980.
20. Auersch, L. Wave propagation in layered soils: theoretical solution in wavenumber domain and experimental results of hammer and railway traffic excitation. *J. Sound Vib.* **1994**, *173*, 233–264. [CrossRef]
21. Lai, C.G.; Wilmanski, K. *Surface Waves in Geomechanics: Direct and Inverse Modelling for Soils and Rocks*; Springer Science & Business Media: Berlin/Heidelberg, Germany 2007; Volume 481.
22. Rix, G.J. Near-surface site characterization using surface waves. In *Surface Waves in Geomechanics: Direct and Inverse Modelling for Soils and Rocks*; Springer: Berlin/Heidelberg, Germany, 2004; pp. 1–46.
23. Song, X.; Tang, L.; Lv, X.; Fang, H.; Gu, H. Application of particle swarm optimization to interpret Rayleigh wave dispersion curves. *J. Appl. Geophys.* **2012**, *84*, 1–13. [CrossRef]
24. Pan, Y.; Schaneng, S.; Steinweg, T.; Bohlen, T. Estimating S-wave velocities from 3D 9-component shallow seismic data using local Rayleigh-wave dispersion curves—A field study. *J. Appl. Geophys.* **2018**, *159*, 532–539. [CrossRef]
25. Tokimatsu, K.; Tamura, S.; Kojima, H. Effects of multiple modes on Rayleigh wave dispersion characteristics. *J. Geotech. Eng.* **1992**, *118*, 1529–1543. [CrossRef]
26. Athanasopoulos, G.; Pelekis, P.; Anagnostopoulos, G. Effect of soil stiffness in the attenuation of Rayleigh-wave motions from field measurements. *Soil Dyn. Earthq. Eng.* **2000**, *19*, 277–288. [CrossRef]
27. Addo, K.; Robertson, P. Shear-wave velocity measurement of soils using Rayleigh waves. *Can. Geotech. J.* **1992**, *29*, 558–568. [CrossRef]
28. IA, V. *Rayleigh and Lamb Waves, Physical Theory and Applications*; Acoustics Institute, Academy of Science of the USSR: Moscow, Russia, 1967.
29. Lamb, H. On the vibrations of an elastic sphere. *Proc. Lond. Math. Soc.* **1881**, *1*, 189–212. [CrossRef]
30. Lamb, H. On waves in an elastic plate. *Proc. R. Soc. Lond.* **1917**, *93*, 114–128.
31. Pilarski, A.; Ditri, J.J.; Rose, J.L. Remarks on symmetric Lamb waves with dominant longitudinal displacements. *J. Acoust. Soc. Am.* **1993**, *93*, 2228–2230. [CrossRef]
32. Lee, B.; Staszewski, W. Modelling of Lamb waves for damage detection in metallic structures: Part I. Wave propagation. *Smart Mater. Struct.* **2003**, *12*, 804. [CrossRef]
33. Mallet, L.; Lee, B.; Staszewski, W.; Scarpa, F. Structural health monitoring using scanning laser vibrometry: II. Lamb waves for damage detection. *Smart Mater. Struct.* **2004**, *13*, 261. [CrossRef]
34. Ng, C.T.; Veidt, M. A Lamb-wave-based technique for damage detection in composite laminates. *Smart Mater. Struct.* **2009**, *18*, 074006. [CrossRef]
35. Hayashi, T.; Inoue, D. Calculation of leaky Lamb waves with a semi-analytical finite element method. *Ultrasonics* **2014**, *54*, 1460–1469. [CrossRef] [PubMed]
36. Su, Z.; Ye, L.; Lu, Y. Guided Lamb waves for identification of damage in composite structures: A review. *J. Sound Vib.* **2006**, *295*, 753–780. [CrossRef]
37. Foti, S. *Multistation Methods for Geotechnical Characterization Using Surface Waves*; Politecnico di Torino: Turin, Italy, 2000.
38. Block, L.; Cheng, C.; Fehler, M.; Phillips, W. Structural and geotechnical mapping. *Geology* **1992**, *3*, 97–101.
39. Aloisio, A.; Totani, F.; Totani, G. Experimental dispersion curves of non-penetrable soils from direct dynamic measurements using the seismic dilatometer (SDMT). *Soil Dyn. Earthq. Eng.* **2021**, *143*, 106616. [CrossRef]
40. Rydén, N.; Park, C. Surface waves in inversely dispersive media. *Near Surf. Geophys.* **2004**, *2*, 187–197. [CrossRef]
41. Quqa, S.; Landi, L.; Diotallevi, P.P. Modal assurance distribution of multivariate signals for modal identification of time-varying dynamic systems. *Mech. Syst. Signal Process.* **2021**, *148*, 107136. [CrossRef]
42. Quqa, S.; Landi, L.; Diotallevi, P.P. Seismic structural health monitoring using the modal assurance distribution. *Earthq. Eng. Struct. Dyn.* **2021**. [CrossRef]
43. Ditommaso, R.; Mucciarelli, M.; Ponzo, F.C. Analysis of non-stationary structural systems by using a band-variable filter. *Bull. Earthq. Eng.* **2012**, *10*, 895–911. [CrossRef]
44. Marchetti, S. In situ tests by flat dilatometer. *J. Geotech. Geoenviron. Eng.* **1980**, *106*, 299–321.
45. Lechowicz, Z.; Fukue, M.; Rabarijoely, S.; Sulewska, M.J. Evaluation of the undrained shear strength of organic soils from a dilatometer test using artificial neural networks. *Appl. Sci.* **2018**, *8*, 1395. [CrossRef]

46. Ferreira, C.; da Fonseca, A.V.; Ramos, C.; Saldanha, A.S.; Amoroso, S.; Rodrigues, C. Comparative analysis of liquefaction susceptibility assessment methods based on the investigation on a pilot site in the greater Lisbon area. *Bull. Earthq. Eng.* **2020**, *18*, 109–138. [CrossRef]
47. Rodrigues, C.; Cruz, N.; Amoroso, S.; Cruz, M. Stiffness Decay in Structured Soils by Seismic Dilatometer. *Geotech. Test. J.* **2020**, *43*, 1003–1021. [CrossRef]
48. Pavithra, E. Dilatometric Methods: Insights of other researchers. In *Proceedings of the IOP Conference Series: Materials Science and Engineering*; IOP Publishing: Bristol, UK, 2020; Volume 923, p. 012041.
49. Hepton, P. Shear wave velocity measurements during penetration testing. In *Penetration Testing in the UK, Proceedings of the Geotechnology Conference Organized by the Institution of Civil Engineers, Birmingham, UK, 6–8 July 1988*; Thomas Telford Publishing: London, UK, 1989; pp. 275–278.
50. ASTM. *Standard Test Method for Performing the Flat Plate Dilatometer*; ASTM: West Conshohocken, PA, USA, 2015.
51. European Committee for Standardization. *Eurocode 7. EN 1997-2:2007 2007: Geotechnical Design—Part 2: Ground Investigation and Testing*; Brussels, Belgium, 2007.
52. Marchetti, S.; Monaco, P.; Totani, G.; Marchetti, D. In situ tests by seismic dilatometer (SDMT). In *From Research to Practice in Geotechnical Engineering, Proceedings of the Symposium Honoring Dr. John H. Schmertmann for His Contributions to Civil Engineering at Research to Practice in Geotechnical Engineering Congress 2008, New Orleans, LA, USA, 9–12 March 2008*; ASCE: Reston, VA, USA, 2008; pp. 292–311.
53. Foti, S.; Lancellotta, R.; Marchetti, D.; Monaco, P.; Totani, G. Interpretation of SDMT tests in a transversely isotropic medium. In Proceedings of the 2nd International Conference on the Flat Dilatometer, Washington, DC, USA, 2–5 April 2006; pp. 2–5.
54. Cavallaro, A.; Grasso, S.; Maugeri, M. Clay soil characterization by the new seismic dilatometer Marchetti test (SDMT). In Proceedings of the 2nd International Conference on the Flat Dilatometer, Washington, DC, USA, 2–5 April 2006; pp. 261–268.
55. Bihs, A.; Long, M.; Marchetti, D.; Ward, D. Interpretation of CPTU and SDMT in organic, Irish soils. In Proceedings of the 2nd International Symposium on Cone Penetration Testing, Huntington Beach, CA, USA. 9–11 May 2010; Volume 2, pp. 257–264.
56. Grasso, S.; Maugeri, M. The Seismic Dilatometer Marchetti Test (SDMT) for evaluating liquefaction potential under cyclic loading. In Proceedings of the Geotechnical Earthquake Engineering and Soil Dynamics Congress IV, Sacramento, CA, USA, 18–22 May 2008; pp. 1–15.
57. Marchetti, S.; Monaco, P. Recent Improvements in the Use, Interpretation, and Applications of DMT and SDMT in Practice. *Geotech. Test. J.* **2018**, *41*, 837–850. [CrossRef]
58. Amoroso, S.; Monaco, P.; Marchetti, D. Use of the Seismic Dilatometer (SDMT) to estimate in situ G-γ decay curves in various soil types. In *Geotechnical and Geophysical Site Characterization, Proceedings of the 4th International Conference on Site Characterization ISC-4, Pernambuco, Brazil, 18–21 September 2012*; Taylor & Francis Books Ltd.: Abingdon, UK, 2013, Volume 1, pp. 489–497.
59. Cavallaro, A.; Capilleri, P.P.; Grasso, S. Site characterization by dynamic in situ and laboratory tests for liquefaction potential evaluation during Emilia Romagna earthquake. *Geosciences* **2018**, *8*, 242. [CrossRef]
60. Bachrach, R.; Avseth, P. Rock physics modeling of unconsolidated sands: Accounting for nonuniform contacts and heterogeneous stress fields in the effective media approximation with applications to hydrocarbon exploration. *Geophysics* **2008**, *73*, E197–E209. [CrossRef]
61. Pasternak, E.; Mühlhaus, H.B. Generalised homogenisation procedures for granular materials. In *Mathematics and Mechanics of Granular Materials*; Springer: Berlin/Heidelberg, Germany, 2005; pp. 199–229.
62. Koedel, U.; Karl, L. Determination of the damping ratio by multi-channel spectral analysis of seismic downhole data. *Soil Dyn. Earthq. Eng.* **2020**, *136*, 106235. [CrossRef]
63. Amoroso, S.; Comina, C.; Marchetti, D. Combined P-and S-Wave Measurements by Seismic Dilatometer Test (SPDMT): A Case History in Bondeno (Emilia Romagna, Italy). *Geotech. Test. J.* **2020**, *43*, 383–393. [CrossRef]

geosciences

MDPI

Review

Biopolymers as Green Binders for Soil Improvement in Geotechnical Applications: A Review

Hadi Fatehi [1,2,*], Dominic E. L. Ong [1,2], Jimmy Yu [1] and Ilhan Chang [3]

1 School of Engineering and Built Environment, Griffith University, Nathan, QLD 4111, Australia; d.ong@griffith.edu.au (D.E.L.O.); jimmy.yu@griffith.edu.au (J.Y.)
2 Cities Research Institute, Griffith University, Nathan, QLD 4111, Australia
3 Department of Civil System Engineering, Ajou University, Suwon-si 16499, Korea; ilhanchang@ajou.ac.kr
* Correspondence: hadi.fatehigelab@griffithuni.edu.au

Abstract: Soil improvement using biopolymers has attracted considerable attention in recent years, with the aim to reduce the harmful environmental effects of traditional materials, such as cement. This paper aims to provide a review on the environmental assessment of using biopolymers as binders in soil improvement, biopolymer-treated soil characteristics, as well as the most important factors affecting the behavior of the treated soil. In more detail, environmental benefits and concerns about the use of biopolymers in soil improvement as well as biopolymer–soil interaction are discussed. Various geotechnical properties are evaluated and compared, including the unconfined compressive strength, shear strength, erosion resistance, physical properties, and durability of biopolymer-treated soils. The influential factors and soil and environmental conditions affecting various geotechnical characteristics of biopolymer-treated soils are also discussed. These factors include biopolymer concentration in the biopolymer–soil mixture, moisture condition, temperature, and dehydration time. Potential opportunities for biopolymers in geotechnical engineering and the challenges are also presented.

Keywords: biopolymer-treated soils; soil improvement; geotechnical properties; biopolymer soil interaction

Citation: Fatehi, H.; Ong, D.E.L.; Yu, J.; Chang, I. Biopolymers as Green Binders for Soil Improvement in Geotechnical Applications: A Review. *Geosciences* **2021**, *11*, 291. https://doi.org/10.3390/geosciences11070291

Academic Editors: Mohamed Shahin and Jesus Martinez-Frias

Received: 4 June 2021
Accepted: 13 July 2021
Published: 15 July 2021

1. Introduction

Ground improvement is the deliberate enhancement of geotechnical properties of marginal soils to obtain the desired and optimal performance. When faced with a challenge, a geotechnical engineer would endeavor to develop the most effective and economical approach of soil treatment considering the underlying soil types, treatment depth, locations required to be treated, desired level of improvement, availability of skilled personnel and materials, environmental friendliness of the method and project costs. These design factors are the most effective in determining the appropriate treatment method [1]. Chemical modification is defined as the addition of additives to soils to alter the soil properties to be acceptable as defined in design criteria. Nowadays, numerous materials are being used for soil treatment. Lime and cement are considered among the most popular options. Often, additives such as sodium sulphate and fly ash are added to cement to enhance stabilization. Bitumen, chemical compounds, and geotextiles are the other materials which have been used for soil stabilization through various methods such as mixing materials and chemical grouting [2–8].

Global warming and climate change are considered among the most concerning phenomena in recent decades. In the case of materials used in civil applications, traditional and calcium-based materials, such as cement and lime, and synthetic products, have led to greenhouse gases (GHG) emissions, which are directly related to global warming. Therefore, alternative and less polluting materials should be encouraged [9–12].

Although cement production has remained unchanged during the past six years, it is still considered one of the main sources of CO_2 emissions, contributing approximately 8% of the global CO_2 emissions [13]. From 2010 to 2018, the global amount of cement production increased from 3310 to 4100 million tons (23.8% increase) [14]. To reduce cement usage, researchers have attempted to replace traditional environmentally harmful materials with sustainable techniques and methods, such as bio-enzymes, CO_2 absorption, microbial-induced calcite precipitation (MICP) method with viable but low-cost medium, and biopolymers [15–20].

In recent years, different solutions have been developed for environmentally friendly soil improvement, some of which are based on microbial activities and biomaterials. MICP is a multi-disciplinary technique combining the fields of microbiology, biology, and geotechnical engineering [21]. MICP is the process of using bacteria to precipitate calcite ($CaCO_3$) in the soil matrix, which leads to soil strengthening [22–24]. The formed $CaCO_3$ network links the soil grains firmly and while the changes in subsurface conditions are not permanent, the longevity of treated soil mass can be adopted based on project needs and requirements [25]. Another advantage of MICP is to reduce the improved soil permeability to mitigate liquefaction by either bioclogging process or allowing the excess pore water pressure to be dissipated [26]. On the other hand, MICP disadvantages are the requirement of highly specialized sets of environmental and growth conditions, applicability to mostly in coarse-grained soil [27–29], producing ammonia as a by-product [30], and difficulty in gaining a uniform distribution of calcite [25].

In addition, enzymes can also be used to obtain calcite through the same biochemical reactions as in MICP. This improvement method is called enzyme-induced calcium carbonate precipitation (EICP) [31]. Another biological method used is stimulating microbes to generate biopolymer (microbial biopolymer accumulation) within the soil matrix, to improve the soil against erosion and permeability reduction [32–34]. Biofilm formation has a beneficial effect on soil stability and the reduction of soil hydraulic conductivity [26,35–39]. Biogas generation, such as nitrogen, is also a biological method by which the liquefaction potential of sand is decreased. The production of insoluble gases in soil pores leads to a reduction in the degree of soil saturation, to mitigate the liquefaction hazard [40,41].

Polymers are macromolecular materials that are composed of repeating subunits (monomers). There are two types of polymer: synthetic and natural. Synthetic polymers are those produced from petroleum products. Biopolymers are natural polymers that are produced naturally in the environment by plants and living organisms [42]. They have an extensive record in different applications in the construction industry [43]. The majority of biopolymers have been employed as adhesive or water retention agents. Lignin and its derivatives are effective as an additive in concrete and oil well drilling. Cellulose and starch derivatives performed well in various applied fields such as tile adhesives, oil well construction, ceramics, and cement and lime plasters [44–46]. Biopolymers have been used in drilling fluids because they require low viscosity during pumping and high viscosity after circulation has stopped; specific features conferred by xanthan to the solution.[47–50].

In geotechnical engineering, biopolymer binders potentially have a notable effect on enhancing various aspects of marginal soil behaviors. The beneficial effects and advantages of biopolymers have been reviewed in recent review papers from different points of view. However, there is still a gap in understanding of the behavior of biopolymer-treated soil under diverse loadings and environmental conditions. An environmental assessment is undertaken in this study which presents the factors that must be considered when using biopolymers during their life cycle. Also, it is necessary to understand how different parameters could influence the biopolymer interaction processes with soil particles before, during, and after mixing, with particular reference to curing conditions, as well as environmental factors. This paper endeavors to update and comprehensively review the studies on biopolymer-treated soil from the perspective of geotechnical engineering, including the evaluation of unconfined compressive strength, shear strength, erosion resistance, Atterberg limits, compaction characteristics, and pavement stabilization. The various parameters influencing the improved soil geotechnical performance were evaluated under the optimal conditions for each type of biopolymer. Moreover, the interaction processes of biopolymers with different soil types were also discussed.

2. Biopolymers in Geotechnical Applications

Biopolymers can be classified based on different terms including biodegradability (biodegradable and non-biodegradable) and the source of raw materials. Three groups can be considered when classifying biopolymers based on their source of origin: plant-based biopolymers, animal-based biopolymers, and biopolymers produced by microorganisms [51]. Figure 1 presents a list of biopolymers, which are mostly utilized for soil improvement purposes, based on the source of production.

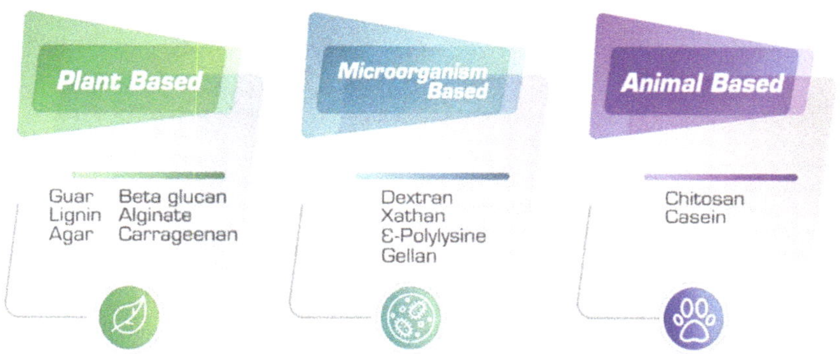

Figure 1. Common biopolymers used for soil improvement based on their source of production.

2.1. Plant-Based Biopolymers

Plant-based biopolymers are produced from plants and agricultural scraps. Biopolymers in geotechnics mainly include plant-based products which are mostly polysaccharides. Guar, lignin, agar, beta-glucan, alginate, and carrageenan fall into this category. Guar gum (guaran) is a polysaccharide composed of two sugars galactose and mannose, extracted from the guar plant. Among water-soluble polysaccharides, guar gum is the one with the highest molecular weight [52]. Guar gum is a naturally charged polysaccharide that can be found in both cationic and anionic states. The world production for the guar gum family is more than 1 million tons each year so that it is commercially available at a reasonable cost. The presence of large hydroxyl groups in guar gum produces a network of hydrogels among soil particles and hydrogen ions through hydrogen bonds [53,54].

Lignin makes an influential contribution to the formation of cell walls, especially in trees and barks. After cellulose, lignin is the most abundant biopolymer existing throughout the world. It is relatively hydrophobic and a recalcitrant biopolymer (resistant against acid- and base-catalyzed hydrolysis) [55]. Due to it being a natural adhesive and its phenolic nature, lignin has been considered a good replacement for wood adhesives [56]. Being cheaper, environment-friendly, and non-toxic has made lignin an appropriate alternative for traditional stabilizers [57]. A stable soil structure is formed when lignin links soil particles together and leads to the reduction of big pores in the soil [58].

Agar gum is a thermogelatin polysaccharide extracted from seaweed (i.e., marine red algae) [59]. This biopolymer tends to dissolve in water at temperatures around 85 °C and to form a gel when cooled to 32–43 °C [60]. The conventional agar production process consists of key stages of pre-treatment, extraction, filtration, concentration, and dehydration [61]. The properties of agar gum are greatly dependent on the seaweed utilized for agar production [61]. Because of having comparatively long molecular structure and low ionic inclination, agar gums act to coat and coagulate soil grains, and direct interactions with soil particles would be difficult for agar gum due to its neutral charged nature [62].

Beta-glucan biopolymer is a natural polymer found in the cell walls of yeast and cellulose in plants. It is a biopolymer of D-glucose monomers linked by glycosidic bonds [63]. The potential of beta-glucan biopolymer to form hydrogen bonds is relatively high due to the hydroxyl and carboxyl (COO-) on the edges, which cause surface polarization, creating a greater capacity for water adsorption by beta-glucan, and resulting in pore filling and volumetric expansion of soil.

Sodium alginate is an anionic polysaccharide with the chemical formula $C_5H_7O_4COONa$, and it is a salt produced from alginic acid [64]. It is mostly produced from the cell walls of marine brown algae [64]. Alginate is a water-soluble biopolymer and forms a viscous gum by linking with water molecules [65]. Biocompatibility, low toxicity, and relatively low cost make alginate a good choice for soil improvement applications. Also, pH can significantly affect the viscosity of alginate solution. A pH reduction results in a decrease in viscosity [64]. Moreover, highly charged molecules, hydrophobic and polar moieties, as well as the side chain carboxylates provide sodium alginate with a distinctive potential additive in soils [66]. Carrageenan is known as a linear polysaccharide produced from red edible seaweeds. Carrageenan is an anionic biopolymer and is soluble in water. In low pH with high temperatures, the stability of carrageenan is expected to decrease [67].

2.2. Microorganism-Based Biopolymers

Biopolymers such as xanthan gum, gellan gum, and dextran, are the products of bacterial fermentation processes, which are considered microorganism-based biopolymers. Xanthan gum is a microorganism-based anionic polysaccharide, produced through glucose/sucrose fermentation by the *Xanthomonas campestris* bacterium. The most well-known characteristics of xanthan gum are pseudo-plasticity [68] and high shear stability [69] even at relatively low concentrations. Moreover, it has several desirable properties, including pH stability, storage stability, and ionic salt compatibility xanthan has found a wide range of applications in cosmetics, oil, paper, paint, pharmaceuticals, and food and textile industries as a gelling, thickening, or suspending agent, and as a flocculent or for viscosity control [70]. Gellan gum is a thermogelatin anionic polysaccharide with a high molecular weight that is produced through bacterial fermentation of *Sphingomonas elodea* [71]. In commercial applications, thermal behavior of gellan gum is used as deacylated polymer which is soluble in water at temperatures higher than 90 °C. It then forms a gel when the temperature is reduced [62]. This biopolymer has a remarkable ability to form high-quality gels, and has also been found to be stable at high temperatures and low pH conditions [72]; these features make this biopolymer a suitable additive for soil improvement.

Polylysine is a lysine homopolymer that includes functional carboxyl and ε-amino groups. It is a cationic polymer [73]. This biopolymer is produced by bacterial fermentation typically in the strains in the genus *Streptomyces* and has both hydrophobic and hydrophilic properties [74]. The positive charges of ε-polylysine are capable of interacting with negatively charged surfaces such as fine clay [73]. Dextran is a natural polymer and a complex branched glucan that is synthetized from sucrose by certain lactic-acid bacteria. Dextran is insoluble in water and pH does not affect its solubility. Dextran solutions behave like Newtonian flow characteristics [75]. In geotechnical purposes, as dextran is produced through a microbial process, a bacterium named *Leuconostoc mesenteroides* was cultured and stimulated in fine sand to generate dextran to reduce the surface erosion rate [34].

2.3. Animal-Derived Biopolymers

Animal-derived biopolymers are the polymers produced from animal sources. Chitin and chitosan are produced from the waste crustacean shells such as shells of shrimps, and protein-based biopolymers are produced from milk and dairy products. Chitosan is a water-soluble biopolymer that tends to bind to negatively charged surfaces [76]. Chitosan is a linear polysaccharide of D-glucosamie and N-acetyl-D-glucosamine, which is usually extracted from crustacean shells of the food processing industry [77]. Chitosan is an animal-based biopolymer with cationic nature, and because of the positive charges of amino groups, it becomes a cationic polyelectrolyte after being dissolved in acetic acid [78]. This cationic feature allows it to interact with chitosan and the negatively charged double layer of clay [79,80]. Casein is a type of protein biopolymer which is negatively charged at a pH of 6.6 with an isoelectric point is around 4.6 [81]. A high-quality casein for soil treatment applications is obtained when extracted from skim milk with the lowest fat amount [82]. Compared with other biopolymers, protein-based biopolymers have a less hydrophilic tendency to water absorption [83]. As the polymeric chains of casein reach into the soil particles, the soil and biopolymer are joined together through electrostatic interaction among protein functional groups [82]. Table 1 presents the most important details of the common biopolymers in soil strengthening.

Table 1. Characteristics of common biopolymers utilized for soil treatment and their potential environmental impacts.

Biopolymer	Source	Charge Type	Water Solubility	Potential Environmental Impact
Xanthan Gum	glucose or sucrose fermentation by the *Xanthomonas campestris* bacterium	anionic	soluble	0.1 kg CO_{2e} for production of 1 kg biopolymer, no land use
Gellan gum	bacterial fermentation of *Sphingomonas elodea*	anionic	soluble at temperatures higher that 90 °C	Low CO_2 record, no land use, and no pesticides
Beta glucan	in cell walls of yeast and cellulose in plants	cationic	soluble	Low carbon footprint, relatively high water footprint
Agar gum	marine algae	neutral	soluble around 85 °C	Negligible carbon footprint and very low water footprint
Lignin and its derivatives	cell walls of trees and barks(produced from the waste by-product of paper manufacturing industry)	mostly anionic	insoluble	Second largest biopolymer in nature, relatively high land use, low carbon footprint
Sodium alginate	brown algae	anionic	soluble	Low energy needed, low acidification by using hydrochloric acid
Chitosan	waste crustacean shells	cationic	soluble	Low greenhouse emissions if it is used as fertilizer; Producible from waste animal products
Carrageenan	red edible seaweeds	anionic	soluble	-
Casein	(waste) dairy products	anionic	insoluble	High water and carbon footprint; Producible from waste dairy product
Guar gum	guar plant	cationic/anionic	soluble	Low carbon footprint
Polylysine	bacterial fermentation	cationic	soluble	-
Dextran	synthetized from sucrose by certain lactic-acid bacteria	anionic/cationic	insoluble	Low CO_2 record, no land use, and no pesticides

3. Assessment of Environmental Impacts of Biopolymers in Geotechnical Engineering

More recently, many industrial policies have been developed towards sustainability, mainly to mitigate the harmful environmental impacts. In this way, the EU commission is set to cut GHG emissions by more than 85% by 2050 compared to the level of 1990 [84]. With rising concerns about planet environmental consequences, soil improvement methods are turning to less harmful methods and materials [85]. Some techniques, such as life-cycle assessment (LCA), have been developed to systematically analyze the environmental performance of processes or materials. LCA investigates the environmental costs and impacts for the entire process from the extraction of the raw materials to final disposal. As an example of a geotechnical work, the life cycle for a road construction project can be summarized in five sections [86,87]:

1. Manufacturing processes for all materials used in the project, including extraction, transportation of raw material from the source to the factory, processing to the final product, and material transportation to the construction site.
2. Construction stage, which is a phase that consists of all the execution processes required for a road to be taken. This includes all the components causing the production of greenhouse gas emissions, even the fuel consumed by the vehicles experiencing delays caused by the project.
3. Use phase, in which most factors are related to the energy and air pollution resulting from road vehicles. Leaching and runoff leading to water pollution is an important parameter to be considered in the use phase.
4. Maintenance and rehabilitation (M&R), a very important section to keep the functionality of a road through its functional life span. Road works to repair possible road damage as well as the traffic occurring due to delays are undertaken in this part. More technically, this includes production and transportation of new materials to the site, and destruction and discarding rubbish materials.
5. End-of-life stage includes the way that disposal materials are to be used, whether recycling or disposing of the materials into the environment. The activities are about demolition and transportation, as well as leaching if the materials are disposed of at the end.

Various factors must be considered when an environmental assessment work is being studied, including acidification potential, aquatic toxicity, human toxicity potential, eutrophication potential (considering the pollution of aquatic ecosystems), global warming potential, non-renewable resource depletion, ozone depletion potential, terrestrial ecotoxicity, and photochemical oxidants creation potential [84]. Among these, global warming, acidification, eutrophication, and human toxicity are the most important ones for geotechnical works [88]. Therefore, the environmental assessment for geotechnical applications is necessary.

Chemical soil treatment is recommended to reduce the transportation cost of virgin materials to the site and disposals in landfills [87]. Lime stabilization decreased the energy consumption by 30% and the CO_2 production by 27% compared with the employment of virgin materials.

Although chemical soil treatment is more preferable to virgin materials, the lime and cement industries contribute to the production of a considerable amount of GHG emissions. The cement industry generates about 2.8 billion tons of CO_2, equal to 8% of total CO_2 amount. More than half of these emissions are produced through the chemical process of calcination for making the clinkers [89], which cannot be mitigated by increasing energy efficiency or using renewable fuel resources. Hydrated lime manufacturing comprises extracting limestone and transporting it to the factory, calcination, and quicklime hydration; this is an energy-intensive procedure (vertical kiln), and this energy is mostly provided from non-renewable resources [90].

For soil improvement using lime, more than 75% of all environmental impacts, including global warming, photochemical oxidation, and embodied energy, are contributed by lime production [90]. Also, treated soil with low density lime binder showed twice larger energy consumption than the treated soil with high-density lime binder, when the goal was reaching a certain strength [90,91]; it can be concluded that chemical soil treatment is preferred to be conducted using binders with high density and low content, which is an advantage for biopolymers as much less content is needed compared with traditional binders.

Recently, the goal of moving towards sustainability has been followed in polymeric materials so that biopolymers have attracted more attention [92]. The biopolymer concept is a polymer that is totally or partly generated from a bio-based source and is somehow biodegradable. The application of biopolymers in various industries helps them to progress in the way of sustainability, however, the presence of some fundamental issues has limited the development of biopolymers in certain industries [93]. Advancement in the biopolymer industry can reduce not only the environmental impacts but also the costs, and strong market potential for biopolymers is estimated in the near future. On the other hand, some synthetic polymers give a better performance in terms of mechanical and thermal resistance [84].

A comprehensive study conducted by the Institute for Energy and Environmental Research (IFEU) concluded that biopolymers perform better than conventional polymers in terms of climate impact, particularly in CO_2 emissions and fossil fuel consumption [94]. Although a bright future can be seen for biopolymers, the potential environmental impacts of the biopolymers cannot be ignored. Few studies have been undertaken so far on the economic and environmental impacts of the biopolymers because they are in an early development stage and have a relatively small market share of the polymers. Most LCA studies on biopolymers are incomplete or are based on a laboratory-scale dataset.

The way biopolymers are degraded when they are used in large volume raises concerns about contamination by entering streams, rivers, and the ocean, as some biopolymers degrade slowly which makes them applicable for use in long-term geotechnical applications [95]. Another concern is the land use for biopolymer production that might occupy the land required to grow crops for food industries. Expanding biopolymer production requires fertilizers for the plants from which biopolymers are extracted, and may lead to a higher level of water eutrophication and soil acidification than to the production of chemical polymers. Also, through the production process and transportation of biopolymers, energy and water must be provided that causes GHG emissions, however, the emission is around 85% less than conventional polymers and much less than traditional adhesives, such as cement and lime.

Environmental impacts for the biopolymers used for soil improvement vary. The amount of GHG emissions generated through biopolymer production cannot be measured accurately, because it is dependent on many factors based on the local conditions and available resources, such as the energy supply pathways, fuel source, and production method. Lignin combustion, which is used as biofuel, decreases GHG emissions by more than 80% relative to gasoline [96]. There are a number of lignin production methods, and climate change impact is variable for each of them, resulting from the highest impact for the cheapest lignin to the least impact for the most expensive lignin. The carbon footprint for xanthan gum is reported to be very low, at 0.1 kg CO_{2e} for production of 1 kg biopolymer, with still unknown water footprint [97]. The carbon footprint comes to a negligible amount for agar biopolymer [98]. Guar is a low-emission crop that produces CO_2 mostly through irrigation, fertilization, and harvesting and threshing process; CO_2 comes from electricity for irrigation, from production and application for fertilization, and the fossil fuels for harvesting and threshing [99].

Beta-glucan is usually produced from oats, and is considered as a low carbon footprint. The carbon emission amount is variable based on the location of source and processing. The largest global warming contribution is from the production and the use of nitrogen fertilizer, as well as the energy used through field activities [100]. The water footprint for beta-glucan is relatively higher with more than 300 gallons needed to produce one kilogram of oats [101,102]. Sodium alginate has low environmental impact, among which electricity accounts for 39% of this followed by the use of chemicals (hydrochloric acid), 26% on average [103].

One of the main concerns of chitosan environmental impacts is that it is the source of ammonia and greenhouse emissions for heat production and when it is used as fertilizer [104]; 2.7% of global GHG emissions are attributed to dairy products, which do not contribute to CO_2 production, significantly; they are dominated by methane (CH4) and nitrous oxide (N2O). On average, 2.4 kg carbon dioxide equivalent is produced per kg of milk, which makes the dairy industry one of the main contributors to climate change.

Land use and water consumption are the other impacts of dairy products [105,106]. Even though producing casein causes serious environmental concerns, casein utilized for soil improvement comes from waste dairy products that are usually dumped into the environment, so that casein cannot only be utilized in soil treatment, but also keeps the environment from being contaminated and brings the recyclability into the dairy products' life cycle [82,107]. Recently, significant advancement in biotechnology has provided the opportunity for biopolymer production, such as xanthan, dextran, and gellan gum, directly by microorganism fermentation of glucose or sucrose or genetically modified crops [108], under controlled conditions; this provides the opportunity to have biopolymers generating less CO_2 and using less land, and there is no need for pesticides.

As mentioned, most of the environmental impacts in road works are related to the material production phase, and low content binders with high efficiency are recommended; so utilizing biopolymers to a large extent meets both conditions by lowering the GHG emissions in their production process and a comparatively low content is needed for soil treatment. For a simple soil improvement project of biopolymer-treated soil, the life cycle can be assumed as shown in Figure 2, showing the process for xanthan biopolymer as the most commonly used biopolymer. Three major parts are included: biopolymer production and transportation, soil extraction and transportation, and blends production.

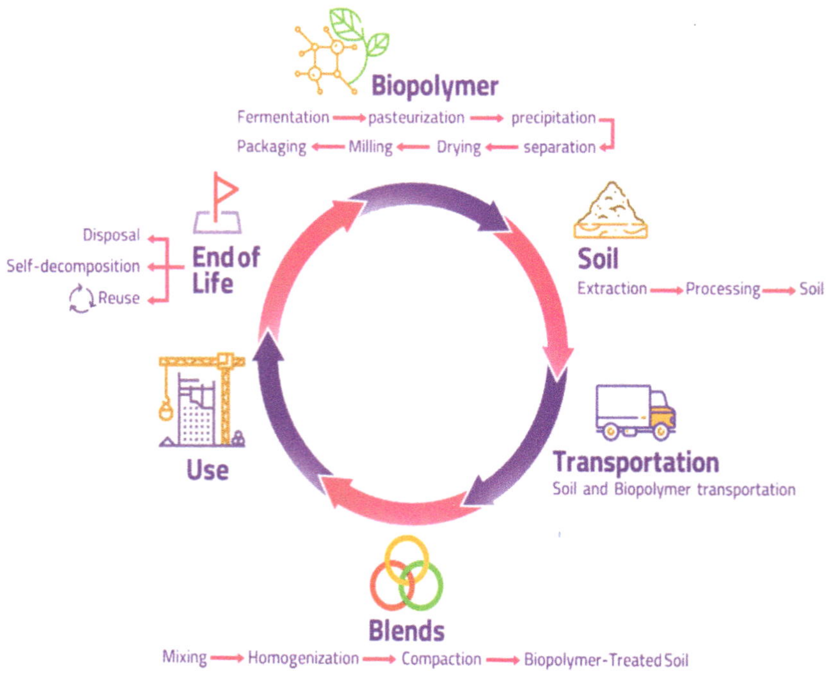

Figure 2. Life-cycle process of a xanthan treated soil.

4. Biopolymer–Soil Interaction

The way that a biopolymer interacts with different kinds of soils has always been considered as an important factor because microstructural study leads to a better understanding of the biopolymer-treated soil behavior. Based on the soil type, the adhesion mechanism of biopolymers is different from one soil to another. Figure 3 shows the schematic form of biopolymer-soil interaction.

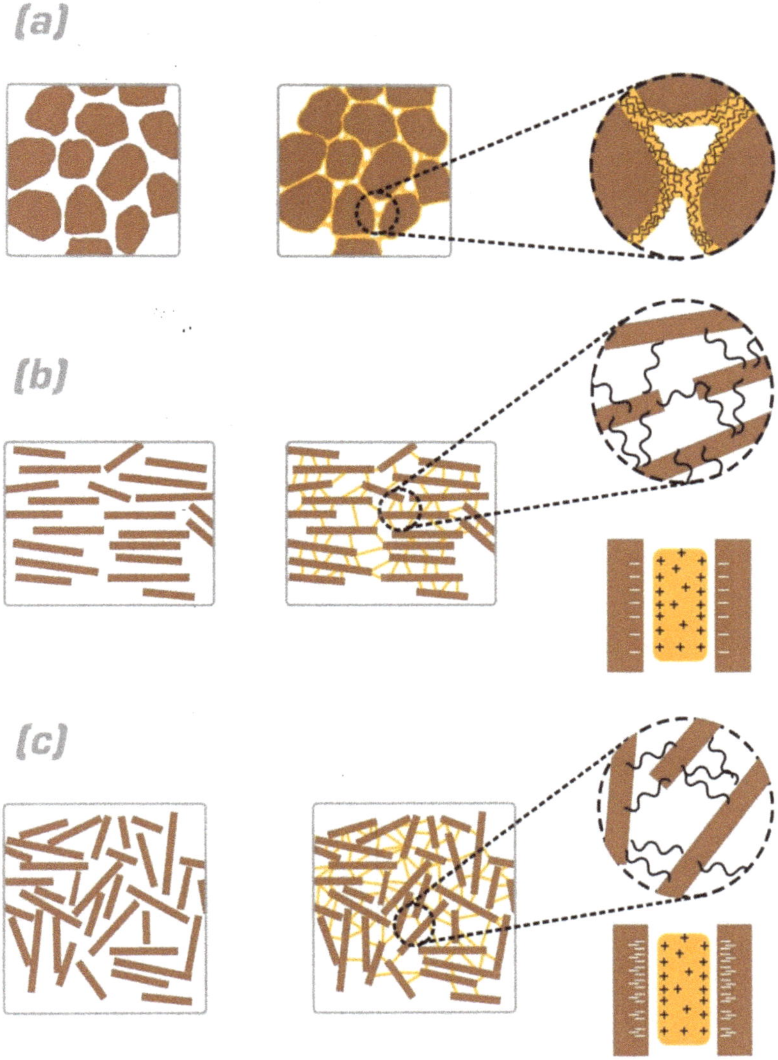

Figure 3. Biopolymer interaction with different kinds of soils (**a**) sand, (**b**) kaolinite clay, (**c**) montmorillonite.

4.1. Biopolymer–Sand Interaction

By adding biopolymer solution to sandy soils, the dissolved bound particles start to permeate into the pores and soak the sand surface. The hydrophilic feature of the sand surface, due to the presence of silica and silicate, as well as the water solubility of biopolymers, quicken the wetting process. When the biopolymer contacts the sand surface, it coats the soil particles and forms a strong film, thus creating the necessary bridges to start the bonding process. By allowing water evaporation over the curing period, the dehydration process of the biopolymer matrix promotes the bonding strength, drawing the particles closer to one another, thus compressing the pore spaces. Having shorter connection chains makes them stronger against external forces so that the geotechnical performance is enhanced during the drying procedure. More biopolymer (up to its optimal content) leads to the formation of higher bonds amongst sand particles and higher shear strength. Figure 3a shows the mechanism of biopolymer adhesion in typical biopolymer-treated sand.

4.2. Biopolymer–Clay Interaction

The interaction mechanism of biopolymer with clayey soils is different from the coarse-grained soils because clay can be structurally transformed over time. In clay minerals, the composition changes within the structures due to substitution of ions. Replacing one structural cation by another of a similar size without changing the basic structure is called isomorphous substitution. This process determines the primary source of electrical charges (negative and positive) in clay minerals [109]. Kaolinite and montmorillonite are the most common clay minerals in the majority of soils. Kaolinite can carry both negative and positive charges, while montmorillonite carries negative charge in most cases [110]. Due to weathering of clay minerals in harsh climate conditions, hydrated oxides of iron (goethite) and aluminum (gibbsite) are generated, thus forming cations over clay sheets. In the case of gibbsite, hydroxyl ions bond with only two positive ions Al^{3+}, so that positive ions remain available in clay structure.

Based on the clay minerals, the inter-layer bonding could have specific characteristics and a tendency to absorb water as well as ions. The layers in kaolinite clay are mostly connected through hydrogen bonds, thus kaolinite has a relatively stable structure with a lower tendency to absorb water and with lower swelling potential. The cation exchange capacity (CEC) of kaolinite is in the range of 0.03–0.1 meg/100 g, which is lower than illite and montmorillonite. This feature makes kaolinite less tendentious to absorb biopolymer particles (see Figure 3b), although new cementitious compounds were produced after adding xanthan to kaolinite clay and subsequently with their pore spaces filled [111].

In montmorillonite, clay sheets are joined by van der Waals bonding, which is a comparatively weak chemical bond that allows water to easily enter between layers. Montmorillonite has a higher expansion potential when it is subjected to water, because of weak chemical bonding. Compared with kaolinite, as can be seen from Figure 3c, higher negative charges exist in montmorillonite with a CEC in the range around 0.8–1.5 meg/100 g. Additionally, the specific surface area of montmorillonite is much larger than that of kaolinite. Consequently, montmorillonite is more likely to absorb a biopolymer solution thus reducing its swelling potential [109,112]. Illite is another silicate clay with a relatively fluctuated structure similar to montmorillonite, the difference is that cations are located among the surfaces causing the formation of ionic bonds. In terms of strength, ionic bonds are stronger than van der Waals bonds, such that illite has a lesser tendency to absorb water than montmorillonite.

In general, clay soils interact with biopolymer particles in a more complex way than the sand. biopolymer can directly interact with clay particles due to the presence of electrical charges. These chemical bonds could be formed through different electrostatic interactions, hydrogen bonding, ionic bonds, or van der Waals bonds. The factors that determine the mechanism of biopolymer-treated interaction include the electrical charges of the biopolymer, natural cations inside the clay, and the amount of charge in the clay sheet surfaces. Moreover, biopolymer coats the soil grains and connects detached particles by forming bridges along with direct chemical bonds. It has also been shown that the specific surface area is reduced when xanthan gum is added to the clay, due to filling the pores [80,111,113–115]. Comparing sandy and clayey soils, biopolymer-treated clay is expected to have a better efficiency because of the stronger interparticle connections between biopolymer and soil particles when compared with sand [113]. A summary of recent studies on biopolymer-treated soil is presented in Table 2.

Table 2. A summary of recent studies on biopolymer-treated soils.

Biopolymer Type	Soil	Assessment	Studied Parameter	Remarks (Ref.)
Xanthan	Silt	Unconfined compressive strength	Biopolymer content	[116–118]
			Dehydration time	[116–119]
			Durability	[117,119]
		Direct Shear	Dehydration time	[117,119]
		Splitting Tensile	Biopolymer content	[117]
		Compaction	Biopolymer content	[116]
		Consistency limits	Biopolymer content	[116]
		Permeability	Dehydration time	[119]
	Sand	Unconfined compressive strength	Biopolymer content	[120–122]
			Dehydration time	[120–122]
			Durability	[120]
		Direct Shear	Biopolymer content	[123,124]
			Moisture content	[123,124]
			Dehydration time	[119,124]
		Triaxial	Biopolymer content	[125]
		Permeability	Biopolymer content	[125]
		Wind erosion	Wind Erosion rate	[120,126]
		Water Erosion	Water erosion rate	[126–128]
		California Bearing Ratio (CBR)	Biopolymer content	[129]
			Durability	[129]
	Clay	Unconfined compressive strength	Biopolymer content	[130–133]
			Dehydration time	[120–122]
			Moisture content	[130]
			Initial moisture content	[131]
			Fatigue loading	[131]
		Direct Shear	Biopolymer content	[119,132,133]
		compaction	Biopolymer content	[131–133]
		Consistency limits	Biopolymer content	[131,132,134]
		Fracture properties	Drying and fracture energy	[135]
		Flexible wall permeameter	Gas permeability	[136]
		Vane shear test	Biopolymer content	[134]
		Consolidation test	Biopolymer content	[133,137]
	Sand-clay mixture	Unconfined compressive strength	Biopolymer content	[138,139]
			Dehydration time	[138]
		Direct Shear	shear strength	[130]
		Consistency limits	Biopolymer content	[138,140]
		compaction	Biopolymer content	[138,140]
		Permeability	Hydraulic Conductivity	[136,138,139]

Table 2. *Cont.*

Biopolymer Type	Soil	Assessment	Studied Parameter	Remarks (Ref.)
		Dispersivity tests	Dispersive properties	[140]
		Water Erosion	Erosion rate	[140,141]
		Consolidation test	Biopolymer content	[140]
Guar	Silt	Unconfined compressive strength	Biopolymer content	[117]
			Dehydration time	[117]
			Durability	[117]
		Direct Shear	Dehydration time	[117]
		Splitting Tensile	Biopolymer content	[117]
	sand	Unconfined compressive strength	Biopolymer content	[120]
			Dehydration time	[120]
			Durability	[120]
		Triaxial	Biopolymer content	[125]
		Permeability	Biopolymer content	[125]
	Clay	Unconfined compressive strength	Biopolymer content	[130,142]
			Dehydration time	[130]
			Moisture content	[130]
		Direct Shear	shear strength	[130]
		Permeability	Biopolymer content	[142]
		Compaction	Biopolymer content	[142]
	Sand-clay mixture	Unconfined compressive strength	Biopolymer content	[143]
			Durability	[143]
		Consistency limits	Biopolymer content	[143]
		Permeability	Hydraulic Conductivity	[136,143]
		Consolidation	Biopolymer content	[143]
Sodium alginate	Silt	Unconfined compressive strength	Biopolymer content	[117,144]
			Dehydration time	[117]
			Durability	[117]
			Moisture effect	[144]
		Splitting Tensile	Biopolymer content	[117]
	Sand	Wind erosion	Erosion rate	[145]
Lignin	Silt	Unconfined compressive strength	Biopolymer content	[146]
			Cuing time	[146]
		Compaction	Biopolymer content	[147]
		Dynamic triaxial	Biopolymer content	[147]
			Curing time	[147]
			Dynamic shear modulus and Damping ratio	[147]
		Consistency limits	Biopolymer content	[146]
		Direct shear	Biopolymer content	[146]

Table 2. *Cont.*

Biopolymer Type	Soil	Assessment	Studied Parameter	Remarks (Ref.)
Beta glucan	Silt	Unconfined compressive strength	Biopolymer content	[117]
			Dehydration time	[117]
			Durability	[117]
		Splitting Tensile	Biopolymer content	[117]
		Direct Shear	Dehydration time	[117]
Chitosan	Silt	Unconfined compressive strength	Biopolymer content	[117]
			Dehydration time	[117]
			Durability	[117]
		Splitting Tensile	Biopolymer content	[117]

5. Geotechnical Properties of Biopolymer Treated Soils

In the current section, different aspects of biopolymer-improved soil are discussed.

5.1. Unconfined Compressive Strength (UCS)

The unconfined compressive strength (UCS) test is the most common test used for evaluating the role of biopolymers in soil treatment [148]. As the interaction way of biopolymer with different soil types could be entirely different, in this section, the UCS is separately discussed for each soil type.

5.1.1. Unconfined Compressive Strength in Biopolymer-Treated Sand

The UCS of sand treated by biopolymers is presented in this section. Figure 4 shows the maximum compressive strength obtained after air curing. Since sand does not have (or it is negligible) the ability to withstand a compressive force. Biopolymer-treated sand (1% and 2%) is shown in Figure 4.

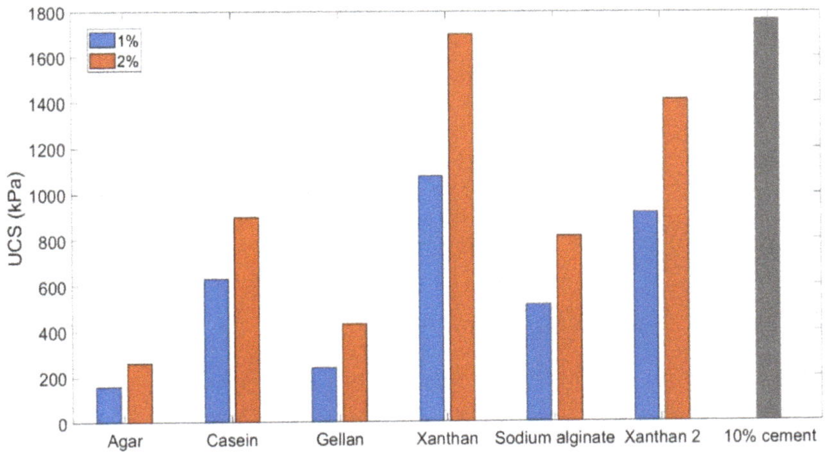

Figure 4. Biopolymer-treated sand [66,82,113,121,149–151].

As seen in Figure 4, the available results show that xanthan biopolymer can increase the UCS values up to 1700 kPa when used to treat sand. The improved UCS value is variable depending on the type of biopolymer used. The lowest UCS value is observed when sand is treated with agar due to the repelling negative charges of both agar and sand. To improve the bond, adding a positively charged material as an intermediate agent would enhance the compressive strength remarkably [149]. Casein and sodium alginate showed almost similar performances, while gellan gum had lower strength compared with casein and alginate. Qureshi et al. (2017) compared the effect of xanthan gum and cement on the UCS variation of treated sand. In terms of peak UCS strength, using 2% xanthan is comparable to the strength obtained by using 10% cement. Xanthan-treated sand gained a less ductile behavior compared to cement-treated sand by showing higher residual strength [151]. Xanthan also demonstrated a notable increase in tensile strength which is in line with compressive strength improvement; the ratio of tensile strength per compressive strength was shown to be constant regardless of the biopolymer content [121].

5.1.2. UCS in Biopolymer-Treated Fine-Graded Soils

Figure 5 shows how biopolymers changed the compressive strength of fine-graded soils. In the case of agar and gellan gum, UCS increased to 2500 (114.8% growth) and 1800 kPa (52.2% growth) for 1% of gellan gum and agar, respectively. This higher amount of strength for gellan gum is due to the presence of multiple hydrogen and hydroxyl groups [62]. Xanthan gum has been comprehensively studied and the results are presented in Figure 5b. In one of the first studies using xanthan gum on clay, the compressive strength increased approximately 470% from 440 kPa for untreated clay and 2540 kPa for 1% xanthan-treated clay [113]. Latifi et al. (2017) showed that xanthan gum is capable of improving different types of clay significantly. For example, both bentonite and kaolinite clays rendered high growth amounts after adding 1% of xanthan gum; however, this increased amount is higher for bentonite [111]. For xanthan gum, through forming hydrogen bonds and cation bridges, monomers could directly interact with electrically charged clay particle surfaces. The added biopolymer, however, lacks a tendency to approach sand surfaces with no electrical charges, thus electrostatic and hydrogen bonding will not be formed [113,152]. Also, different clay types exert different behavior when they are subject to an external material because of the various structural configurations related to the layer lattice group of clays in which oxygen ions sheets are placed on top of one another and the space between these sheets is filled with metallic cations such as Al^{3+} and Si^{4} [153].

Biopolymers improved the compressive strength of silty soils by adding a cohesive component in addition to the silt's frictional strength. Figure 5b shows the compressive strength of silt stabilized by 2% of alginate, beta-glucan, guar gum, and xanthan gum. The same silt has been used for all these biopolymers. Xanthan gum gave a better performance in comparison with others, although all four biopolymers led to the enhancement in the compressive strength. Also, alginate and beta-glucan increased the UCS to almost the same level [117]. It has also been indicated that beta glucan effectively increased the compressive strength of *hwangtoh* soil (MH) up to 200% for 0.496% of biopolymer. The properties of beta glucan-treated soil were governed by the available cations in soil (Na^{+}, K^{+}, Ca^{2+}, Mg^{2+}), charges in the soil surface as well as electrical interactions between the biopolymer [154]. Other biopolymers such as lignin have also presented good results on improving the geotechnical behavior of silt [58].

5.2. Shear Strength

Shear strength of biopolymer-treated soils in their dried state and cured in the environmental conditions have been presented in Figure 6. The most noticeable point is that all biopolymers enhanced the shear strength, and this improvement continued by the increase of biopolymer content. The highest growth rate is for xanthan-treated sand in terms of cohesion, which is in line with the UCS results.

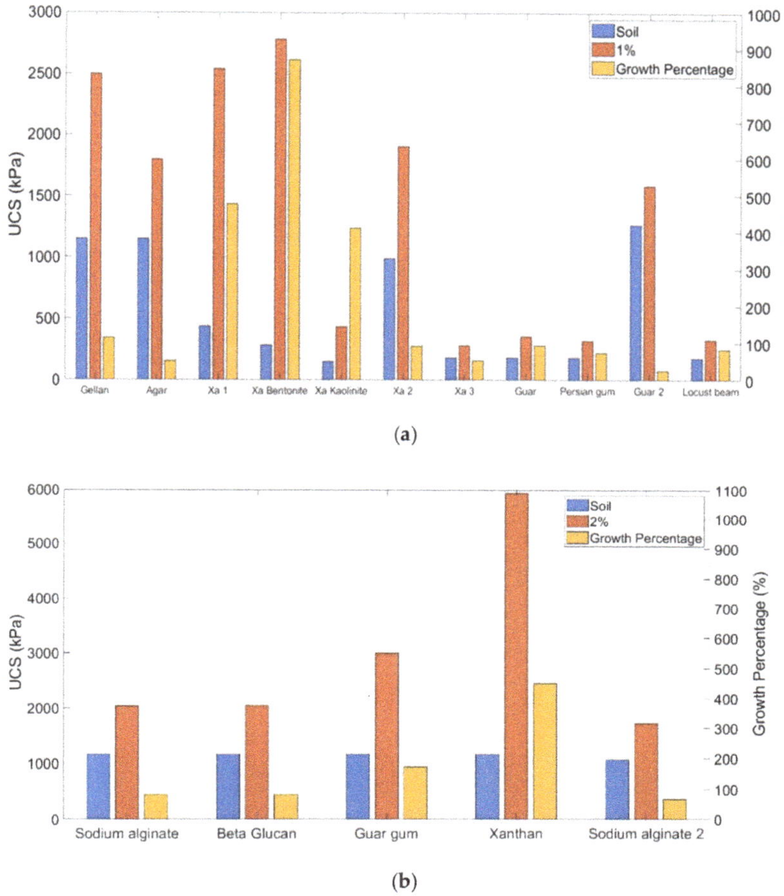

Figure 5. The effect of various biopolymers on unconfined compressive strength of fine-graded soils, (**a**) clay, (**b**) silt [62,111,113,118,130,143,155].

The shear behavior of biopolymer-treated sand is interpreted differently from the cement-treated sand; gellan-treated sand is a compound of solid grains and gellan hydrogels while the chemical structure of cemented sand is changed after the hydration process (sand and C-S-H hydrates) [156]. No chemical reaction was found between agar and sand grains, and the shear strength enhancement is due to the hydrogel generated by agar gum [157]. Also, the reason for the increase in cohesion by increasing biopolymer content is directly dependent on the biopolymer gel strength that is defined as the force needed to fracture the gel [158]; 2% xanthan gum-treated sand rendered a considerably better performance compared with gypsum-treated sand; however, 20% of gypsum led to a much larger shear strength than 2.5% xanthan [159]. For gellan-treated clay, it has been shown that gellan gum contributed to increasing both friction and cohesion values. As such, gellan gum plays an important role in the conglomeration of fine particles, this behavior is almost similar to the cement-treated clay [156].

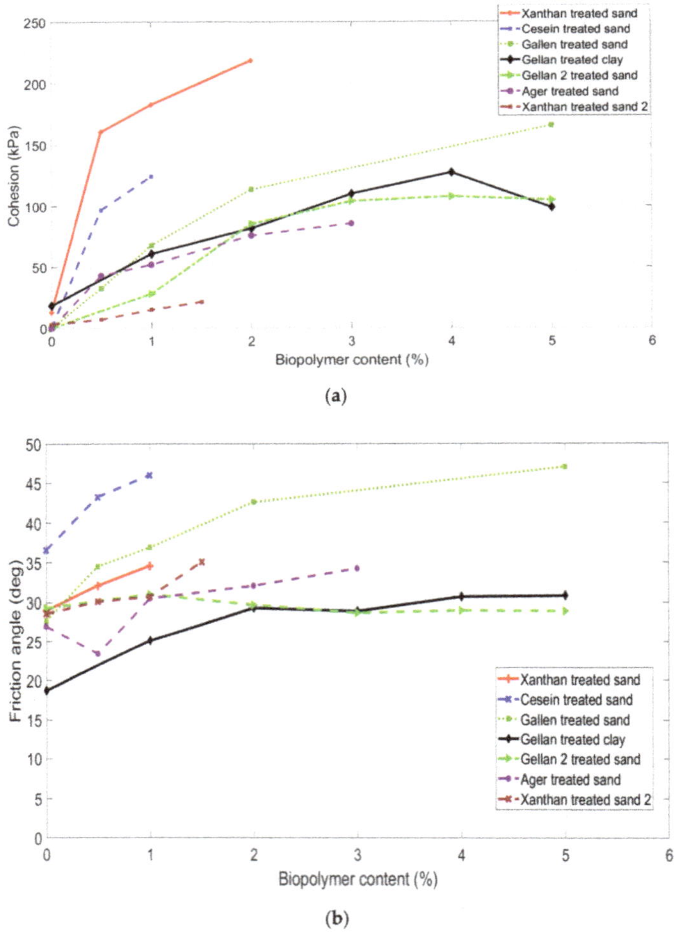

Figure 6. Variation of shear strength parameters of biopolymer-treated soils: (**a**) cohesion, (**b**) friction angle [82,123,150,156,157,159].

Residual strength is the lowest amount of stress that a material can tolerate: so that, if the applied stress is lower than residual resistance, no failure will occur [160]. The residual strength of sand treated by xanthan gum was also studied [159]; the results showed that the residual strength of dried samples experienced growth for both cohesion and friction angle. This growth could be attributed to the presence of van der Waals forces between xanthan hydrogel and sand aggregates. As biotreated sand samples remain ductile after failure, their residual strength was higher than sand samples improved by 10% and 20% gypsum. Unconsolidated-undrained triaxial tests also showed that xanthan and guar increased cohesion and frictional angle of coalmine waste. Xanthan indicated better enhancement compared to guar gum after 7 days of curing [125]. Also, Zhang et al. (2021) studied dynamic characteristics of lignin-treated silt. It was reported that dynamic shear strength and modulus were effectively increased using lignin, and a decrease was observed in damping ratio [147]. The schematic form of soil with and without biopolymer treatment under direct shear testing has been presented in Figure 7. The natural soil has a low shear strength due to the lack of cohesion among soil particles. The adhesive feature of biopolymer increases the cohesion among sand particles and leads to the formation of larger aggregation [124].

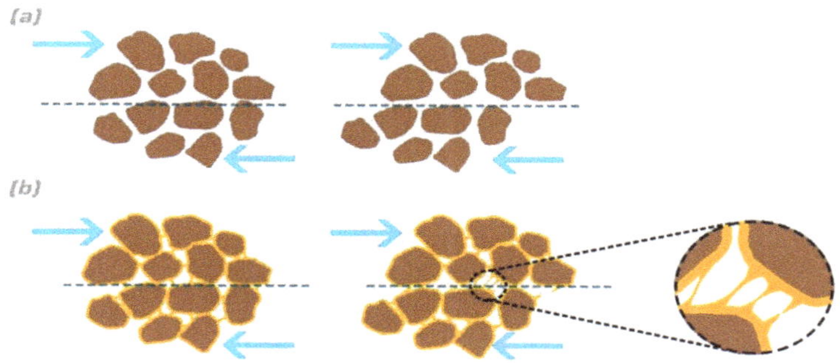

Figure 7. Schematic patterns of sand motion under direct shear testing: (**a**) natural sand state, (**b**) biopolymer-treated sand.

5.3. Erosion Resistance

Through the interaction between fluids (both water and wind) and soil, if the drag force of fluids overcomes both soil shear strength and gravity, soil erosion happens [161,162]. Biological clogging using microbial biopolymers plays a crucial role in reducing soil erosion by blocking the pathways of water within the pore spaces. Bioclogging leads to the formation of a less permeable layer and the reduction in water flow rate throughout the soil space [59]. It has been proven that using dextran improves the erosion resistance of soil and critical shear strength. This high erosion strength is because of increasing the cohesion and lowering the soil permeability and void ratio [34,163]. Under moderate and heavy rain simulation, the water erosion content decreased from 60% for untreated soil to under 2% for both beta-glucan and xanthan-treated soils, respectively [163]. Moreover, levee structures can be stabilized by employing xanthan gum against flood conditions, where stiffer soil surface with higher shear strength can be produced, thus excess pore water pressure is prevented from building up within the levee structures. This stabilizing process keeps the levee structures stable even after partial damage [127]. Water drip resistance test and water runoff test demonstrated that biopolymers can effectively change the surface properties of the sand and loose soils. However, carrageenan cannot stabilize the soil surface, and it should be used along with other biopolymers [164]. Hydraulic flume testing was utilized in order to simulate the hydraulic flow and evaluate the surface erosion resistance of biopolymer modified soil; it has been shown that adding xanthan gum resulted in 80% reduction in hydraulic erosion by making a protective jelly layer on the sand surface, and is even more effective than adding 10% of kaolinite clay to sand [127,128,165].

Another serious environmental problem occurring in dry sandy soils or anywhere covered with loose soil is wind erosion. Fugitive dust, as a major consequence of wind erosion, inevitably covers roads and crops leading to the reduction in farming productivity [166]. Biopolymers potentially improve the soil resistance against wind erosion by forming a firm crust over the sand surface, not even one crack was observed after a wind tunnel test on biopolymer treated samples (Figure 8). Xanthan gum and guar gum gave better performance compared to carrageenan and modified starch [120,166,167]. Moreover, the durability of xanthan and chitosan-treated soils is expected to be effective, at least up to two weeks against wind erosion [166,168]. In a relatively large-scale project aimed at dust mitigation for 90 days, synthetic polymers demonstrated better efficiency in the reduction of dust amount in comparison to guar gum as a biopolymer [169]. In terms of mine tailing stabilization, the weight loss of mine tailings treated by both xanthan and guar gums was smaller than water-treated sample; however, after five cycles of wetting and drying, the reduction was greater than the first cycle due to microstructural damage, wet-dry cycles, and ultraviolet (UV) radiation [170].

Figure 8. Crust formation of sand treated by sodium alginate biopolymer [145].

Figure 9 shows how soil resists wind erosion in both natural state and biopolymer-treated soil. As seen, a set of forces including resistant forces and driving forces exist during wind erosion. Driving forces consist of aerodynamic drag (F_d) and aerodynamic lift (F_L), which try to disconnect the soil grains from the soil mass. Resistant forces include the gravity force (F_g) and the interparticle cohesive force (F_i), that prevent soils from being eroded [167,171]. In the natural state of the sand, F_g is defined as the weight of one soil grain. Typically, F_i has a small value because of the low amount of interaction among soil particles so that the airflow is easily capable of eroding the soil. When the biopolymer is added to the soil mass and after drying, the biopolymer hydrogel directly interacts with the soil by coating the soil particles and forming strong chemical bonds, which leads to the formation of a network consisting of soil grains and biopolymer. In this condition, the F_g is the weight of the whole soil mass so that it has an amount much larger than that in the natural state. Also, the inter-particle force is significantly increased because of the biopolymer network. The additional resistant force results in a higher total restraining force than the driving force, which remains unchanged in both biotreated soil and natural state.

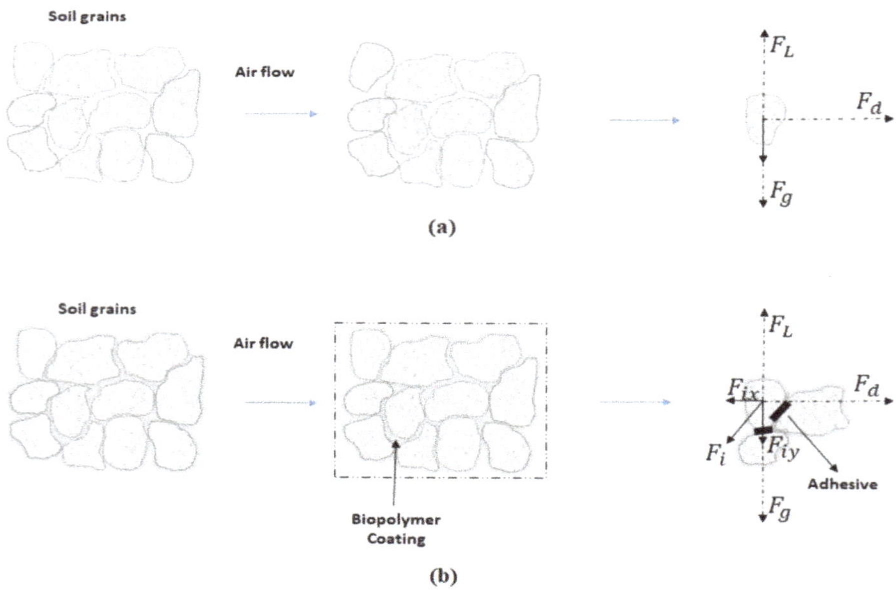

Figure 9. Mechanical trend of soil resistance against wind erosion: (**a**) sand in natural state, (**b**) biopolymer treated sand [154].

5.4. Atterberg Limits

Atterberg limits are considered a fundamental concept of soil consistency in soil classification. Atterberg limit tests define the boundaries where clay and silt go through four different physical changes (solid, semi-solid, plastic, and liquid) by moisture content variation [172]. Figure 10 demonstrates how different biopolymers can affect the Atterberg limits of soils. As seen, almost all biopolymers have an increasing effect on the liquid limit (LL) (Figure 9a) and plastic limit (PL) (Figure 10b). Also, the more the biopolymer content, the higher the PL and LL. However, the growth trends of LL and PL vary based on the biopolymer type and soil characteristics; this difference of growth rates results in having both decreasing and incremental trends for PI via the increase in biopolymer content. The enhancement of LL for beta-glucan treated clay is due to the high-water adsorption of beta-glucan, a linear correlation to approximate the PI based on beta-glucan content has also been presented [154]. Lignin-treated silt experienced a reduction in PI after rising the amount of lignin, while PL and LL were increased; due to the quick cation exchange between silt and lignin, the lignin coated soil particles and formed a strong network by filling the pores [173].

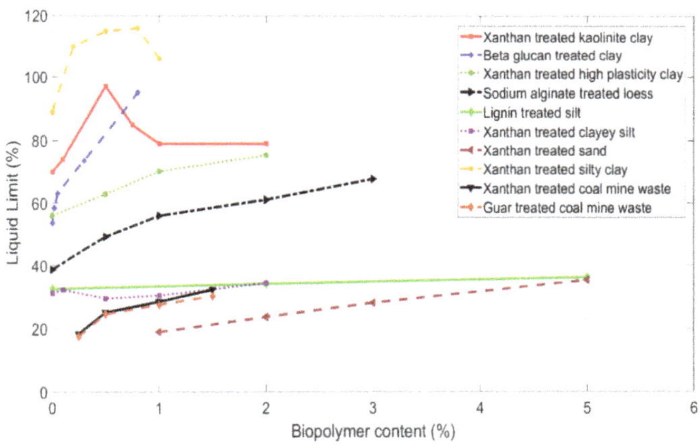

(a)

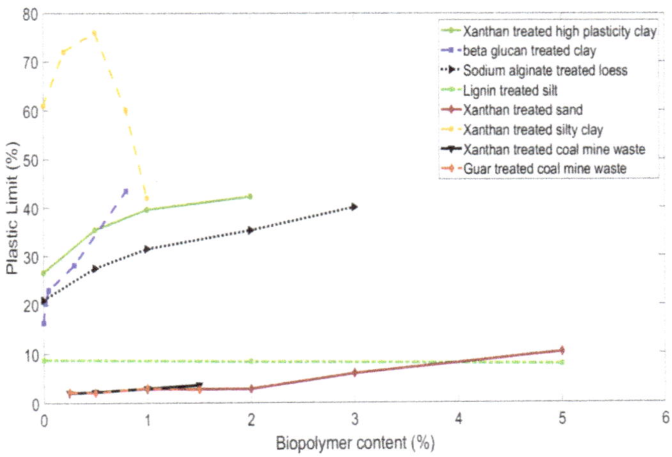

(b)

Figure 10. *Cont.*

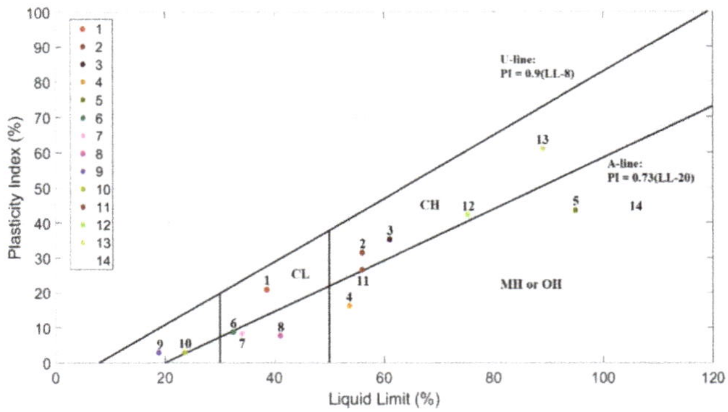

(c)

Details of the numbered coordinates

1	Loess	8	12% lignin-treated silt
2	1% sodium alginate-treated loess	9	1% xanthan-treated sand
3	2% sodium alginate-treated loess	10	2% xanthan-treated sand
4	Korean residual soil (KRS)	11	High-plasticity clay
5	0.8% beta glucan treated KRS	12	2% xanthan treated high-plasticity clay
6	Silt	13	Silty clay
7	2% Lignin-treated silt	14	1% xanthan treated silty clay

Figure 10. Variation of Atterberg limits by adding biopolymer, (**a**) liquid limit, (**b**) plasticity index, (**c**) plasticity chart [102,111,120,123,130,138,141,161].

For soft marine soil, it was shown that LL and PL had a tendency to increase with the increase in biopolymer to soil content, which is a result of forming hydrogel in pore space and direct chemical interaction with charged clay surfaces [134]. For sand treatment by xanthan gum, the hydrophilic feature of xanthan adsorbs water which provides the conditions to form xanthan hydrogel around and between sand particles, so that the behavior of soil becomes plastic and cohesive. This good property of treated sand improved the shear strength and wind erosion resistance of cohesionless sand [151]. Chang et al. (2019) studied the soil consistency characteristics for different kinds of clay mixed with sand employing xanthan biopolymer. The presence of sand in the clay reduced the value of LL and plasticity. Xanthan gum is able to transform the clay state from liquid to plastic for the contents more than 8% [134]. Also, trends for LL variation were not the same for different clay types. Xanthan-treated montmorillonite clay indicated a decreasing variation by the increase in xanthan content which is a consequence of the higher xanthan tendency to react with montmorillonite particles instead of pore fluid. The kaolinite-sand mixture initially experienced an increase in LL, then it decreased to an almost constant value; this shows xanthan simultaneously interacts with kaolinite particles and forms a hydrogel in the pores [137].

The relationships of the LL and PI for biomodified soils are presented in the plasticity chart (Figure 10c). It can be seen that adding sodium alginate into loess resulted in moving from the area of medium plasticity to high plasticity. This graph also shows that other biopolymer-treated soils tend to generate greater plasticity. This variation is directly related to the amount of surface charge and the size of clay particles in which biopolymers can be effective, in terms of hydrogel formation and electrical interactions. The cohesionless sand develops low plasticity after the addition of 1% xanthan gum. This improvement for desert sand is in accordance with achieving high erosion resistance as well as controlling soil degradation and desertification.

5.5. Compaction Test

The compaction behavior was different for various biopolymer-treated soils. In the case of beta glucan-treated KRS (Korean residual soil), both optimal moisture content (OMC) and maximum dry density (MDD) increase with increasing biopolymer content [154], whereas the MDD of biotreated collapsible soil by guar and xanthan gum reduced compared with untreated soil; this reduction was related to the lightweight collapsible soil grains which allowed them to move around because of the effect of solution viscosity [174]. For lignin-treated silt, the MDD tends to increase with increasing lignin content, but for lignin contents higher than 12%, the MDD reduced because 12% was reported to be the optimal biopolymer content, thus the biopolymer did not act as a stabilizer. In other words, it can be said that all the pores were filled by the biopolymer, and the larger biopolymer content leads to the accumulation of lignin and the reduction in soil density. OMC was also affected by the lignin in such a way that OMC would slightly decrease compared with natural silt: it should be noted that this variation can change depending on the biopolymer and soil characteristics [58,173].

For sand treated by xanthan gum, it has been reported that the MDD and OMC behaviors were symmetrical such that 1% biopolymer caused the highest MDD and lowest OMC to be reached, then by increasing the biopolymer content, MDD reduced and OMC started to increase. The initial growth in MDD was due to the lubrication effect of biopolymer until 1%, while for higher content, xanthan showed a negative effect so that a swelling behavior of the treated soil was observed [151,175]. The xanthan-treated clay showed different behavior in comparison to the biomodified sand. It has been shown that MDD reduced when 3% xanthan was added to the clay and OMC increased [114]. As most of the biopolymers used for soil improvement purposes are hydrophilic and tend to absorb water, the moisture content of the biopolymer-treated soil should be strictly controlled to achieve the appropriate degree of compaction, especially at the construction site. To have a comparative outlook, it should be mentioned that increasing the content of traditional and calcium-based traditional stabilizers typically leads to the reduction in MDD and the increase in OMC [173].

5.6. Durability

The variation of geotechnical properties of soil treated by different polymers under different environmental conditions including cycles of wetting and drying as well as freeze-thaw, and decomposition has always been a concern [176]. As most biopolymers employed in soil improvement are soluble in water, exposure to water would surely affect their performance during wet–dry cycles. After the first dehydration of gellan-treated sand, once the dried samples were subjected to water, the dehydrated gel began to absorb water and undergo volumetric expansion. Then, the outer parts of the gel are slightly detached from the main gellan gel. This process leads to a reduction in the compressive strength for the wet sample compared to the dried state. During redrying, dissociated gellan gum fibrils attach to the main body again due to a lowering in moisture content. The wet and dry cycle leads to the detachment and attachment of some minor parts of the gel which will result in weakening of the biotreated sample and incomplete recovery of all hydrogel. Repeating more cycles causes the compressive strength to gradually reduce and converge to a specific level. Moreover, the stiffness follows a similar trend and shows a more ductile behavior than traditional binders such as cement and lime which experience a brittle collapse after a limited number of wetting and drying cycles [177].

The durability of treated soils has also been studied using xanthan gum through various experiments. Slake durability tests indicate that xanthan gum is capable of improving slake durability index to a value that is comparable with cement-treated sand. A higher amount of biopolymer led to better performance against wet–dry cycles. Although xanthan gum adsorbed water which subsequently reduces the strength, the hydrogel remained available for the subsequent cycles, that is why some proportions of xanthan biopolymer performed better than cement [151,178]. As for decomposition, 750 days of curing the Red Yellow soil enhanced by xanthan gum without wetting and drying slightly increased the compressive strength and stiffness in comparison to 28-day cured samples. This procedure was also the same for curing time of 21 and 63 days for sand and clay treated by xanthan gum which could be attributed to the high stability of xanthan gum under temperature variation and dehydration [113]. In addition, keeping samples for 30 days under real conditions with different temperature, humidity, as well as wet-dry cycles indicated a considerably better performance for xanthan treated sample compared with untreated silt [117]. However, it has been presented that guar gum rendered a greater efficiency than xanthan gum in water erosion after five wet-dry cycles, whereas carrageenan showed poor performance in this regard and disappeared after the second cycle [167]. Furthermore, xanthan gum-treated silt performed much better than virgin high plasticity silt after 12 days of freezing and thawing cycles [118]. The results in the literature demonstrate that the durability of biopolymers in long-term applications requires more attention in order to explain how they can be affected by various environmental and destructive circumstances such as wetting and drying cycles, freezing and thawing cycles, microorganisms, and UV attack [179,180].

5.7. Pavement Application

Currently, there are limited studies on the role of biopolymers in the enhancement of pavement layers. The California bearing ratio (CBR) and resilient modulus tests are parameters of evaluation for pavement layer stabilization. A considerable increase has been reported in the resilient modulus of silt and clay by the addition of sodium alginate. Up to 2% for clay and 4% for silt, the trend was increasing, then a declining trend was observed for larger proportions. The better performance of sodium alginate in clay enhancement is due to the existence of sufficient free cations tending to react with an anionic biopolymer and forming a gelling structure, and filling the pores; while the available cations may not be enough to turn biopolymer into the gel form [181]. Sodium alginate also has significant efficiency in enhancing the unsoaked CBR of dune sand [66]. A very low amount of xanthan (0.4%) reduced 36% of the required thickness for the unstabilized sand

Resilient modulus (RM) also increased when lignin, up to 12% (optimal content), was added in silt which was in line with the UCS results [58,182–184]. This coincidence was the same for adding protein-based biopolymers including casein and sodium caseinate into the dune sand [82]. Thus, it can be seen that the variation of pavement parameters (RM and CBR) for biopolymer-treated soil is in line with the UCS results. The criteria of pavement design consider a minimum amount for maximum compressive strength obtained from the UCS test; this standard value is determined based on specific conditions in different countries. The feasibility of using xanthan gum for sub-base stabilization as well as road shoulder construction has been investigated and compared with cement and geopolymers. The results showed that the UCS of xanthan-treated silty sand passed all design criteria in terms of the required UCS. Also, the higher ductility of xanthan-treated soil was a positive impact if compared against cement and fly ash [122]. However, more studies are still required to comprehensively investigate how different types of biopolymer behave under various environmental and loading conditions.

6. Factors Affecting Geotechnical Properties of Biopolymer-Treated Soils

Using biopolymer binders for soil improvement should be applied in optimal conditions because various parameters can influence the biomodified soil characteristics. In this section, the important factors in geotechnical aspects of stabilized soils have been critically evaluated. As the experimental conditions for these results vary in many aspects, such as the soil types and curing conditions, they cannot simply be judged by the same standards, hence specific understanding of each test condition or consideration is important.

6.1. Biopolymer Concentration

Biopolymer content is usually the first parameter investigated in many studies of biopolymer-treated soils to develop an understanding of how the soil would eventually react to the additives. Concentration is mostly demonstrated based on the biopolymer ratio to the soil mass. In the dry state, the required biopolymer content for reaching a specified strength is much lower than that of cement or other traditional stabilizers [150]; however, lignin needs a higher quantity compared with other biopolymers [58,185]. The untreated soil strength has often been less than biopolymer-treated soil regardless of biopolymer content. The effect of biopolymer content on the unconfined compressive strength for biomodified soils in their dry states has been presented in Figure 11. As is the case for treated sands, an increasing trend can be observed for almost all samples when the biopolymer content increased, while this observation does not seem to apply to xanthan that tends to reduce by adding biopolymer higher than 2% by weight. When an excessive amount of xanthan is added to the soil, the interaction of sand and hydrogel is reduced by concentrating in localized places. Protein based biopolymers (casein and sodium caseinate) continuously enhanced the sand compressive strength significantly up to 5% biopolymer content. Casein-treated sand indicated a higher growth rate in compressive strength in comparison to agar and gellan-treated sands [66,82,149–151,186].

The comparative results of silt-treated soils are presented in Figure 11b. The procedure is also similar to coarse-grained soil improvement and it was observed that all biopolymers improved the UCS. It was shown that UCS increases significantly when biopolymer, such as xanthan or beta-glucan, is added to the soil. It was observed that considerable improvement takes place until an optimal level is reached, after which its rate of improvement reduces e.g., lignin-treated silt. It is noteworthy that the respective optimal amounts for (a) lignin to treat low-plasticity silt is 12%, (b) xanthan to treat bentonite and kaolinite clays is 1% and 1.5%, respectively, (c) guar gum to treat silt with low plasticity is 1%, and (d) xanthan on low-plasticity silt is 2%. Higher lignin contents led to the excessive local concentration within the soil pores so that the reduction in frictional resistance would be a consequence of this effect [173]. Also, Soldo et al. (2020) observed that the increase in the treated soil's Young's moduli increases in a similar fashion as the optimal concentration for xanthan and guar gum for improving UCS [117]. However, for biopolymers including beta-glucan and casein, the optimal content cannot be accurately specified due to the variable trends of results. These optimum values for some mixtures of biopolymer–soil indicate that the optimized quantities should be rigorously investigated for more types of biopolymer-treated soils to pave the way for using biopolymers as alternative sources to traditional materials.

Sodium alginate and xanthan reduced the initial hydraulic conductivity of silty sand, and higher biopolymer contents caused lower permeability, because of the high tendency of xanthan gum to adsorb water and clay minerals [187]. Using a sufficient concentration of casein has also satisfied the required UCS for typical (non-fired) adobe blocks (UCS $\geq$ 1.32–1.56 MPa), as well as traditional adobe block construction (UCS $\geq$ 1.18 MPa) [107,188–190]. Although, the minimum UCS required for fired block (UCS $\geq$ 10.3 MPa) is larger than the UCS obtained for casein-treated soil [191]. Besides, it has been indicated that the higher the biopolymer content, the higher the shear strength. In terms of improving erosion resistance, the increase in adhesive content also resulted in better performance.

6.2. Moisture Content

Water susceptibility is one of the main problems of using biopolymers in soil improvement because dehydrated biopolymer tends to absorb water and then swell. This is due to the hydrophilic property of most biopolymers that results in weakening the geotechnical behavior of treated soils [192]. Nevertheless, the mechanical strengths of bio-improved soil in saturated conditions are mostly higher than untreated soil samples. Agar and gellan-treated sand under submerged conditions showed behavior that was independent of time, with the compressive strength remaining constant with time. This fixed strength can be explained by the hypothesis that water content governs the mechanical properties of these thermogelatin-treated soils rather than time. Therefore, if high UCS values are desired in dry conditions, reduction of water is the best option [62].

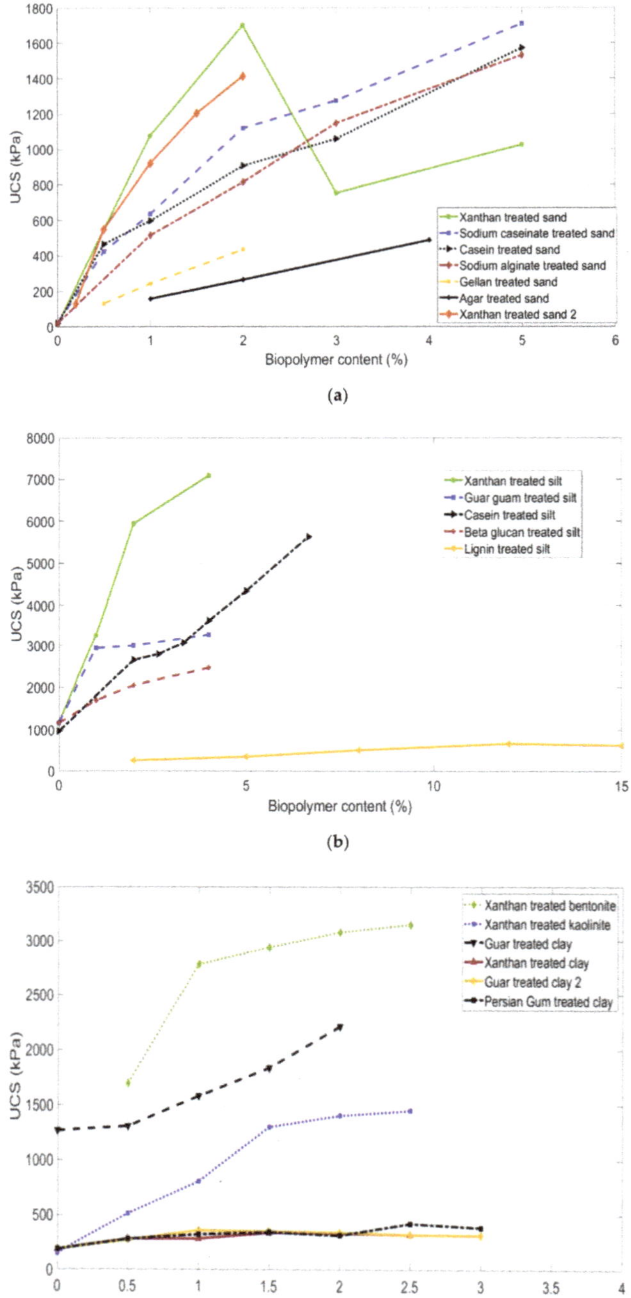

Figure 11. Effect of biopolymer content on compressive strength of soils (**a**) biopolymer-treated sand, (**b**) biopolymer-treated silt, (**c**) biopolymer-treated clay [56,70,94,97,103,135,136,138,161].

The geotechnical features of treated soils with the same amount of moisture could be different in initial and submerged states. Resubmerging the samples after drying provides the conditions for biopolymers to absorb water and return to their hydrogel state, and the water content would often be less than the initial state [62,107]. For gellan gum, the lower results after resubmergence can be explained by the fact that gellan gum is not recoverable after one dehydration [150]. Casein performed better in the wet state than some biopolymers such as beta-glucan, xanthan gum, gellan, and agar. The reason for the poor efficiency of these biopolymers is that the strengthening procedure of polysaccharides is mostly governed by hydrogen bonding and hydrogel behavior [15,107].

The most fundamental concern of collapsible soils is related to being saturated. Adding xanthan gum improved this drawback significantly by increasing the wet shear strength [174]. In addition, loess soil, which is unstable for undergoing the effect of water and rainfall, was improved by adding sodium alginate, which helped the loess to stay totally in shape after a long period of soaking, while the non-treated loess was lost completely after 30 s. Therefore, sodium alginate could prevent a loess slope from becoming saturated and enhance slope stability [144]. In terms of shear strength, 2% xanthan gum increased the cohesion and total shear strength of the soil, while the friction angle was reduced compared to untreated sand. At the initial condition, the hydrogel bond strength is responsible for the increase of shear resistance because, at this stage, there is still no direct chemical interaction between biopolymer and the host soil. Meanwhile, in the resubmerged state, the residual biofilm close to sand particles along with hydrogel helps to increase the shear strength of the sand [159].

Figure 12 demonstrates how moisture delays the formation of direct contact between the clay surface and biopolymer. The presence of water and biopolymer together means more time will be needed for biopolymer-treated soil to dry, which can be attributed to the hydrophilicity and water retention of biopolymers [193]. More importantly, the initial moisture content is usually taken as OMC obtained from the compaction test that is not necessarily the best choice for the highest compressive strength. For a clay treated by 1% xanthan gum, 4% greater water content than OMC of treated samples led to a more prominent stress-strain behavior. Given this, there could be an ideal initial moisture content for each biopolymer–soil mixture [131].

6.3. Temperature

Elevated temperature is a parameter that can affect the mechanical behavior of a biopolymer. The most affected biopolymer group used in soil improvement is related to thermogelatin biopolymers including agar and gellan. Thermo-gelation biopolymers tend to dissolve and are suspended in hot water (between 80 and 90 °C); following that, they coagulate into a gel state after the temperature decreases to below 50 °C [194]. This thermal process provides the conditions to obtain the highest efficiency of these biopolymers, which is why agar and gellan resulted in higher compressive strength of the soil compared to the biomodified soil without thermal treatment, as well as higher stability in submerged conditions [62]. Air-dried curing (20 °C) caused the formation of a solid layer in the external part, which then led to a delay in curing the inner part of the sample. A similar sample cured at 40 °C experienced a uniform cementitious process on both external and internal sections. Therefore, drying at 40 °C gave a greater performance compared with air-dried conditions [124].

Other types of biopolymer have also presented a similar behavior when the biopolymer-treated soil is subjected to curing at different thermal conditions, in the way that raising the temperature to an optimal value would be quite beneficial for improving the strengthening process. Then, temperatures higher than the optimal value led to a loose structure, and sometimes higher temperatures led to decomposition of the biopolymers because of weakening the bonds and connections amongst biopolymers and soil grains became weaker [66,82,115]. For example, temperatures more than 60 °C for casein-treated sand caused the compressive strength to fall to a low of 48% which was similar to the consequence of decomposing casein in amino acid [82]. According to the literature review, the optimal temperature for different biopolymer-treated soils can be expressed as follows: 45 °C for sodium alginate-treated sand, 60 °C for both casein and sodium caseinate-treated sand, and 60 °C for beta glucan-treated silt [66,82,115].

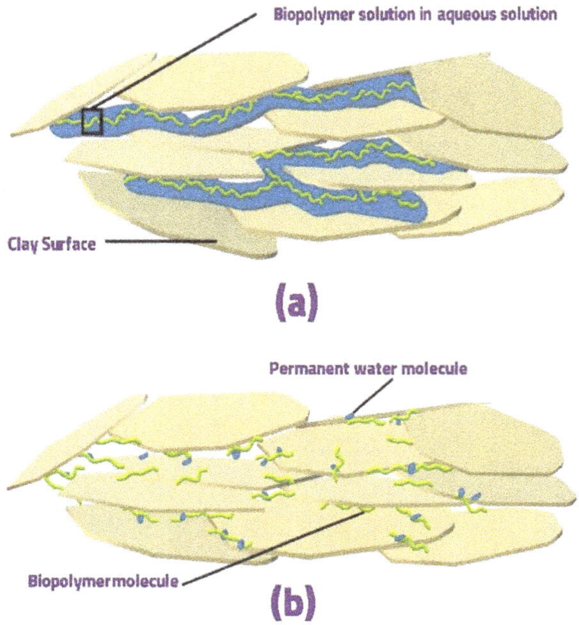

Figure 12. Moisture effect in interaction of biopolymer with clay sheets (**a**) moist biotreated sample, (**b**) dried biotreated sample [121].

6.4. Dehydration Time

After preparation of the biopolymer–soil mixture, hydrophilic biopolymers try to absorb water and form hydrogel networks within the soil pores; thus, it is the inherent geotechnical properties of the host soil that govern the eventual strength of the improved soil. However, it has also been shown that some biopolymers can play a role in increasing the initial compressive strength of soil [80,177]. Moreover, it is also not feasible to perform unconfined compressive tests on biotreated sand when it is still wet because the biopolymer has not effectively attached to the soil particles [66,82]. It is thought that curing duration is directly related to the moisture content, therefore, elapsed time provides the hydrogel the opportunity to lose water and dehydrate.

Figure 13 demonstrates the optimal curing time needed for each type of biopolymer-treated soil. In this chart, the optimal time is considered the number of days that are needed for a biopolymer-treated soil to obtain greater than 80% of its final strength. As shown, based on the biopolymer, the optimal curing time could be different for various cases because of the biopolymer type as well as the soil type. For instance, the optimal time for different types of clay treated by xanthan gum is usually around 28 days, while this is less than five days for xanthan-treated silt [111,114,117,118,133]. Moreover, the higher amount of biopolymer could also increase the dehydration process, because more water would be absorbed by the biopolymer [195].

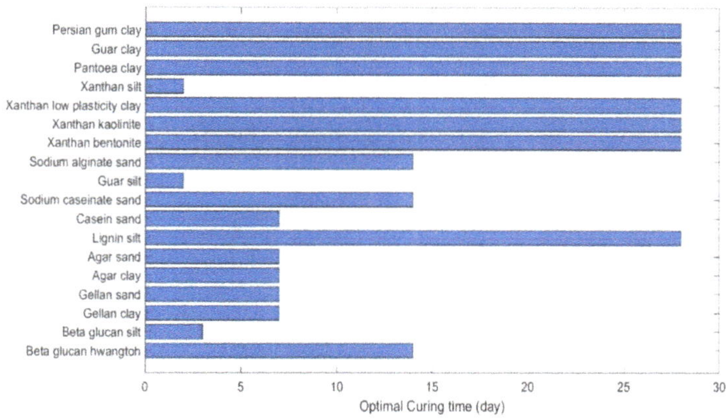

Figure 13. Optimal time required for each biopolymer-treated soil [58,62,66,82,111,115,130,133,138,158,174,196,197].

6.5. *Effect of Adding Fiber*

Adding fiber to soils offers a solution to increasing the ductility (in case of brittle behavior) and enhancing the swelling behavior of clay. This subject has still not been comprehensively investigated, however, some research indicated that adding fiber was effective in soil treatment. Both plant-based and animal-based fibers can be applied regarding this matter. Date palm fibers along with xanthan gum were used to enhance dune sand. Results showed that fiber increased the vane shear strength of the treated sand at a low concentration of xanthan, while high xanthan contents (2%) experienced almost the same strength with and without adding fiber [175]. The addition of sheep wool along with sodium alginate showed significant growth in the compressive strength performance of cohesive soil from 2.23 MPa for natural soil to 4.44 MPa for reinforced biotreated soil. It should be noted that wool amount should be limited to keep the high strength. Also, the flexural strength was increased by 30% using a mixture of wool and alginate [196].

7. Potential Geotechnical Applications of Biopolymer-Treated Soil

As shown, biopolymers can improve different soil characteristics so that they can be useful in various geotechnical applications, some of which are discussed in this part of the paper.

7.1. *Base and Sub-Base Stabilization*

A successful pavement design satisfies the minimum structural criteria in terms of shearing resistance, excessive deflections, and reduction of permanent deformations, such that soil binders are used to ameliorate problematic soil properties and reduce the required thickness of layers. When it comes to selecting a stabilizer, the following parameters are of the highest importance: soil type, soil improvement type, required durability and strength level, cost, and environmental considerations. General employable materials for pavement improvement are cement, lime, lime-cement blends, bitumen, granular materials, and chemical products. A comprehensive geotechnical investigation must be undertaken before and after stabilization to assess how a binder changes soil property. These geotechnical tests include CBR, UCS, repeated load triaxial test, wheel-tracking test, indirect tensile modulus, flexural modulus, capillary rise test, vertical saturation test, erodibility test, and leaching test [197].

Currently, there are limited studies on the role of biopolymers in the enhancement of pavement layers. CBR and resilient modulus tests are parameters of evaluation for pavement layer stabilization. A considerable increase has been reported in the resilient modulus of silt and clay by adding sodium alginate. Up to 2% for clay and 4% for silt, the trend was increasing, then a declining trend was observed for larger proportions. The better performance of sodium alginate in clay enhancement is due to the existence of sufficient free cations tending to react with an anionic biopolymer and form a gelling structure and filling the pores; while the available cations may not be enough to turn biopolymer into the gel form [181]. Sodium alginate also has significant efficiency in enhancing the unsoaked CBR of dune sand [66]. A very low amount of xanthan (0.4%) reduced 36% of the required thickness for the unstabilized sand.

Resilient modulus (RM) also increased when lignin, up to 12% (optimal content), was added in silt which was in line with the UCS results [58,182]. This coincidence was the same for adding protein-based biopolymers including casein and sodium caseinate into the dune sand [82]. Thus, it can be seen that the variation of pavement parameters (RM and CBR) for biopolymer-treated soil is in line with the UCS results. The criteria of pavement design consider a minimum amount for maximum compressive strength obtained from the UCS test; this standard value is determined based on specific conditions in different countries. The feasibility of using xanthan gum for sub-base stabilization as well as road shoulder construction has been investigated and compared with cement and geopolymers. The presented results showed that the UCS of xanthan-treated silty sand passed all design criteria in terms of the required UCS. Also, the higher ductility of xanthan-treated soil had a positive impact compared to cement and fly ash [122].

7.2. Increasing Soil Bearing Capacity for Shallow Foundations

Shallow foundations are designed based on an allowable bearing capacity that must meet the safety factor for the prevention of shear failure.. When the applied load exceeds the soil bearing capacity, the footing penetrates the soil either in horizontal or vertical directions so that it could lead to a bearing capacity failure. Soil bearing capacity is variable for different soil conditions and is dependent on several factors [198]. A number of works have already been studied the shear parameters of biopolymer treated soils especially for cohesion increment. A good result is expected in the case of biopolymers as soil binders. In a numerical model by PLAXIS 2D, it was shown that low amounts of guar and xanthan increased the soil bearing capacity, attenuated the negative effect of saturation degree, and decreased the settlement during and after saturation [174].

7.3. Landfills

One of the most common methods of waste disposal and management is landfill. The surface system of a landfill needs to be sufficiently resistant against wind and water erosion, promote vegetation if present by storing water, and satisfy aesthetic and ecological criteria. The cover system prevents the underlying layers from exposure to damaging environments, such as freeze-thaw, desiccation, and ultraviolet light. Landfills comprise a very low-permeable barrier system to delay the contamination of leachate and toxic components into the soil and groundwater system. The permeability of liners in a landfill system should not exceed 10^{-7} cm/s, and the thickness must be more than 600 mm [199]. Clay liners with low hydraulic conductivity are used in landfill systems, but their drawbacks, mostly because of high volume change and cracking as well as non-renewability, limit their use [200]. Stabilizing the host soil with environment-friendly binders can be considered as an alternative to clay liners.

Biopolymer treated soils considerably improved the erodibility resistance by maintaining the soil ability of vegetation growth. Also, absorbing water due to the hydrophilicity of biopolymers enables the cover system to store water. The ability of vegetation promotion growth for biopolymer treated soil was investigated in a large-scale study for a one year period; biopolymer increased the vegetation density and higher sprouts were observed compared to the untreated soil [43]. Around 2.72×10^5 times decrease in hydraulic conductivity was recorded when guar gum was added to the soil, while the shear strength was also increased by more than 70%. Furthermore, durability indexes under wet-dry cycles were placed in the permissible range. It is also notable that biopolymer treated soil is capable of enhancing the heavy metal attenuation capacity of the soil, causing less harmful leachate containing heavy metal to pass through the soil to groundwater [143]. Therefore, with the great potential of biopolymer-treated soil in landfill application, further use of biopolymers in research and practical applications is expected in the future.

7.4. Biopolymers in Slurry Barrier Trench Excavations

Bentonite slurry is used to stabilize the sides for trench excavation, which is called the slurry trench method of excavation. This technique enables excavation where a high groundwater profile exists, and the trench needs to be extended below the water level [201]. During the excavation process, the slurry is pumped into the trench to prevent the sides from collapse by applying hydrostatic pressure towards the walls, penetrating inside the pores, and forming a filter layer over the side surface. Usually, bentonite is the main material in this technique, but recently biopolymer slurry has shown a potential to be utilized as an alternative. The high gel strength of the biopolymer and low tendency to lose water make it capable of being used in this technique [202]. To stop internal erosion in an embankment dam, a biopolymer slurry trench can be excavated with partial penetration to reduce seepage occurring on the downstream slope in case the excavation should be continued beneath the water level. In 1997, guar gum slurry was utilized to stabilize sidewalls during the excavation process of installing a permeable iron barrier at Oak Ridge, Tennessee, USA. The biopolymer slurry trench method has also been used in several other projects in different parts of the world [203].

8. Prospects and Limitations of Biopolymer in Soil Treatment

It can be said that the most important reason for introducing biopolymers in soil strengthening is to mitigate the harmful effects of other environmentally unfriendly materials. Biopolymers would provide the opportunity of using sustainable materials with low impact on the environment as well as groundwater. Moreover, it has been shown that a smaller content of biopolymers would be able to obtain the same or higher strength compared with cement. Unlike some other soil improvement methods, the presence of fine soil is not considered an obstacle to using biopolymers, while using the MICP technique is also limited to coarse-grained soils, to reach a uniform distribution of precipitated calcite. Introducing biopolymer into the soil can be performed through various methods such as grouting, injection, mixing, and spraying [9,15,204]. In comparison to the MICP method, less time and complexity are needed in terms of the application and productivity of using biopolymers. In addition, biopolymers have the potential of mass production as well as relatively immediate reaction with soil particles, which enable them to be employed in temporary and rapid applications [15]. Furthermore, artificial intelligence methods and algorithms could be readily developed and deployed to optimize the performance of biopolymers as a viable soil improvement method. For example, genetic programming neural networks or Bayesian network have been successfully used in the study of geopolymer concrete and brick, which could be emulated for biopolymers [205,206].

More research needs to be performed to resolve some limitations of biopolymer usage. The characteristics of biopolymer-treated soil can change under real environmental conditions such as pH, temperature, wet-dry cycles, as well as ultraviolet radiation [207]. Additionally, the durability of biopolymer mixed with soil should be more comprehensively investigated, especially for degradation because of microorganisms. The types of microorganisms can differ completely from one soil to another, which means that each soil might need to be specifically evaluated in terms of biodegradability and duration. Also, a biopolymer–soil mixture may work well at a certain time of the year, while it has poor workability in a different season or atmospheric condition. Biopolymer-treated soil performance under saturated conditions should be studied further, as most biopolymers are quite sensitive to water.

Few studies have been performed to evaluate the effect of biopolymers in field conditions. Recently, a biopolymer was used to control the soil erosion in the field. Xanthan, casein, and xanthan-starch compound were utilized. Monitoring was carried out after 100 days to evaluate the vegetation density and shear strength. Biopolymer-treated soils exhibited an increase in shear strength compared to the untreated soil. In addition, the vegetation density increased considerably due to the addition of biopolymers [43]. The feasibility of utilizing biopolymers for pavement stabilization was studied in the treatment of a 50 m-long pedestrian trail. The effectiveness was observed in enhanced surface stiffness and erosion resistance [43].

In terms of economic feasibility, using biopolymers in geotechnical applications is currently less feasible in comparison to traditional materials. The trend towards more commercialization during the last few decades has led to the cost reduction of some biopolymers, such as xanthan gum, the price of which has decreased from 30,000 US$/ton to 3000 US$/ton over 20 years [150]. Furthermore, the existing prices are usually for the high-quality grade of biopolymers, which are mostly used in food and medical applications. In recent years, lower-grade biopolymers have been successfully used in soil improvement work [18]. Successful identification of new products or bacteria strains [208] would enhance productivity as well as applying a wider use of technological tools [209] incorporating soil improvement methods [210] and would thus provide another major step towards mass application on real-life projects [211–215].

9. Conclusions

Adding biopolymer to act as a binder into the host soils is one of the most effective applications for soil treatment. In this study, a technical review has been conducted on the role of biopolymers in soil improvement and how they change the geotechnical properties of soils.

- Biopolymers from various sources have been employed. The most common source is biopolymers produced from plants. Animal-based and microorganism-based biopolymers have demonstrated great potential for geotechnical purposes. The ocean is also a valuable resource for extracting biopolymers from marine plants and animals.
- Sand particles are neutral in charge, such that the biopolymer adhesion mechanism is achieved via coating the soil particles and forming bridges amongst the particles. On the other hand, clay sheets tend to directly interact with biopolymer particles through different electrostatic interactions such as hydrogen bonding, ionic bonds, as well as van der Waals bonds. This behavior is due to the existence of negative charges and cations in clay minerals.

Important points of the geotechnical characteristics of biopolymer-treated soils are as follows:

- The unconfined compressive test is usually considered the most common test method to evaluate the behavior of biopolymer-treated soils. The general conclusion is that biopolymers increase the UCS values; however, the UCS improvement could be variable depending on the biopolymer and the soil testing conditions. Xanthan-treated sand achieved a compressive strength that was comparable to cement.
- The reported results of direct shear tests have also shown improvements in shear strengths, in particular on increasing the cohesion. Internal friction amongst soil particles experiences relatively lower strength improvement during treatment.
- By adding biopolymers, stiffer soil surfaces with improved strength could be achieved to resist wind and water erosion. Using biopolymers to control wind erosion is an effective short-term technique that could be resistant for at least one or two weeks.
- The physical soil properties have also been affected by adding biopolymer such that higher additive content led to higher liquid and plastic limits. However, this trend for the plasticity index (PI) was not the same, and both decreasing and increasing trends were observed, respectively.
- Variation of both the optimal moisture contents and maximum dry densities could vary due to various biopolymer and soil types. As many of the biopolymers employed for soil improvement are soluble in water, the moisture content is an important

parameter for further research so that the improved soils achieve the desirable compaction conditions.

- The lack of research can be observed in the role of biopolymers used in pavement applications. However, the effective performance of biopolymers in increasing the CBR strength, resilient modulus (RM), and UCS has been reported. However, further research is still needed to assess its long-term cyclic durability under traffic or seismic loadings.

Various factors can affect the biotreated soil behavior. The most important factors are presented below:

- One advantage of using a biopolymer as a binder is the relatively low content needed to achieve a compressive strength that is comparable to traditional materials such as cement and lime. For some biopolymers, it can be seen that an optimal amount must be ascertained otherwise reduction in soil strength may be expected. Compared with other biopolymers, greater lignin content is needed to reach the optimal content.
- Sensitivity to water is considered to be a fundamental issue of biopolymers in soil treatment; strength loss and apparent swelling/shrinkage behavior are the drawbacks. The maximum strength of biopolymer-treated soil is obtained in the dry state and higher moisture content will lead to lower strength. Susceptibility to water is also different from one biopolymer to another.
- The durability of the biopolymer-treated soil requires more attention because various environmental conditions including wetting–drying cycles, freeze–thaw cycles, microorganisms, and ultraviolet radiation could significantly decrease the strength. For gellan gum, exposure to wet–dry cycles resulted in major damage to the biopolymer matrix and thus loss of strength. However, xanthan gum offered a relatively better efficiency under wet–dry cycles.
- Biopolymer-treated soil properties could change significantly under different thermal variations. Thermogelatin biopolymers, such as agar and gellan gums, are among the biopolymers most susceptible to temperature. Some biopolymers (casein, alginate, and beta-glucan) perform better at their optimal temperatures. Generally, high temperatures may lead to the decomposition of the biopolymers.
- The moisture content of the biopolymer soil mixture has a close relationship with dehydration time, in such a way that time causes the biotreated sample to lose its moisture and form the linked network. If the samples are under cured in normal atmospheric conditions, it takes time for the specimens to achieve more than 90% of their maximum strength value, which could be considered as the optimal dehydration time. This optimal value could vary from 3 to 28 days, depending on soil and biopolymer types.
- The brittle behavior of biopolymer-treated soils could be modified by adding fiber into the mixture to increase their flexibility. Along with synthetic products, natural fibers such as date palm fibers and animal wools have demonstrated a positive impact on the biotreated soil's characteristics.

Author Contributions: Conceptualization, H.F., D.E.L.O., J.Y. and I.C.; investigation, H.F., D.E.L.O., J.Y. and I.C.; writing—original draft preparation, H.F.; writing—review and editing, D.E.L.O., J.Y. and I.C.; supervision, D.E.L.O., J.Y. and I.C. All authors have read and agreed to the published version of the manuscript.

Funding: This research received no external funding.

Conflicts of Interest: The authors declare no conflict of interest.

References

1. Nicholson, P.G. *Soil Improvement and Ground Modification Methods*; Butterworth-Heinemann: Oxford, UK, 2014.
2. Leong, H.Y.; Ong, D.E.L.; Sanjayan, J.G.; Nazari, A. Strength development of soil–fly ash geopolymer: Assessment of soil, fly ash, alkali activators, and water. *J. Mater. Civ. Eng.* **2018**, *30*, 04018171. [CrossRef]

3. Saberian, M.; Li, J.; Boroujeni, M.; Law, D.; Li, C.-Q. Application of demolition wastes mixed with crushed glass and crumb rubber in pavement base/subbase. *Resour. Conserv. Recycl.* **2020**, *156*, 104722. [CrossRef]
4. Saberian Boroujeni, M.; Li, J.; Nguyen, B.; Saberian Boroujeni, M. Experimental and analytical study of dynamic properties of UGM materials containing waste rubber. *Soil Dyn. Earthq. Eng.* **2019**, *130*, 1–12. [CrossRef]
5. Correia, A.G.; Winter, M.; Puppala, A. A review of sustainable approaches in transport infrastructure geotechnics. *Transp. Geotech.* **2016**, *7*, 21–28. [CrossRef]
6. Cola, S.; Schenato, L.; Brezzi, L.; Tchamaleu Pangop, F.C.; Palmieri, L.; Bisson, A. Composite anchors for slope stabilisation: Monitoring of their in-situ behaviour with optical fibre. *Geosciences* **2019**, *9*, 240. [CrossRef]
7. Toghroli, A.; Mehrabi, P.; Shariati, M.; Trung, N.T.; Jahandari, S.; Rasekh, H. Evaluating the use of recycled concrete aggregate and pozzolanic additives in fiber-reinforced pervious concrete with industrial and recycled fibers. *Constr. Build. Mater.* **2020**, *252*, 118997. [CrossRef]
8. Jahandari, S.; Saberian, M.; Tao, Z.; Mojtahedi, S.F.; Li, J.; Ghasemi, M.; Rezvani, S.S.; Li, W. Effects of saturation degrees, freezing-thawing, and curing on geotechnical properties of lime and lime-cement concretes. *Cold Reg. Sci. Technol.* **2019**, *160*, 242–251. [CrossRef]
9. Chang, I.; Lee, M.; Cho, G.-C. Global CO_2 Emission-Related Geotechnical Engineering Hazards and the Mission for Sustainable Geotechnical Engineering. *Energies* **2019**, *12*, 2567. [CrossRef]
10. Etim, R.; Eberemu, A.; Osinubi, K. Stabilization of black cotton soil with lime and iron ore tailings admixture. *Transp. Geotech.* **2017**, *10*, 85–95. [CrossRef]
11. Rahgozar, M.A.; Saberian, M.; Li, J. Soil stabilization with non-conventional eco-friendly agricultural waste materials: An experimental study. *Transp. Geotech.* **2018**, *14*, 52–60. [CrossRef]
12. Afshar, A.; Jahandari, S.; Rasekh, H.; Shariati, M.; Afshar, A.; Shokrgozar, A. Corrosion resistance evaluation of rebars with various primers and coatings in concrete modified with different additives. *Constr. Build. Mater.* **2020**, *262*, 120034. [CrossRef]
13. Bauer, V. Global Cement Production, Responsible for 8% of the World's CO_2 Emissions. Available online: https://www.timber-online.net/uebrige_wirtschaft/2018/12/global-cement-production.html (accessed on 31 May 2021).
14. Wang, T. Cement Production Globally and in the U.S. from 2010 to 2018. Available online: https://www.statista.com/statistics/219343/cement-production-worldwide/ (accessed on 31 May 2021).
15. Chang, I.; Im, J.; Cho, G.-C. Introduction of microbial biopolymers in soil treatment for future environmentally-friendly and sustainable geotechnical engineering. *Sustainability* **2016**, *8*, 251. [CrossRef]
16. Ong, D.E.L.; Choo, C. Sustainable construction of a bored pile foundation system in erratic phyllite. In Proceedings of the ASEAN Australian Engineering Congress, Kuching, Malaysia, 25–27 July 2011.
17. Ngu, L.H.; Song, J.W.; Hashim, S.S.; Ong, D.E.L. Lab-scale atmospheric CO_2 absorption for calcium carbonate precipitation in sand. *Greenh. Gases Sci. Technol.* **2019**, *9*, 519–528. [CrossRef]
18. Omoregie, A.I.; Ngu, L.H.; Ong, D.E.L.; Nissom, P.M. Low-cost cultivation of Sporosarcina pasteurii strain in food-grade yeast extract medium for microbially induced carbonate precipitation (MICP) application. *Biocatal. Agric. Biotechnol.* **2019**, *17*, 247–255. [CrossRef]
19. Liu, S.; Wang, R.; Yu, J.; Peng, X.; Cai, Y.; Tu, B. Effectiveness of the anti-erosion of an MICP coating on the surfaces of ancient clay roof tiles. *Constr. Build. Mater.* **2020**, *243*, 118202. [CrossRef]
20. Ghasemzadeh, H.; Mehrpajouh, A.; Pishvaei, M. Laboratory analyses of Kaolinite stabilized by vinyl polymers with different monomer types. *Eng. Geol.* **2021**, *280*, 105938. [CrossRef]
21. DeJong, J.; Soga, K.; Kavazanjian, E.; Burns, S.; Van Paassen, L.; Al Qabany, A.; Aydilek, A.; Bang, S.; Burbank, M.; Caslake, L.F. Biogeochemical processes and geotechnical applications: Progress, opportunities and challenges. *Geotechnique* **2013**, *63*, 287–301. [CrossRef]
22. Omoregie, A.I.; Palombo, E.A.; Ong, D.E.L.; Nissom, P.M. Biocementation of sand by Sporosarcina pasteurii strain and technical-grade cementation reagents through surface percolation treatment method. *Constr. Build. Mater.* **2019**, *228*, 116828. [CrossRef]
23. Omoregie, A.I.; Palombo, E.A.; Ong, D.E.L.; Nissom, P.M. A feasible scale-up production of Sporosarcina pasteurii using custom-built stirred tank reactor for in-situ soil biocementation. *Biocatal. Agric. Biotechnol.* **2020**, *24*, 101544. [CrossRef]
24. Bahmani, M.; Fatehi, H.; Noorzad, A.; Hamedi, J. Biological soil improvement using new environmental bacteria isolated from northern Iran. *Environ. Geotech.* **2019**, *40*, 1–13. [CrossRef]
25. Mujah, D.; Shahin, M.A.; Cheng, L. State-of-the-art review of biocementation by microbially induced calcite precipitation (MICP) for soil stabilization. *Geomicrobiol. J.* **2017**, *34*, 524–537. [CrossRef]
26. Choi, S.-G.; Chang, I.; Lee, M.; Lee, J.-H.; Han, J.-T.; Kwon, T.-H. Review on geotechnical engineering properties of sands treated by microbially induced calcium carbonate precipitation (MICP) and biopolymers. *Constr. Build. Mater.* **2020**, *246*, 118415. [CrossRef]
27. Tang, C.-S.; Yin, L.-y.; Jiang, N.-J.; Zhu, C.; Zeng, H.; Li, H.; Shi, B. Factors affecting the performance of microbial-induced carbonate precipitation (MICP) treated soil: A review. *Environ. Earth Sci.* **2020**, *79*, 1–23. [CrossRef]
28. Dhami, N.K.; Reddy, M.S.; Mukherjee, A. Significant indicators for biomineralisation in sand of varying grain sizes. *Constr. Build. Mater.* **2016**, *104*, 198–207. [CrossRef]
29. Sharma, M.; Satyam, N.; Reddy, K.R. Effect of freeze-thaw cycles on engineering properties of biocemented sand under different treatment conditions. *Eng. Geol.* **2021**, *284*, 106022. [CrossRef]

30. Achal, V.; Mukherjee, A. A review of microbial precipitation for sustainable construction. *Constr. Build. Mater.* **2015**, *93*, 1224–1235. [CrossRef]
31. Chu, J.; Ivanov, V.; Stabnikov, V.; Li, B. Microbial method for construction of an aquaculture pond in sand. *Géotechnique* **2013**, *63*, 871–875. [CrossRef]
32. Kwon, T.-H.; Ajo-Franklin, J.B. High-frequency seismic response during permeability reduction due to biopolymer clogging in unconsolidated porous media. *Geophysics* **2013**, *78*, EN117–EN127. [CrossRef]
33. Noh, D.H.; Ajo-Franklin, J.B.; Kwon, T.H.; Muhunthan, B. P and S wave responses of bacterial biopolymer formation in unconsolidated porous media. *J. Geophys. Res. Biogeosci.* **2016**, *121*, 1158–1177. [CrossRef]
34. Ham, S.-M.; Chang, I.; Noh, D.-H.; Kwon, T.-H.; Muhunthan, B. Improvement of Surface Erosion Resistance of Sand by Microbial Biopolymer Formation. *J. Geotech. Geoenviron. Eng.* **2018**, *144*, 06018004. [CrossRef]
35. Baveye, P.; Vandevivere, P.; Hoyle, B.L.; DeLeo, P.C.; de Lozada, D.S. Environmental impact and mechanisms of the biological clogging of saturated soils and aquifer materials. *Crit, Rev. Environ. Sci. Technol.* **1998**, *28*, 123–191. [CrossRef]
36. Dunsmore, B.C.; Bass, C.J.; Lappin-Scott, H.M. A novel approach to investigate biofilm accumulation and bacterial transport in porous matrices. *Environ. Microbiol.* **2004**, *6*, 183–187. [CrossRef]
37. Matsubara, H. Stabilisation of weathered limestone surfaces using microbially enhanced calcium carbonate deposition. *Eng. Geol.* **2021**, *284*, 106044. [CrossRef]
38. Abdullah, H.H.; Shahin, M.A.; Walske, M.L. Review of fly-ash-based geopolymers for soil stabilisation with special reference to clay. *Geosciences* **2020**, *10*, 249. [CrossRef]
39. Sadeghian, F.; Haddad, A.; Jahandari, S.; Rasekh, H.; Ozbakkaloglu, T. Effects of electrokinetic phenomena on the load-bearing capacity of different steel and concrete piles: A small-scale experimental study. *Can. Geotech. J.* **2021**, *58*, 741–746. [CrossRef]
40. Rebata-Landa, V.; Santamarina, J.C. Mechanical effects of biogenic nitrogen gas bubbles in soils. *J. Geotech. Geoenviron. Eng.* **2012**, *138*, 128–137. [CrossRef]
41. He, J.; Chu, J.; Ivanov, V. Mitigation of liquefaction of saturated sand using biogas. *Geotechnique* **2013**, *63*, 267–275. [CrossRef]
42. Stal, L.J. Biopolymer. In *Encyclopedia of Astrobiology*; Gargaud, M., Amils, R., Quintanilla, J.C., Cleaves, H.J., Irvine, W.M., Pinti, D.L., Viso, M., Eds.; Springer: Berlin/Heidelberg, Germany, 2011; pp. 199–200.
43. Chang, I.; Lee, M.; Tran, A.T.P.; Lee, S.; Kwon, Y.-M.; Im, J.; Cho, G.-C. Review on biopolymer-based soil treatment (BPST) technology in geotechnical engineering practices. *Transp. Geotech.* **2020**, *24*, 100385. [CrossRef]
44. Liu, X.; Wang, C.; Wang, A.; Qu, J.; Wen, Y.; Wei, B. Application of Cellulose and Cellulose Nanofibers in Oil Exploration. *Pap. Biomater.* **2019**, *4*, 69.
45. Mays, G.C.; Hutchinson, A.R. *Adhesives in Civil Engineering*; Cambridge University Press: Cambridge, UK, 2005.
46. Zhang, Y.; Zhao, Q.; Liu, C.; Zhou, M. Properties comparison of mortars with welan gum or cellulose ether. *Constr. Build. Mater.* **2016**, *102*, 648–653. [CrossRef]
47. Plank, J. Applications of biopolymers in construction engineering. *Biopolym. Online Biol. Chem. Biotechnol. Appl.* **2005**, *10*. [CrossRef]
48. Khatami, H.; O'Kelly, B.C. Prevention of bleeding of particulate grouts using biopolymers. *Constr. Build. Mater.* **2018**, *192*, 202–209. [CrossRef]
49. Li, M.; Xie, D.; Shu, Q.; Ou, H.; Guo, X. Study on sodium fatty alcohol polyoxyethyleneether sulfate relieve the contamination of oil well cement with mineral oil-based drilling fluids. *Constr. Build. Mater.* **2018**, *163*, 450–459. [CrossRef]
50. Rempel, A.W.; Rempel, A.R. Frost resilience of stabilized earth building materials. *Geosciences* **2019**, *9*, 328. [CrossRef]
51. Niaounakis, M. *Biopolymers: Applications and Trends*; William Andrew: Norwich, NY, USA, 2015.
52. Gupta, S.C.; Hooda, K.; Mathur, N.; Gupta, S. Tailoring of guar gum for desert sand stabilization. *Indian J. Chem. Technol.* **2009**, *16*, 507–512.
53. Chudzikowski, R. Guar gum and its applications. *J. Soc. Cosmet. Chem.* **1971**, *22*, 43.
54. Chen, R.; Zhang, L.; Budhu, M. Biopolymer stabilization of mine tailings. *J. Geotech. Geoenviron. Eng.* **2013**, *139*, 1802–1807. [CrossRef]
55. Martone, P.T.; Estevez, J.M.; Lu, F.; Ruel, K.; Denny, M.W.; Somerville, C.; Ralph, J. Discovery of lignin in seaweed reveals convergent evolution of cell-wall architecture. *Curr. Biol.* **2009**, *19*, 169–175. [CrossRef]
56. Hemmilä, V.; Trischler, J.; Sandberg, D. Lignin: An adhesive raw material of the future or waste of research energy? In Proceedings of the Northern European Network for Wood Science and Engineering (WSE. Proceedings of the 9th Meeting), Hannover, Germany, 11–12 September 2013; pp. 98–103.
57. Ta'negonbadi, B.; Noorzad, R. Stabilization of clayey soil using lignosulfonate. *Transp. Geotech.* **2017**, *12*, 45–55. [CrossRef]
58. Zhang, T.; Liu, S.; Cai, G.; Puppala, A.J. Experimental investigation of thermal and mechanical properties of lignin treated silt. *Eng. Geol.* **2015**, *196*, 1–11. [CrossRef]
59. Ivanov, V.; Chu, J. Applications of microorganisms to geotechnical engineering for bioclogging and biocementation of soil in situ. *Rev. Environ. Sci. Biol. Technol.* **2008**, *7*, 139–153. [CrossRef]
60. McHugh, D.J. *A Guide to the Seaweed Industry*; Food and Agriculture Organization of the United Nations: Rome, Italy, 2003.
61. Hernandez-Carmona, G.; Freile-Pelegrín, Y.; Hernández-Garibay, E. Conventional and alternative technologies for the extraction of algal polysaccharides. In *Functional Ingredients from Algae for Foods and Nutraceuticals*; Elsevier: Amsterdam, The Netherlands, 2013; pp. 475–516.

62. Chang, I.; Prasidhi, A.K.; Im, J.; Cho, G.-C. Soil strengthening using thermo-gelation biopolymers. *Constr. Build. Mater.* **2015**, *77*, 430–438. [CrossRef]
63. Bacic, A.; Fincher, G.B.; Stone, B.A. *Chemistry, Biochemistry, and Biology of 1–3 Beta Glucans and Related Polysaccharides*; Academic Press: Cambridge, MA, USA, 2009.
64. Ouwerx, C.; Velings, N.; Mestdagh, M.; Axelos, M. Physico-chemical properties and rheology of alginate gel beads formed with various divalent cations. *Polym. Gels. Netw.* **1998**, *6*, 393–408. [CrossRef]
65. Lee, K.Y.; Mooney, D.J. Alginate: Properties and biomedical applications. *Prog. Polym. Sci.* **2012**, *37*, 106–126. [CrossRef]
66. Fatehi, H.; Bahmani, M.; Noorzad, A. Strengthening of Dune Sand with Sodium Alginate Biopolymer. In Proceedings of the Geo-Congress 2019: Soil Improvement, Philadelphia, Pennsylvania, 24–27 March 2019; pp. 157–166.
67. Kulkarni, V.S.; Shaw, C. *Essential Chemistry for Formulators of Semisolid and Liquid Dosages*; Elsevier Science: Amsterdam, The Netherlands, 2015.
68. Milas, M.; Rinaudo, M. Properties of xanthan gum in aqueous solutions: Role of the conformational transition. *Carbohydr. Res.* **1986**, *158*, 191–204. [CrossRef]
69. Chen, C.S.H.; Sheppard, E. Conformation and shear stability of xanthan gum in solution. *Polym. Eng. Sci.* **1980**, *20*, 512–516. [CrossRef]
70. Garcıa-Ochoa, F.; Santos, V.; Casas, J.; Gómez, E. Xanthan gum: Production, recovery, and properties. *Biotechnol. Adv.* **2000**, *18*, 549–579. [CrossRef]
71. Jansson, P.-E.; Lindberg, B.; Sandford, P.A. Structural studies of gellan gum, an extracellular polysaccharide elaborated by Pseudomonas elodea. *Carbohydr. Res.* **1983**, *124*, 135–139. [CrossRef]
72. Chang, I.; Jeon, M.; Cho, G.-C. Application of microbial biopolymers as an alternative construction binder for earth buildings in underdeveloped countries. *Int. J. Polym. Sci.* **2015**, *2015*, 326745. [CrossRef]
73. Kwon1a, Y.-M.; Im1b, J.; Chang, I.; Cho, G.-C. ε-polylysine biopolymer for coagulation of clay suspensions. *Geomech. Eng.* **2017**. [CrossRef]
74. Shima, S.; MATSUOKA, H.; IWAMOTO, T.; SAKAI, H. Antimicrobial action of ε-poly-L-lysine. *J. Antibiot.* **1984**, *37*, 1449–1455. [CrossRef]
75. Hemmings, H.C.; Egan, T.D. *Pharmacology and Physiology for Anesthesia: Foundations and Clinical Application: Expert Consult—Online and Print*; Elsevier/Saunders: Amsterdam, The Netherlands, 2013.
76. Lee, D.W.; Lim, C.; Israelachvili, J.N.; Hwang, D.S. Strong adhesion and cohesion of chitosan in aqueous solutions. *Langmuir* **2013**, *29*, 14222–14229. [CrossRef]
77. Kumar, M.N.R. A review of chitin and chitosan applications. *React. Funct. Polym.* **2000**, *46*, 1–27. [CrossRef]
78. Jayasuriya, A. Production of micro-and nanoscale chitosan particles for biomedical applications. In *Chitosan Based Biomaterials Volume 1*; Elsevier: Amsterdam, The Netherlands, 2017; pp. 185–209.
79. Wilson, M.; Wilson, L. Clay mineralogy and shale instability: An alternative conceptual analysis. *Clay Miner.* **2014**, *49*, 127–145. [CrossRef]
80. Hataf, N.; Ghadir, P.; Ranjbar, N. Investigation of soil stabilization using chitosan biopolymer. *J. Clean. Prod.* **2018**, *170*, 1493–1500. [CrossRef]
81. Southward, C. Manufacture and applications and edible casein products. I. Manufacture and properties. *N. Z. J. Dairy Sci. Technol.* **1985**, *20*, 79–101.
82. Fatehi, H.; Abtahi, S.M.; Hashemolhosseini, H.; Hejazi, S.M. A novel study on using protein based biopolymers in soil strengthening. *Constr. Build. Mater.* **2018**, *167*, 813–821. [CrossRef]
83. Némethy, G.; Scheraga, H.A. Structure of water and hydrophobic bonding in proteins. I. A model for the thermodynamic properties of liquid water. *J. Chem. Phys.* **1962**, *36*, 3382–3400. [CrossRef]
84. La Rosa, A. Life cycle assessment of biopolymers. In *Biopolymers and Biotech Admixtures for Eco-Efficient Construction Materials*; Elsevier: Amsterdam, The Netherlands, 2016; pp. 57–78.
85. Girod, B.; van Vuuren, D.P.; de Vries, B. Influence of travel behavior on global CO_2 emissions. *Transp. Res. Part A Policy Pract.* **2013**, *50*, 183–197. [CrossRef]
86. Zhang, R.; Long, M.; Zheng, J. Comparison of environmental impacts of two alternative stabilization techniques on expansive soil slopes. *Adv. Civ. Eng.* **2019**, *2019*, 9454929. [CrossRef]
87. Celauro, C.; Corriere, F.; Guerrieri, M.; Casto, B.L. Environmentally appraising different pavement and construction scenarios: A comparative analysis for a typical local road. *Transp. Res. Part D Transp. Environ.* **2015**, *34*, 41–51. [CrossRef]
88. Marcelino-Sadaba, S.; Kinuthia, J.; Oti, J.; Meneses, A.S. Challenges in Life Cycle Assessment (LCA) of stabilised clay-based construction materials. *Appl. Clay Sci.* **2017**, *144*, 121–130. [CrossRef]
89. Uwasu, M.; Hara, K.; Yabar, H. World cement production and environmental implications. *Environ. Dev.* **2014**, *10*, 36–47. [CrossRef]
90. da Rocha, C.G.; Passuello, A.; Consoli, N.C.; Samaniego, R.A.Q.; Kanazawa, N.M. Life cycle assessment for soil stabilization dosages: A study for the Paraguayan Chaco. *J. Clean. Prod.* **2016**, *139*, 309–318. [CrossRef]
91. Saberian, M.; Jahandari, S.; Li, J.; Zivari, F. Effect of curing, capillary action, and groundwater level increment on geotechnical properties of lime concrete: Experimental and prediction studies. *J. Rock Mech. Geotech. Eng.* **2017**, *9*, 638–647. [CrossRef]

92. Kabir, E.; Kaur, R.; Lee, J.; Kim, K.-H.; Kwon, E.E. Prospects of biopolymer technology as an alternative option for non-degradable plastics and sustainable management of plastic wastes. *J. Clean. Prod.* **2020**, *258*, 120536. [CrossRef]
93. Stoica, M.; Antohi, V.M.; Zlati, M.L.; Stoica, D. The financial impact of replacing plastic packaging by biodegradable biopolymers-A smart solution for the food industry. *J. Clean. Prod.* **2020**, *277*, 124013. [CrossRef]
94. Detzel, A.; Kauertz, B.; Derreza-Greeven, C.; Kirsch, A. *Untersuchung der Umweltwirkungen von Verpackungen aus biologisch abbaubaren Kunststoffen*; Umweltbundesamt: Dessau-Roßlau, Germany, 2012.
95. Song, J.; Murphy, R.; Narayan, R.; Davies, G. Biodegradable and compostable alternatives to conventional plastics. *Philos. Trans. R. Soc. B Biol. Sci.* **2009**, *364*, 2127–2139. [CrossRef]
96. Scown, C.D.; Gokhale, A.A.; Willems, P.A.; Horvath, A.; McKone, T.E. Role of lignin in reducing life-cycle carbon emissions, water use, and cost for United States cellulosic biofuels. *Environ. Sci. Technol.* **2014**, *48*, 8446–8455. [CrossRef]
97. Healable. Is Xanthan Gum Good or Bad? Available online: https://healabel.com/x-ingredients/xanthan-gum (accessed on 31 May 2021).
98. Healable. Is Agar Agar Good or Bad? Available online: https://healabel.com/a-ingredients/agar-agar (accessed on 31 May 2021).
99. Gresta, F.; De Luca, A.I.; Strano, A.; Falcone, G.; Santonoceto, C.; Anastasi, U.; Gulisano, G. Economic and environmental sustainability analysis of guar (Cyamopsis tetragonoloba L.) farming process in a Mediterranean area: Two case studies. *Ital. J. Agron.* **2014**, *9*, 20–24. [CrossRef]
100. Heusala, H.; Sinkko, T.; Sözer, N.; Hytönen, E.; Mogensen, L.; Knudsen, M.T. Carbon footprint and land use of oat and faba bean protein concentrates using a life cycle assessment approach. *J. Clean. Prod.* **2020**, *242*, 118376. [CrossRef]
101. Allwood, J.W.; Martinez-Martin, P.; Xu, Y.; Cowan, A.; Pont, S.; Griffiths, I.; Sungurtas, J.; Clarke, S.; Goodacre, R.; Marshall, A.; et al. Assessing the impact of nitrogen supplementation in oats across multiple growth locations and years with targeted phenotyping and high-resolution metabolite profiling approaches. *Food Chem.* **2021**, *355*, 129585. [CrossRef]
102. Healable. Are Oats Good or Bad? Available online: https://healabel.com/o-ingredients/oats (accessed on 31 May 2020).
103. Langlois, J.; Fréon, P.; Delgenès, J.-P.; Steyer, J.-P.; Helias, A. Life cycle assessment of alginate production. In Proceedings of the 8th International Conference on LCA in the Agri-Food Sector, Saint-Malo, France, 2 October 2012.
104. Muñoz, I.; Rodríguez, C.; Gillet, D.; Moerschbacher, B.M. Life cycle assessment of chitosan production in India and Europe. *Int. J. Life Cycle Assess.* **2018**, *23*, 1151–1160. [CrossRef]
105. Gerber, P.; Vellinga, T.; Opio, C.; Henderson, B.; Steinfeld, H. *Greenhouse Gas Emissions from the Dairy Sector: A Life Cycle As-sessment*; Food and Agricultural Organisation of the United Nations: Rome, Italy, 2010.
106. Flysjö, A.M. Greenhouse Gas Emissions in Milk and Dairy Product Chains: Improving the Carbon Footprint of Dairy Products. Available online: http://pure.au.dk/portal/files/45485022/Anna_20Flusj_.pdf (accessed on 31 May 2021).
107. Chang, I.; Im, J.; Chung, M.-K.; Cho, G.-C. Bovine casein as a new soil strengthening binder from diary wastes. *Constr. Build. Mater.* **2018**, *160*, 1–9. [CrossRef]
108. Niaounakis, M. *Biopolymers: Reuse, Recycling, and Disposal*; William Andrew: Norwich, NY, USA, 2013.
109. Barton, C. Clay minerals. In *Encyclopedia of Soil Science*; Rattan Lal, ed. Marcel Dekker: New York, NY, USA, 2002; pp. 187–192.
110. Tombácz, E.; Szekeres, M. Surface charge heterogeneity of kaolinite in aqueous suspension in comparison with montmorillonite. *Appl. Clay Sci.* **2006**, *34*, 105–124. [CrossRef]
111. Latifi, N.; Horpibulsuk, S.; Meehan, C.L.; Abd Majid, M.Z.; Tahir, M.M.; Mohamad, E.T. Improvement of problematic soils with biopolymer—an environmentally friendly soil stabilizer. *J. Mater. Civ. Eng.* **2017**, *29*, 04016204. [CrossRef]
112. Strawn, D.G.; Bohn, H.L.; O'Connor, G.A. *Soil Chemistry*; John Wiley & Sons: Hoboken, NJ, USA, 2019.
113. Chang, I.; Im, J.; Prasidhi, A.K.; Cho, G.-C. Effects of Xanthan gum biopolymer on soil strengthening. *Constr. Build. Mater.* **2015**, *74*, 65–72. [CrossRef]
114. Cabalar, A.F.; Awraheem, M.H.; Khalaf, M.M. Geotechnical properties of a low-plasticity clay with biopolymer. *J. Mater. Civ. Eng.* **2018**, *30*, 04018170. [CrossRef]
115. Chang, I.; Cho, G.-C. Strengthening of Korean residual soil with β-1, 3/1, 6-glucan biopolymer. *Constr. Build. Mater.* **2012**, *30*, 30–35. [CrossRef]
116. Singh, S.P.; Das, R.; Seth, D. Plasticity and Strength Characteristics of Expansive Soil Treated with Xanthan Gum Biopolymer. In *Problematic Soils and Geoenvironmental Concerns*; Springer: Berlin/Heidelberg, Germany, 2021; pp. 649–663.
117. Soldo, A.; Miletić, M.; Auad, M.L. Biopolymers as a sustainable solution for the enhancement of soil mechanical properties. *Sci. Rep.* **2020**, *10*, 267. [CrossRef]
118. Singh, S.P.; Das, R. Geo-engineering properties of expansive soil treated with xanthan gum biopolymer. *Geomech. Geoeng.* **2019**, *15*, 107–112. [CrossRef]
119. Soldo, A.; Miletić, M. Study on Shear Strength of Xanthan Gum-Amended Soil. *Sustainability* **2019**, *11*, 6142. [CrossRef]
120. Toufigh, V.; Ghassemi, P. Control and Stabilization of Fugitive Dust: Using Eco-Friendly and Sustainable Materials. *Int. J. Geomech.* **2020**, *20*, 04020140. [CrossRef]
121. Chen, C.; Peng, Z.; Gu, J.; Peng, Y.; Huang, X.; Wu, L. Exploring Environmentally Friendly Biopolymer Material Effect on Soil Tensile and Compressive Behavior. *Int. J. Environ. Res. Public Health* **2020**, *17*, 9032. [CrossRef] [PubMed]
122. Lee, S.; Chung, M.; Park, H.M.; Song, K.-I.; Chang, I. Xanthan Gum Biopolymer as Soil-Stabilization Binder for Road Construction Using Local Soil in Sri Lanka. *J. Mater. Civ. Eng.* **2019**, *31*, 06019012. [CrossRef]

123. Lee, M.; Im, J.; Cho, G.-C.; Ryu, H.H.; Chang, I. Interfacial Shearing Behavior along Xanthan Gum Biopolymer-Treated Sand and Solid Interfaces and Its Meaning in Geotechnical Engineering Aspects. *Appl. Sci.* **2021**, *11*, 139. [CrossRef]
124. Chen, C.; Wu, L.; Perdjon, M.; Huang, X.; Peng, Y. The drying effect on xanthan gum biopolymer treated sandy soil shear strength. *Constr. Build. Mater.* **2019**, *197*, 271–279. [CrossRef]
125. Bonal, N.; Prasad, A.; Verma, A. Use of biopolymers to enhance the geotechnical properties of coal mine overburden waste. *Géotech. Lett.* **2020**, *10*, 179–185. [CrossRef]
126. Kwon, Y.-M.; Ham, S.-M.; Kwon, T.-H.; Cho, G.-C.; Chang, I. Surface-erosion behaviour of biopolymer-treated soils assessed by EFA. *Géotech. Lett.* **2020**, *10*, 106–112. [CrossRef]
127. Lee, S.; Chung, M.; Kwon, Y.; Cho, G.; Chang, I. Investigation of erosion behavior of biopolymer treated soil using laboratory hydraulic flume testing. In Proceedings of the 16th Asian Regional Conference on Soil Mechanics and Geotechnical Engineering, Taipei, Taiwan, 14–18 October 2019.
128. Movasat, M.; Tomac, I. Post-Fire Mudflow Prevention by Biopolymer Treatment of Water Repellent Slopes. In Proceedings of the Geo-Congress, Minneapolis, MN, USA, 25–28 February 2020; pp. 170–178.
129. Elkafoury, A.; Azzam, W. Utilize Xanthan gum for enhancing CBR value of used cooking oil-contaminated fine sand subgrade soil for pavement structures. *Innov. Infrastruct. Solut.* **2021**, *6*, 1–10. [CrossRef]
130. Ghasemzadeh, H.; Modiri, F. Application of novel Persian gum hydrocolloid in soil stabilization. *Carbohydr. Polym.* **2020**, *246*, 116639. [CrossRef]
131. Ni, J.; Li, S.-S.; Ma, L.; Geng, X.-Y. Performance of soils enhanced with eco-friendly biopolymers in unconfined compression strength tests and fatigue loading tests. *Constr. Build. Mater.* **2020**, *263*, 120039. [CrossRef]
132. Rashid, A.S.A.; Tabatabaei, S.; Horpibulsuk, S.; Yunus, N.Z.M.; Hassan, W.H.W. Shear Strength Improvement of Lateritic Soil Stabilized by Biopolymer Based Stabilizer. *Geotech. Geol. Eng.* **2019**, *37*, 5533–5541. [CrossRef]
133. Joga, J.R.; Varaprasad, B. Sustainable Improvement of Expansive Clays Using Xanthan Gum as a Biopolymer. *Civ. Eng. J.* **2019**, *5*, 1893–1903. [CrossRef]
134. Kwon1a, Y.-M.; Chang2b, I.; Lee1c, M.; Cho, G.-C. Geotechnical engineering behavior of biopolymer-treated soft marine soil. *Geomech. Eng.* **2019**, *17*, 453–464.
135. Barani, O.R.; Barfar, P. Effect of Xanthan Gum Biopolymer on Fracture Properties of Clay. *J. Mater. Civ. Eng.* **2021**, *33*, 04020426. [CrossRef]
136. Biju, M.; Arnepalli, D. Effect of biopolymers on permeability of sand-bentonite mixtures. *J. Rock Mech. Geotech. Eng.* **2020**, *12*, 1093–1102. [CrossRef]
137. Chang, I.; Kwon, Y.-M.; Im, J.; Cho, G.-C. Soil consistency and interparticle characteristics of xanthan gum biopolymer–containing soils with pore-fluid variation. *Can. Geotech. J.* **2019**, *56*, 1206–1213. [CrossRef]
138. Sujatha, E.R.; Atchaya, S.; Sivasaran, A.; Keerdthe, R. Enhancing the geotechnical properties of soil using xanthan gum—an eco-friendly alternative to traditional stabilizers. *Bull. Eng. Geol. Environ.* **2021**, *80*, 1157–1167. [CrossRef]
139. Khosravi, M.; Tabarsa, A.; Osouli, A.; Latifi, N. A biopolymer-based waterproofing mortar for irrigation channel joints. In Proceedings of the Geo-Congress 2020, Minneapolis, MN, USA, 25–28 February 2020; pp. 159–169.
140. Joga, J.R.; Varaprasad, B. Effect of xanthan gum biopolymer on dispersive properties of soils. *World J. Eng.* **2020**, *17*, 563–571. [CrossRef]
141. Reddy, J.J.; Varaprasad, B.; Reddy, P.V. Application of Grey Taguchi Method to optimize the internal erosion parameters of stabilized soil. *Multiscale Multidiscip. Model. Exp. Des.* **2020**, *4*, 99–108. [CrossRef]
142. Sujatha, E.R.; Sivaraman, S.; Subramani, A.K. Impact of hydration and gelling properties of guar gum on the mechanism of soil modification. *Arab. J. Geosci.* **2020**, *13*, 1–12. [CrossRef]
143. Kumar, S.A.; Sujatha, E.R.; Pugazhendi, A.; Jamal, M.T. Guar gum-stabilized soil: A clean, sustainable and economic alternative liner material for landfills. *Clean Technol. Environ. Policy* **2021**. [CrossRef]
144. Zhao, Y.; Zhuang, J.; Wang, Y.; Jia, Y.; Niu, P.; Jia, K. Improvement of loess characteristics using sodium alginate. *Bull. Eng. Geol. Environ.* **2019**, *79*, 1879–1891. [CrossRef]
145. Lemboye, K.; Almajed, A.; Alnuaim, A.; Arab, M.; Alshibli, K. Improving sand wind erosion resistance using renewable agriculturally derived biopolymers. *Aeolian Res.* **2021**, *49*, 100663. [CrossRef]
146. Singh, S.P.; Palsule, P.S.; Anand, G. Strength Properties of Expansive Soil Treated with Sodium Lignosulfonate. In *Problematic Soils and Geoenvironmental Concerns*; Springer: Berlin/Heidelberg, Germany, 2021; pp. 665–679.
147. Zhang, J.; Han, Y.; Wang, X.; Bian, H. Experimental Investigation of the Dynamic Characteristics of Treated Silt Using Lignin: Case Study of Yellow River Flood Basin. *Int. J. Geomech.* **2021**, *21*, 04021056. [CrossRef]
148. QABANY, A.A.; Soga, K. Effect of chemical treatment used in MICP on engineering properties of cemented soils. *Géotechnique* **2013**, *63*, 331–339. [CrossRef]
149. Khatami, H.R.; O'Kelly, B.C. Improving mechanical properties of sand using biopolymers. *J. Geotech. Geoenviron. Eng.* **2013**, *139*, 1402–1406. [CrossRef]
150. Chang, I.; Im, J.; Cho, G.-C. Geotechnical engineering behaviors of gellan gum biopolymer treated sand. *Can. Geotech. J.* **2016**, *53*, 1658–1670. [CrossRef]
151. Qureshi, M.U.; Chang, I.; Al-Sadarani, K. Strength and durability characteristics of biopolymer-treated desert sand. *Geomech. Eng.* **2017**, *12*, 785–801. [CrossRef]

152. Nugent, R.A.; Zhang, G.; Gambrell, R.P. Effect of exopolymers on the liquid limit of clays and its engineering implications. *Transp. Res. Rec.* **2009**, *2101*, 34–43. [CrossRef]
153. Sarkar, A. *Mould and Core Material for the Steel Foundry*; Elsevier: Amsterdam, The Netherland, 1967.
154. Chang, I.; Cho, G.-C. Geotechnical behavior of a beta-1, 3/1, 6-glucan biopolymer-treated residual soil. *Geomech. Eng.* **2014**, *7*, 633–647. [CrossRef]
155. Albayrak, Z.N.K.; Gencer, G. The Usability of Clay/Pumice Mixtures Modified with Biopolymer as an Impermeable Liner. *KSCE J. Civ. Eng.* **2021**, *25*, 28–36. [CrossRef]
156. Chang, I.; Cho, G.-C. Shear strength behavior and parameters of microbial gellan gum-treated soils: From sand to clay. *Acta Geotech.* **2019**, *14*, 361–375. [CrossRef]
157. Smitha, S.; Sachan, A. Use of agar biopolymer to improve the shear strength behavior of sabarmati sand. *Int. J. Geotech. Eng.* **2016**, *10*, 387–400. [CrossRef]
158. Stephen, A.M.; Phillips, G.O. *Food Polysaccharides and Their Applications*; CRC press: Boca Raton, FL, USA, 2016.
159. Lee, S.; Chang, I.; Chung, M.-K.; Kim, Y.; Kee, J. Geotechnical shear behavior of xanthan gum biopolymer treated sand from direct shear testing. *Geomech. Eng.* **2017**, *12*, 831–847. [CrossRef]
160. Gilbert, R.; Najjar, S.; Shields, M. Importance of residual strengths in factors of safety and reliability. In *Geosynthetics Research and Development in Progress*; ASCE: Reston, VA, USA, 2005; pp. 1–6.
161. Morgan, R. *Soil Erosion and Conservation*; Blackwell Publ: Oxford, UK, 2005.
162. Mehdizadeh, A.; Disfani, M.M.; Evans, R.; Arulrajah, A.; Ong, D.E.L. Mechanical consequences of suffusion on undrained behaviour of a gap-graded cohesionless soil-an experimental approach. *Geotech. Test. J.* **2017**, *40*, 1026–1042. [CrossRef]
163. Chang, I.; Im, J.; Cho, G.-C. An environmentally-friendly geotechnical approach for soil erosion reduction using microbial biopolymers. In *Geo-Chicago 2016*; ASCE: Reston, VA, USA, 2016; pp. 17–24.
164. Donayre, A.; Sanchez, L.; Kim, S.; Aguilar, R.; Nakamatsu, J. Eco-friendly Improvement of Water Erosion Resistance of Unstable Soils with Biodegradable Polymers. In Proceedings of the IOP Conference Series: Materials Science and Engineering, Kuala Lumpur, Malaysia, 13–14 August 2018; p. 012044.
165. Lee, S.; Kwon, Y.-M.; Cho, G.-C.; Chang, I. Investigation of Biopolymer Treatment Feasibility to Mitigate Surface Erosion Using a Hydraulic Flume Apparatus. In Proceedings of the Geo-Congress 2020, Minneapolis, MN, USA, 25–28 February 2020; pp. 46–52.
166. Alsanad, A. *Novel Biopolymer Treatment for Wind Induced Soil Erosion*; Arizona State University: Tempe, AZ, USA, 2011.
167. Ayeldeen, M.; Negm, A.; El Sawwaf, M.; Gädda, T. Laboratory study of using biopolymer to reduce wind erosion. *Int. J. Geotech. Eng.* **2018**, *12*, 228–240. [CrossRef]
168. Kavazanjian, E., Jr.; Iglesias, E.; Karatas, I. Biopolymer soil stabilization for wind erosion control. In Proceedings of the 17th International Conference on Soil Mechanics and Geotechnical Engineering, Alexandria, Egypt, 5–9 October 2009; pp. 881–884.
169. Rushing, J.F.; Harrison, A.; Tingle, J.S.; Mason, Q.; McCaffrey, T. Evaluation of dust palliatives for unpaved roads in arid climates. *J. Perform. Constr. Facil.* **2006**, *20*, 281–286. [CrossRef]
170. Chen, R.; Lee, I.; Zhang, L. Biopolymer stabilization of mine tailings for dust control. *J. Geotech. Geoenviron. Eng.* **2015**, *141*, 04014100. [CrossRef]
171. Morgan, R.P.C.; Nearing, M. *Handbook of Erosion Modelling*; John Wiley & Sons: Hoboken, NJ, USA, 2016.
172. Wong, S.T.; Ong, D.E.L.; Robinson, R.G. Behaviour of MH silts with varying plasticity indices. *Geotech. Res.* **2017**, *4*, 118–135. [CrossRef]
173. Cai, G.; Zhang, T.; Liu, S.; Li, J.; Jie, D. Stabilization mechanism and effect evaluation of stabilized silt with lignin based on laboratory data. *Mar. Georesour. Geotechnol.* **2016**, *34*, 331–340. [CrossRef]
174. Ayeldeen, M.; Negm, A.; El-Sawwaf, M.; Kitazume, M. Enhancing mechanical behaviors of collapsible soil using two biopolymers. *J. Rock Mech. Geotech. Eng.* **2017**, *9*, 329–339. [CrossRef]
175. Qureshi, M.U.; Al-Hilly, A.; Al-Zeidi, O.; Al-Barrami, A.; Al-Jabri, A. Vane shear strength of bio-improved sand reinforced with natural fibre. In Proceedings of the E3S Web of Conferences, Dalian, China, 29–31 October 2019; p. 12004.
176. Shogren, R.; Doane, W.; Garlotta, D.; Lawton, J.; Willett, J. Biodegradation of starch/polylactic acid/poly (hydroxyester-ether) composite bars in soil. *Polym. Degrad. Stab.* **2003**, *79*, 405–411. [CrossRef]
177. Chang, I.; Im, J.; Lee, S.-W.; Cho, G.-C. Strength durability of gellan gum biopolymer-treated Korean sand with cyclic wetting and drying. *Constr. Build. Mater.* **2017**, *143*, 210–221. [CrossRef]
178. Qureshi, M.; Al-Qayoudhi, S.; Al-Kendi, S.; Al-Hamdani, A.; Al-Sadrani, K. The effects of slaking on the durability of bio-improved sand. *Int. J. Sci. Eng. Res.* **2015**, *6*, 486–490.
179. Kirk, T.K. Effects of microorganisms on lignin. *Annu. Rev. Phytopathol.* **1971**, *9*, 185–210. [CrossRef]
180. Teaca, C.A.; Roşu, D.; Mustaţă, F.; Rusu, T.; Roşu, L.; Roşca, I.; Varganici, C.-D. Natural Bio-Based Products for Wood Coating and Protection against Degradation: A Review. *BioResources* **2019**, *14*, 4873–4901. [CrossRef]
181. Arab, M.G.; Mousa, R.; Gabr, A.; Azam, A.; El-Badawy, S.; Hassan, A. Resilient Behavior of Sodium Alginate–Treated Cohesive Soils for Pavement Applications. *J. Mater. Civ. Eng.* **2019**, *31*, 04018361. [CrossRef]
182. Goren, S.; Alagha, O. Soil treatment with lignin sulphide chemical stabilizer: Environmental and structural assessment. *J. Residuals Sci. Technol.* **2008**, *5*, 189–194.
183. Liu, Y.; Chang, M.; Wang, Q.; Wang, Y.; Liu, J.; Cao, C.; Zheng, W.; Bao, Y.; Rocchi, I. Use of sulfur-free lignin as a novel soil additive: A multi-scale experimental investigation. *Eng. Geol.* **2020**, *269*, 105551. [CrossRef]

184. Pérez, I.P.; Pasandín, A.M.R.; Pais, J.C.; Pereira, P.A.A. Use of lignin biopolymer from industrial waste as bitumen extender for asphalt mixtures. *J. Clean. Prod.* **2019**, *220*, 87–98. [CrossRef]
185. Gopalakrishnan, K.; Ceylan, H.; Kim, S. Renewable biomass-derived lignin in transportation infrastructure strengthening applications. *Int. J. Sustain. Eng.* **2013**, *6*, 316–325. [CrossRef]
186. Ni, J.; Hao, G.-L.; Chen, J.-Q.; Ma, L.; Geng, X.-Y. The Optimisation Analysis of Sand-Clay Mixtures Stabilised with Xanthan Gum Biopolymers. *Sustainability* **2021**, *13*, 3732. [CrossRef]
187. Bouazza, A.; Gates, W.P.; Ranjith, P.G. Hydraulic conductivity of biopolymer-treated silty sand. *Géotechnique* **2009**, *59*, 71–72. [CrossRef]
188. Wu, F.; Li, G.; Li, H.-N.; Jia, J.-Q. Strength and stress–strain characteristics of traditional adobe block and masonry. *Mater. Struct.* **2013**, *46*, 1449–1457. [CrossRef]
189. Silveira, D.; Varum, H.; Costa, A.; Martins, T.; Pereira, H.; Almeida, J. Mechanical properties of adobe bricks in ancient constructions. *Constr. Build. Mater.* **2012**, *28*, 36–44. [CrossRef]
190. Sorsa, A.; Senadheera, S.; Birru, Y. Engineering Characterization of Subgrade Soils of Jimma Town, Ethiopia, for Roadway Design. *Geosciences* **2020**, *10*, 94. [CrossRef]
191. Mamlouk, M.S.; Zaniewski, J.P. *Materials for Civil and Construction Engineers*; Pearson Prentice Hall: Upper Saddle River, NJ, USA, 2006.
192. Hatakeyama, H.; Hatakeyama, T. Interaction between water and hydrophilic polymers. *Thermochim. Acta* **1998**, *308*, 3–22. [CrossRef]
193. Maghchiche, A.; Haouam, A.; Immirzi, B. Use of polymers and biopolymers for water retaining and soil stabilization in arid and semiarid regions. *J. Taibah Univ. Sci.* **2010**, *4*, 9–16. [CrossRef]
194. Jeong, B.; Kim, S.W.; Bae, Y.H. Thermosensitive sol–gel reversible hydrogels. *Adv. Drug Deliv. Rev.* **2012**, *64*, 154–162. [CrossRef]
195. Dehghan, H.; Tabarsa, A.; Latifi, N.; Bagheri, Y. Use of xanthan and guar gums in soil strengthening. *Clean Technol. Environ. Policy* **2019**, *21*, 155–165. [CrossRef]
196. Galán-Marín, C.; Rivera-Gómez, C.; Petric, J. Clay-based composite stabilized with natural polymer and fibre. *Constr. Build. Mater.* **2010**, *24*, 1462–1468. [CrossRef]
197. Jameson, G. *Guide to Pavement Technology Part 4D: Stabilised Materials*; Austroads: Sydney, Australia, 2019; ISBN 1925854124.
198. Kimmerling, R. *Geotechnical Engineering Circular No. 6 Shallow Foundations*; Federal Highway Administration. Office of Bridge Technology: Washington, NJ, USA, 2002.
199. Epa, U. *Solid Waste Disposal Facility Criteria*; EPA: Washington, DC, USA, 1993.
200. Shankar, M.U.; Muthukumar, M. Comprehensive review of geosynthetic clay liner and compacted clay liner. In Proceedings of the IOP Conference Series: Materials Science and Engineering, Busan, Korea, 25–27 August 2017; p. 032026.
201. Chen, R.; Ding, X.; Ramey, D.; Lee, I.; Zhang, L. Experimental and numerical investigation into surface strength of mine tailings after biopolymer stabilization. *Acta Geotech.* **2016**, *11*, 1075–1085. [CrossRef]
202. Day, S.R.; Ryan, C.R. State of the art in bio-polymer drain construction. In *Slurry Walls: Design, Construction, and Quality Control*; ASTM International: West Conshohocken, PA, USA, 1992.
203. Aminpour, M.; O'Kelly, B.C. Applications of biopolymers in dam construction and operation activities. In Proceedings of the Proceedings of the 2nd International Dam World Conference, Lisbon, Portugal, 2015; pp. 937–946.
204. Cheng, L.; Shahin, M.A.; Chu, J. Soil bio-cementation using a new one-phase low-pH injection method. *Acta Geotech.* **2019**, *14*, 615–626. [CrossRef]
205. Leong, H.Y.; Ong, D.E.L.; Sanjayan, J.G.; Nazari, A.; Kueh, S.M. Effects of Significant Variables on Compressive Strength of Soil-Fly Ash Geopolymer: Variable Analytical Approach Based on Neural Networks and Genetic Programming. *J. Mater. Civ. Eng.* **2018**, *30*, 04018129. [CrossRef]
206. Leong, H.; Ong, D.E.L.; Sanjayan, J.; Nazari, A. A genetic programming predictive model for parametric study of factors affecting strength of geopolymers. *RSC Adv.* **2015**, *5*, 85630–85639. [CrossRef]
207. Castellane, T.C.L.; Persona, M.R.; Campanharo, J.C.; de Macedo Lemos, E.G. Production of exopolysaccharide from rhizobia with potential biotechnological and bioremediation applications. *Int. J. Biol. Macromol.* **2015**, *74*, 515–522. [CrossRef]
208. Omoregie, A.I.; Ong, D.E.L.; Nissom, P.M. Assessing ureolytic bacteria with calcifying abilities isolated from limestone caves for biocalcification. *Lett. Appl. Microbiol.* **2019**, *68*, 173–181. [CrossRef]
209. Ong, D.E.L.; Yang, D.; Phang, S. Comparisons of finite element modeling of a deep excavation using SAGE-CRISP and PLAXIS. In Proceedings of the International Conferenceon Deep Excavations, Singapore, 28–30 June 2006; pp. 28–30.
210. Ong, D.E.L. Benchmarking of FEM Technique Involving Deep Excavation, Pile-soil Interaction and Embankment Construction. In Proceedings of the 12th International Conference of IACMAG, Goa, India, 1–6 October 2008.
211. Ong, D.E.L.; Leung, C.; Chow, Y. Piles subject to excavation-induced soil movement in clay. In Proceedings of the Proceedings of the 13th European Conference on Soil Mechanics and Geotechnical Engineering, Prague, Czech Republic, 25–28 August 2003; pp. 777–782.
212. Ong, D.E.L.; Leung, C.F.; Chow, Y.K. Behavior of pile groups subject to excavation-induced soil movement in very soft clay. *J. Geotech. Geoenviron. Eng.* **2015**, *135*, 1462–1474. [CrossRef]
213. Ong, D.E.L.; Leung, C.; Chow, Y. Pile behavior due to excavation-induced soil movement in clay. I: Stable wall. *J. Geotech. Geoenviron. Eng.* **2006**, *132*, 36–44. [CrossRef]

214. Cheng, W.-C.; Li, G.; Ong, D.E.L.; Chen, S.-L.; Ni, J.C. Modelling liner forces response to very close-proximity tunnelling in soft alluvial deposits. *Tunn. Undergr. Space Technol.* **2020**, *103*, 103455. [CrossRef]
215. Mehdizadeh, A.; Disfani, M.M.; Evans, R.; Arulrajah, A.; Ong, D.E.L. Discussion of "development of an internal camera-based volume determination system for triaxial testing" by se Salazar, a. Barnes and Ra Coffman. The technical note was published in geotechnical testing journal, vol. 38, no. 4, 2015. *Geotech. Test. J.* **2016**, *39*, 165–168. [CrossRef]

geosciences

MDPI

Article

Strength and Microstructural Assessment of Reconstituted and Stabilised Soft Soils with Varying Silt Contents

Yaxu Liu *, Zhuang Liu, Erwin Oh and Dominic Ek Leong Ong

School of Engineering and Built Environment, Griffith University, Southport, QLD 4215, Australia; zhuang.liu@griffithuni.edu.au (Z.L.); y.oh@griffith.edu.au (E.O.); d.ong@griffith.edu.au (D.E.L.O.)
* Correspondence: yaxu.liu@griffithuni.edu.au; Tel.: +61-(0)-452533616

Abstract: The study of the strength of reconstituted and stabilised soft soils is very important in geotechnical engineering. The soil particles, such as clay, sand, and silt play important roles in determining the behaviour of soils. The behaviour of clay and sand particles are unique; however, the behaviour of silt particles lie in a transitional form between sand and clay. Therefore, this paper seeks to investigate (a) the effect of silt contents on the strength of soft soils; (b) the effect of silt content on the strength of cement-stabilised soft soils; and (c) the microstructure of the soft soil specimens stabilised by cement with varying particle size distribution. A series of tests consisting in consolidated, isotropic undrained (CIU) triaxial tests, unconfined compressive strength (UCS) tests, and scanning electron microscope (SEM) images were conducted in this study to achieve these objectives. In conclusion, the relationship between the silt content and critical state behaviour of soft soils (both clay and silt particles) are proposed. For the cement-stabilised specimens, the unconfined compressive strength increases with the increase in silt content when the cement content is 10%. However, the UCS decreases with the increase in silt content when cement content is 30%. With cement content ranging from 15–25%, the UCS increases at first with the increase of silt content but decreases once the silt content reaches a 'saturation' point.

Keywords: soft soil behaviour; soft soil stabilisation; silt content; unconfined compressive strength; consolidated undrained strength; microstructure analysis

Citation: Liu, Y.; Liu, Z.; Oh, E.; Ong, D.E.L. Strength and Microstructural Assessment of Reconstituted and Stabilised Soft Soils with Varying Silt Contents. *Geosciences* **2021**, *11*, 302. https://doi.org/10.3390/geosciences11080302

Academic Editor: Mohamed Shahin

Received: 18 May 2021
Accepted: 19 July 2021
Published: 21 July 2021

Publisher's Note: MDPI stays neutral with regard to jurisdictional claims in published maps and institutional affiliations.

1. Introduction

Soft soil deposits, such as mine tailings, hydraulic fills, soft estuarine or marine soils exist globally on the planet. Available land for construction has decreased due to rapid urbanisation and population growth around the globe. It is crucial to develop or reclaim land with poor geotechnical conditions in order to build on it to solve the issue of land shortage. Advancements in construction material science considering sustainability [1–6] and the characterisation of soils form the basis of geoenvironmental and geotechnical design but not without challenges faced in laboratory tests [7,8], computer simulations [9,10] and inferior in situ geomaterial conditions requiring conditioning and improvement [11–16]. For example, many projects, such as the construction of coastal facilities, such as shipyard [17] and quay walls, deep excavation [18–20] or foundations for building basements [21,22], road embankments on flood plains [23,24] and tunnelling [25–28] require infrastructure to be built on areas with widespread soft soil deposits [29–32]. In Brisbane, Queensland, Australia, dubbed the 'river city', soft estuarine clays can be found in embankments supporting main motorways, the Brisbane airport and the Port of Brisbane. Hence, it is key that the latest understanding of the behaviour of reconstituted and stabilised estuarine soft soil deposits with varying silt contents is developed for future potential, large-scale ground improvement projects.

It is widely recognised that the behaviour of soil in a triaxial compression test is highly related to confining pressures, particle size distribution, and soil minerals, as well as the composition of the soil particles [33]. It is also widely known that the clay and sand phases have dominating influences on the properties and behaviour of soil [34,35]. For example, the friction angle, stiffness, and permeability decrease with increased clay percentage, while these parameters increase with increased sand content. The behaviour of soils with a higher percentage of clay particles shows more stress softening behaviour, while the soils with a higher percentage of sand particles show more stress hardening behaviour [35,36].

Soft marine and estuarine soils commonly exist in many coastal areas. Both have similar weaknesses, such as high water content, low shear strength, and high compressibility. These soils must be stabilised before construction projects are carried out on such soft soils. The behaviour of the soft soils has been extensively investigated by many studies. From previous studies, it can be concluded that the main components of those soils are soft clay or soft silt deposits admixed with some sand particles. Furthermore, soft clay is commonly found in most marine deposits, such as soft marine soils in Bangkok; Shanghai; the coast of Cyprus; the Port of Brisbane, Australia; the Port of Lianyungang, China; the ports of Tokuyama, Mizushima, Hibiki, and Moji, Japan; and Singapore [37–46]. Meanwhile, soft silt is commonly distributed in most of the estuarine deposits and few of the marine deposits as well, such as Rance estuarine silt; Coode Island silt; Limavady silt (Northern Ireland); Yellow River alluvial silt; Scotland Kinnegar estuarine silt and New South Wales silt [37–50]. Therefore, to investigate the behaviours of soft soils, the study of the behaviour of the clay and silt particles is very significant.

In general, the behaviours of clay and sand particles are unique and easy to identify. However, the behaviour of silt particles lies in between the behaviours of clay and sand particles. It is important that some previous studies found that the behaviour of silt is not in accordance with the critical-state framework adopted for clay and sand [51,52], and suggested that the behaviour of silt is more complex than that of sand and clay, as the behaviour of silts is in a transitional form between clay and sand. Consequently, it is important to understand the physical, mineralogical, strength and microstructural behaviour of silt, as it is presently recognised that gaps in understanding its fundamental behaviour exist.

Many soft soils need to be stabilised by suitable ground improvement techniques before constructing infrastructure on them. Cement is one of the commonly used cementitious materials to treat soft soils. It can increase the strength and decrease the water content of stabilised soil through the hydration of cement and a pozzolanic reaction. The use of cement to stabilise the soft soils and the behaviour of cement-stabilised soils have been extensively investigated in many previous studies [31,37,40–42,46,53–55]. It has been proven by those previous studies that the addition of cement can efficiently increase the strength of stabilised soft soils, such as sandy-silty soft soil, by significantly reducing the water content and creating a cementation bond amongst soil particles. However, there is limited research that has fully investigated the influence of silt particles on the efficiency of cement stabilisation. For example, [56] conducted a series of experimental tests with soft soils stabilised by cement. To investigate the effect of silt and clay fractions on the reaction between cement and soil, soft clay was admixed with kaolin soil and then stabilised by cement with a cement additive content of 20% in that study. Unconfined compressive strength tests were then conducted on stabilised samples to find out the differences between cement-stabilised soft clay and the cement-stabilised soft clay—kaolin soil mixture. However, that study is limited since only a 20% additive content of cement was taken into consideration [56]. The high moisture content in most of the soft estuarine or marine deposits might result in a required amount of cement, which is likely to exceed 20%. In addition, the mechanism of the reaction between silt, clay particles and cement may result in that the influence of silt particles on the stabilisation may vary with the additive content of cement. Therefore, a broad range of silt and cement contents are necessary to be taken into consideration to further understand the influence and behaviour of cement-stabilised soft soils. Meanwhile, it is essential to study the microstructural development of soil as considerable scientific publications have proven that the macroscopic physical properties of soil are closely related to those microstructural characteristics and can be affected by the soil microstructure to a certain extent [57,58].

To overcome the shortcomings stated above, this study examines: (a) the properties, such as stress-strain behaviour, stress paths, and critical state parameters of reconstituted soft marine clay with varying silt contents based on the experimental results from consolidated undrained triaxial tests (CIU); (b) the influence of silt content on the unconfined compressive strength of cement-stabilised soft soils by performing unconfined compression tests (UCTs) on the specimens with varying silt and cement contents; and (c) the influence of silt content on the micro-structure aspects of cement-stabilised soft soils by conducting scanning electron microscope (SEM) tests for stabilised specimens.

In particular, five types of soft soil samples with different clay and silt fractions, and plasticity indexes, were adopted in this study. The soil samples were prepared by mixing soft soil collected from the Port of Brisbane (PoB) (Southeast Queensland, Australia) with kaolin soil at 5 different mixing ratios. Consolidated isotropic undrained (CIU) tests were then conducted for each group of the PoB soft soil-kaolin soil mixture at varying confining pressures. Consequently, the influence of silt fraction on the properties of soft soil is investigated. Then, each group of soil was stabilised by cement with a wide range of additive contents. Unconfined compressive strength (UCS) and scanning electron microscope (SEM) tests were then conducted on the stabilised specimens. Therefore, the influence of silt fraction in strength and micro-structure behaviour of stabilised and unstabilised soft soils can be further investigated. The effect of the presence and the variation in sand contents can be minimised as both the PoB soft soil and the kaolin soil used in this study have relatively low contents of sand.

2. Materials and Methods

The soils used in this study are summarised in Table 1. Soft marine clay collected from a depth of 1 to 2 m from the Port of Brisbane (PoB) in Southeast Queensland, Australia was used as the base soil. This marine clay was of dark grey in colour and was a mixture of clay and silt. The main source of this soil was the deposit from the Holocene age [59].

Table 1. Basic properties of the soils.

Soils	Clay (%)	Silt (%)	Sand (%)	LL (%)	PL (%)	PI (%)
Port of Brisbane soft soil	79.8	18.2	2.0	91.6	36.7	54.9
Kaolin soil	19.5	79.6	0.9	59.2	48.1	11.1

Commercial kaolin soil, which is called KM25 and produced by Kaolin SDN, Selangor, Malaysia, was mixed with the Port of Brisbane soft soil to form 5 types of reconstituted silt- clay samples with varying silt contents. In general, 5 types of soils with varying silt contents were adopted, with 0%, 25%, 50%, 75%, and 100% of Kaolin soil, respectively. Kaolin soil was used in this study due to the main components of kaolin being silt, so it can be mixed with the PoB soft soil to form reconstituted samples with varying silt contents. In a study conducted by [56], a same mixing methodology was adopted, in which 50% of a marine clay was mixed with 50% of kaolin soil. In this study, the 5 types of PoB soft soil-Kaolin soil mixtures are named as S1, S2, S3, S4, and S5, with the following components in particular: S1—100% Port of Brisbane (PoB) soil; S2—75% PoB soil mixed with 25% Kaolin soil (KM25); S3—50% PoB soil mixed with 50% Kaolin soil (KM25); S4—25% PoB soft soil mixed with 75% Kaolin soil (KM25); and S5—100% Kaolin soil (KM25). After the mixing, dead weights were placed gradually on the top plate with a pressure of 20 kPa. The samples were prepared for CIU testing by one-dimensional compression from a slurry to a state where they could be handled without disturbance. Figure 1a–c are the SEM images of the Port of Brisbane (PoB) soft soil, kaolin soil, and Ordinary Portland cement, respectively. The Atterberg limit of 3 groups of mixed soils was also tested (S2, S3, and S4).

The Atterberg limit of these 3 groups of soils are as follows: S2: PL is 37.6%, LL is 75.6%; S3: PL is 38.5%, LL is 68.4%; S4: PL is 42.7%, LL is 61.7%.

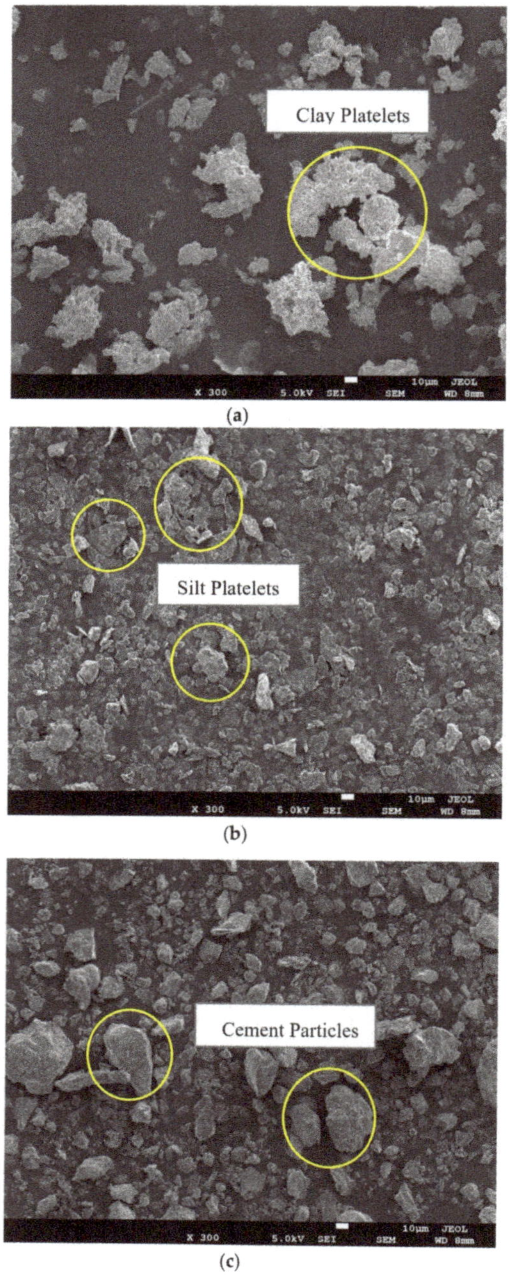

Figure 1. SEM images of soils and cement used in this study (**a**) Port of Brisbane soft soil; (**b**) Kaolin Soil (KM25); (**c**) Ordinary Portland Cement (300 times magnification).

2.1. Soil Properties

The particle size distribution of each sample was obtained by performing wet sieving analysis and a hydrometer test in accordance with ASTM standards. The initial moisture content, plastic limit, liquid limit, and percentages of clay, silt, and sand in each type of soil are obtained and summarised in Table 1. The chemical composition of the Port of Brisbane (PoB) soft soil, Kaolin soil and cement are shown in Table 2. It shows that both PoB soft soil and Kaolin soil contain very small amounts of sand, which can be eliminated during soil mixing. Kaolin soil has a much higher silt content of 79.6% and PoB soft soil has 79.8% of clay content. The SEM images of two kinds of soil particles and cement are shown in Figure 1. It can be seen that the particles of PoB soft soil are much smaller than those of Kaolin soil and cement, which generally conglutinate together into groups of aggregates and cluster with various sizes. The cement particles are generally irregular in shape. However, clay and silt particle shape usually present as aggregates of platelets with different contact modes in which clay platelets generally contact by surface-surface mode and silt particles usually contact by edge-to-surface and surface-to-surface mode.

Table 2. Chemical composition of the materials.

Oxide	PoB Soft Soil (%)	Kaolin Soil (%)	Cement (%)
LOI	14.84	11.81	2.42
Fe_2O_3	8.48	1.38	3.20
CaO	1.44	-	65.21
K_2O	1.61	4.02	0.45
SO_3	1.47	0.02	2.98
SiO_2	48.00	49.34	19.85
Al_2O_3	17.12	32.71	5.06
MgO	2.10	1.07	1.09
Na_2O	3.36	0.04	0.22

2.2. Consolidated Isotropic Undrained Triaxial Tests

The triaxial testing specimens were prepared by following an odometer-based reconstituted clay preparation method as developed and suggested by previous studies [60–63]. These previous studies tested reconstituted soft clay samples and proposed a reliable preparation method. In particular, water was first added to the soft soils to form a soil slurry ranging from 1.25 to 2 times of its liquid limit. A study conducted by [60–63] suggested water contents of the reconstitution of clay samples should be 1.25 and 1.5 times of its liquid limit, respectively. In a different work by [64], triaxial compression tests were conducted on both undisturbed and reconstituted specimens prepared by this method. The testing results indicated that the undisturbed and reconstituted specimens behaved similarly in terms of deviator stresses and stress paths. Therefore, this preparation method was adopted in this study to ensure the accuracy of the results of reconstituted specimens.

In this study, all the soil samples were prepared by adding water to achieve a water content of 1.5 times its liquid limit to eliminate the air bubbles. The amount of water added to the soil was calculated by the initial moisture content of each sample. The soil slurry was then poured into a preparation mould made of PVC tube with a 50 mm diameter, and 300 mm height. Silicone grease was applied on the inner surfaces of the preparation mould to reduce the friction effect. A dead weight was placed on top of the specimen to provide a vertical pressure on the soil sample through a loading cap and a porous stone. Consequently, all the specimens were consolidated under a preloading pressure of 20 kPa. The preparation mould can provide a double-drainage path during the consolidation process. The settlement of each specimen was recorded every day until the primary consolidation was completed. After the consolidation was completed, the specimen was extruded out from the preparation mould. Figure 2 shows some soft soil specimens after the one-dimension consolidation preparation. These specimens were then trimmed to 50 mm in diameter, and 100 mm in height for triaxial testing. The trimming end cut from each specimen was used to measure the water content of the specimen before the triaxial test. The trimmed specimen was then immediately placed in the triaxial apparatus.

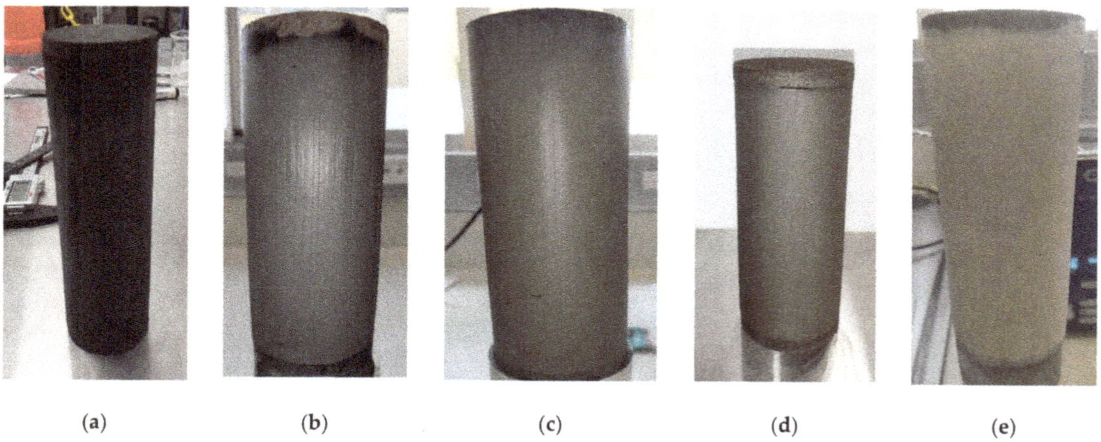

Figure 2. Some soft soil specimens for consolidated undrained triaxial testing: (**a**) S1—100% PoB soft soil; (**b**)S2—75% PoB soft soil + 25% kaolin soil; (**c**) S3—50% PoB soft soil + 50% kaolin soil; (**d**) S4—25% PoB soft soil + 75% kaolin soil; and (**e**) S5—100% kaolin soil.

Consolidated isotropic undrained (CIU) triaxial tests for unstabilised soil samples were carried out in accordance with ASTM standards. Filter paper strips were placed around each specimen to provide radical drainage and therefore, to accelerate the consolidation and the saturation phases during the triaxial test. It is necessary to flush water through the drainage path to remove trapped air bubbles. The back pressure saturation method was used to saturate the specimens until the Skempton's pore pressure parameter (B) value reached 0.98. The specimens were then tested at effective confining pressure of 50, 100, and 200 kPa, respectively. The shear rate of each specimen was calculated based on the time required to achieve t_{100} consolidation in accordance with the method proposed by [65]. During the shear phase, the specimens were sheared under undrained conditions, and all the tests were stopped when an axial displacement of 20% was achieved. At the end of the test, the specimens were removed from the triaxial cell and the water content of each specimen was then measured. The testing results were calibrated by eliminating the effect of the filter paper strips and membrane stiffness in accordance with the method suggested by [65]. Five types of soil samples with varying ratios between the Port of Brisbane soft soil and Kaolin soil were tested. For each kind of sample, specimens were tested under 3 different effective confining pressures as mentioned above. In total, 15 consolidated undrained triaxial tests were conducted in this study, as summarised in Table 3. Figure 2a–e presents some of the tested specimens under consolidated undrained condition.

2.3. *Unconfined Compressive Strength Tests*

Four types of soil slurry were prepared by mixing the Port of Brisbane soft soil and kaolin soil at different mixing ratio stated above. The cement additive content was 10, 15, 20, 25, and 30%. As mentioned in previous, the study conducted by [56] considered one mixing ratio between kaolin soil and marine clay, and a cement additive content of 20%. This study considered a more comprehensive range of cement additive content to further develop the effect of silt particle on the stabilisation of soft soil.

Table 3. Summary of all tested triaxial specimens under consolidated undrained conditions.

Testing No	Percentage of Port of Brisbane Soft Soil	Percentage of Kaolin Soil	Confining Pressure (kPa)
S1—50	100%	0%	50
S1—100	100%	0%	100
S1—200	100%	0%	200
S2—50	75%	25%	50
S2—100	75%	25%	100
S2—200	75%	25%	200
S3—50	50%	50%	50
S3—100	50%	50%	100
S3—200	50%	50%	200
S4—50	25%	75%	50
S4—100	25%	75%	100
S4—200	25%	75%	200
S5—50	0%	100%	50
S5—100	0%	100%	100
S5—200	0%	100%	200

The amount of cement was calculated to achieve each targeting additive content for each type of soil. The additive content of cement is defined as the mass ratio of dry cement powder (M_c): dry soil solid (M_s). De-aired distilled water was added into the cement powder to form cement slurry. The slurry was then mixed with the soil slurry for approximately 10 min. The soil-cement mixture was then placed into a cylindrical PVC with 50 mm in diameter and 100 mm in height. The specimens were vibrated to get rid of the air bubbles trapped in the specimens. Each PVC mould was sealed by wax in both ends. All of the specimens were cured in a curing room with a constant temperature of 22 °C. After reaching the targeting 28-day curing period, the specimens were extruded from the PVC moulds for testing. The specimens were weighed and checked for consistency to within +3 g of the mass of the matrix. Figure 3 presents the photos of some of the stabilised specimens for unconfined compressive strength test.

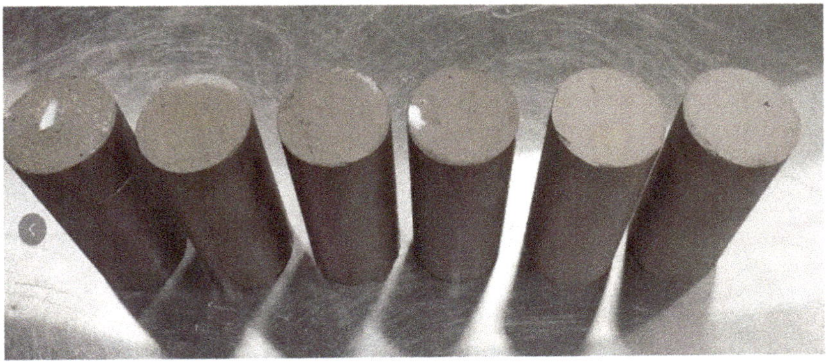

Figure 3. Examples of some cement-stabilised specimens for unconfined compressive strength tests, Port of Brisbane Soft Soil + 5, 10, 15, 20, 25, and 30% cement, respectively.

The Unconfined Compressive Strength tests for all the stabilised specimens were carried out in accordance with ASTM D-2166 [66]. A shearing rate of 1.00 mm/min was adopted as suggested by the standard. All the tests were terminated when a 15% axial strain was achieved, or when a peak deviator stress was observed, whichever one came first. All the specimens for UCS tests have a 50 mm height and 100 mm diameter.

2.4. Scanning Electron Microscope (SEM) Tests

Scanning Electron Microscope Tests were performed to investigate the change of the microstructure of the cement-stabilised PoB soft soil with different kaolin soil replacement ratio (0%, 25%, 50%, 75%) and cement additive content (10%, 15%, 20%, 25%, 30%). The SEM samples used in this study were obtained from the sheared specimens of the Unconfined Compression Test. Firstly, the tested UCS specimens were cut into columns of approximately 10 × 10 × 30 mm and then immersed in liquid nitrogen (−196 °C) for quick freezing. Afterwards, the vacuum freeze dryer produced by John Morris Scientific Pty Ltd. was utilised to sublimate and evacuate the solid-state water in the specimen at −100 °C to −50 °C for 5 days, and then the freeze-dried specimens were broken gloved hands at the middle part to ensure the fresh cross-sections are not out of shape or contaminated. Finally, 10 nm of gold were coated on the sample surface to ensure a good electrical conductivity. The SEM images of stabilised samples are acquired by means of JEOL JSM-7001f in Central Analytical Research Facility at Queensland University of Technology, QLD, Australia, which is a Field Emission Scanning Electron Microscope (FE-SEM) with a wide range of magnification up to 1,000,000 times.

3. Results and Discussion

The experimental results obtained from the experimental tests mentioned previously are presented and discussed in this chapter.

3.1. Consolidated Undrained Triaxial Tests

The CIU testing results of all the specimens are summarised in this chapter. For clarity, the component and confining pressure of each tested specimen are summarised in Table 3. For example, S1—50 refers to the specimen with 100% Port of Brisbane soft soil and tested under 50 kPa confining pressure, and S3—100 refers to the specimen with 50% Port of Brisbane soft soil admixed with 50% kaolin soil and tested under a confining pressure of 100 kpa.

3.1.1. Stress-Strain Behaviour

The effective stress paths, excess pore pressure-axial strain behaviour, and stress-axial strain behaviour of each specimen is illustrated in Figures 4–6, respectively. The mean effective stress and deviator stress are defined by the following two equations:

$$p' = (\sigma_{1'} + 2\sigma_{3'})/3 \quad (1)$$

$$q = \sigma_{1'} - \sigma_{3'} \quad (2)$$

From Figure 4, it can be noted that stress-strain behaviour shows strain-softening for all the specimens, because the changes in mean effective stress were all negative values for all the specimens. However, the effective stress path shifts to the right as the silt content increases under the same confining pressure. This can be verified from the effective stress paths and deviator stress-strain behaviour of these specimens. For specimens in groups S1, S2 and S3, the mean effective stress decreased during the shear phase. For specimens in groups S5, and S4—50, S4—200, the mean effective stress decreased at the beginning of the shear phase, but then increases in mean effective stress were observed from those specimens.

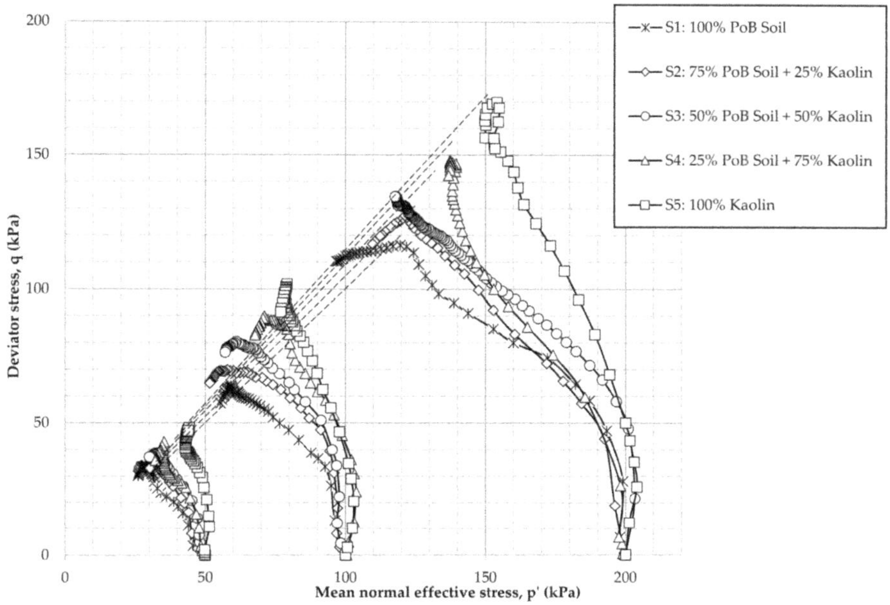

Figure 4. Effective stress paths for different confining pressures with varying silt contents (S1 represents 100% PoB soft soil; S2 represents 75% PoB soft soil + 25% kaolin soil; S3 represents 50% PoB soft soil + 50% kaolin soil, S4 represents 25% PoB soft soil + 75% kaolin soil; and S5 represents 100% kaolin soil; while 50,100, and 200 refer to the confining pressure.).

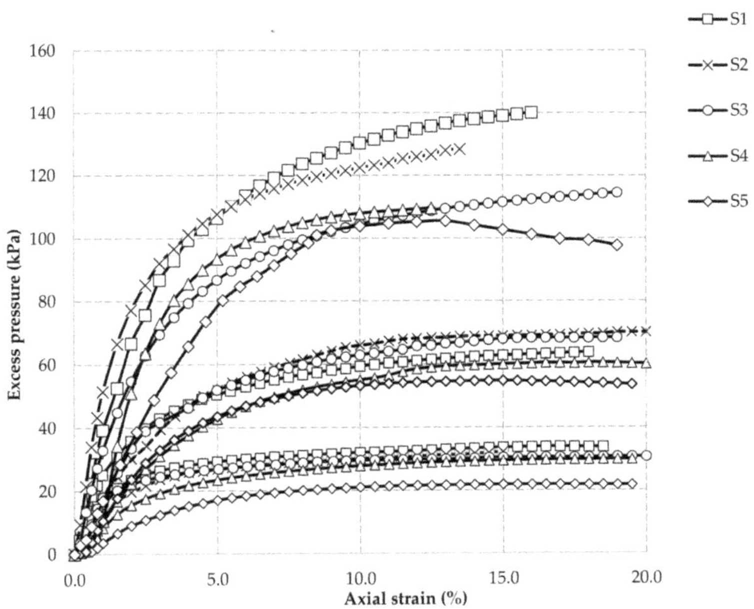

Figure 5. Excess pore pressure versus axial strain with different confining pressures for different silt contents. (S1 represents 100% PoB soft soil; S2 represents 75% PoB soft soil + 25% kaolin soil; S3 represents 50% PoB soft soil + 50% kaolin soil, S4 represents 25% PoB soft soil + 75% kaolin soil; and S5 represents 100% kaolin soil; while 50, 100, and 200 refer to the confining pressure.).

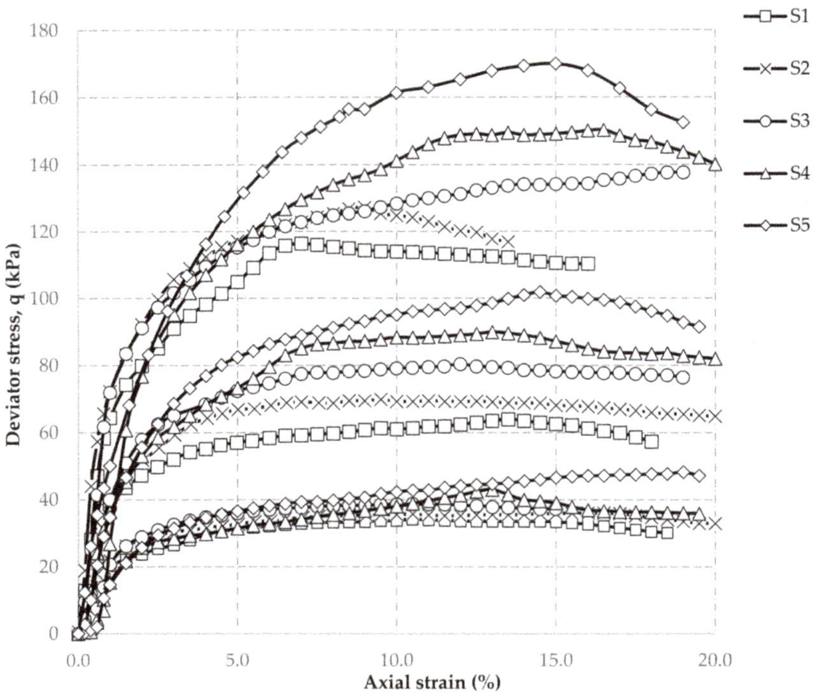

Figure 6. Deviator stress versus axial strain with different confining pressures for different silt contents. (S1 represents 100% PoB soft soil; S2 represents 75% PoB soft soil + 25% kaolin soil; S3 represents 50% PoB soft soil + 50% kaolin soil, S4 represents 25% PoB soft soil + 75% kaolin soil; and S5 represents 100% kaolin soil; while 50, 100, and 200 refer to the confining pressure.).

Dilation or slight dilation behaviour can also be found for specimens S5—50, S5—100, S5—200, and S4—100, with negative values in the change of excess pore pressure as shown in Figure 5. Consequently, this resulted in the stress paths of those specimens turning direction in the p-q space. Based on these observations, for the S1, S2, and S3 specimens (100% Port of Brisbane soft soil), the effective stress paths are typical of normally consolidated clay. For the S5 specimens (100% kaolin soil), the effective stress paths are typical of loose to medium-dense silty sand. Behaviour of the specimens in the S4 group show transitional form, as the stress paths of S4—50, and S4—100 show normal consolidated clay behaviour, however, that of the S4—200 shows slight loose to medium-dense sand behaviour. Silt content is also a key factor determining the maximum deviator stress of the specimens. Figure 6 shows the relationship between stress and axial strain of the specimens under each confining pressure. Figure 5 summarises the change in excess pore pressure for each group of specimens. Most of the testing results show that the maximum deviator stress of the specimens and the silt content increased simultaneously, with the same confining pressure. The maximum excess pore pressure, however, decreased as the silt content increased, as shown in Figure 5.

3.1.2. Critical State Behaviour

Figure 7 is plotted based on the deviator stress and mean effective stress of each specimen at failure condition. According to the published literature [65], the deviator stress at failure can be determined when either of the following conditions are achieved: (a) maximum principal stress, and (b) constant deviator stress and pore pressure. The corresponding values of strain and pore pressure are axial strain at failure and pore pressure at failure, respectively [65]. In this study, the maximum principal stress ratio method has been adopted to identify deviator stress at failure. Hence, the mean effective stress and deviator stress at failure refer to the mean effective stress and deviator stress at the condition when the maximum principal ratio was observed.

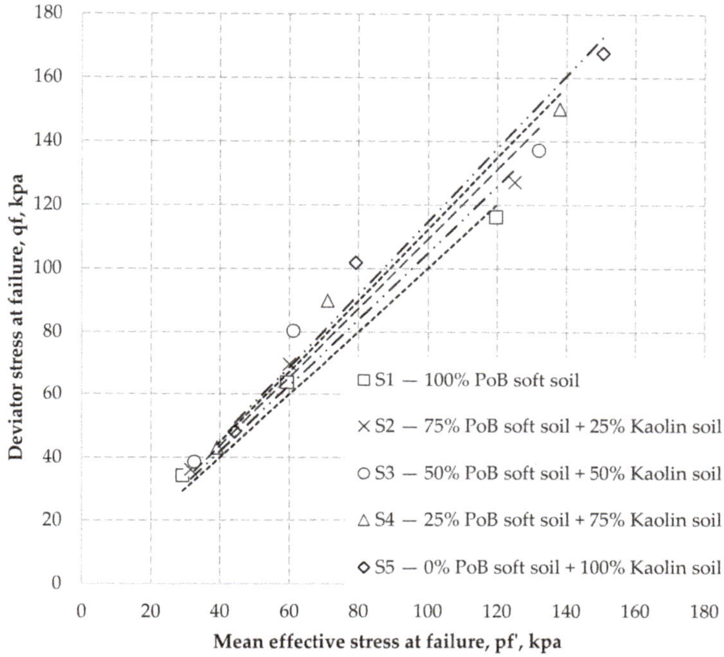

Figure 7. CSLs of the specimens at different confining pressures with varying kaolin contents. The dotted-point lines represent the best fitted critical state line for each group of soil.

The slope M has been used as a fitting parameter to indicate the critical state behaviour of soils. It can also be used to calculate the effective friction angles of the soil. The fitting parameter M was determined by the gradient of CSL for each group of specimens. Then the effective friction angle for each group of soil was back calculated by the following equation proposed by [67]:

$$\sin \phi\prime\ (^{\circ}) = \frac{3M}{6+M} \tag{3}$$

A summary of the fitting factors M and the effective internal friction angle of the tested specimens are summarised in Table 4. It can be observed that the CSL of S5 group specimens shows the highest M value of 1.15, in which the specimens are 100% kaolin soil without any Port of Brisbane soft soil added into the specimen. On the other hand, the S1 group specimens shows the lowest M value of 1.01, where the specimens are 100% Port of Brisbane soft soil without any addition of kaolin soil. In general, M of the fitting line increases as the silt content increases or the clay content decreases.

Table 4. Summary of critical state parameters.

Soil Groups	Clay Content (%)	Silt Content (%)	*M*	Friction Angle °	Cohesion (kPa)
S1	79.80	18.00	1.01	25.43	5.12
S2	64.73	33.41	1.05	26.54	4.58
S3	49.65	48.81	1.09	27.53	2.52
S4	34.58	64.22	1.12	28.25	1.87
S5	19.50	79.62	1.14	28.80	1.86

It is also apparent that the effective friction angle and the silt content increase simultaneously. For clarification, the calculated values of the friction angle are plotted against the silt content in percentage in Figure 8. The relationship between the silt content and effective angles of friction is inversely proportional, as indicated in Figure 8. The correlation between silt content and effective internal friction angle for the tested soils is:

$$\phi' (^\circ) = 0.0549 \text{ Silt } (\%) + 24.631 \quad (4)$$

with a coefficient of determination (R^2) equal to 0.98.

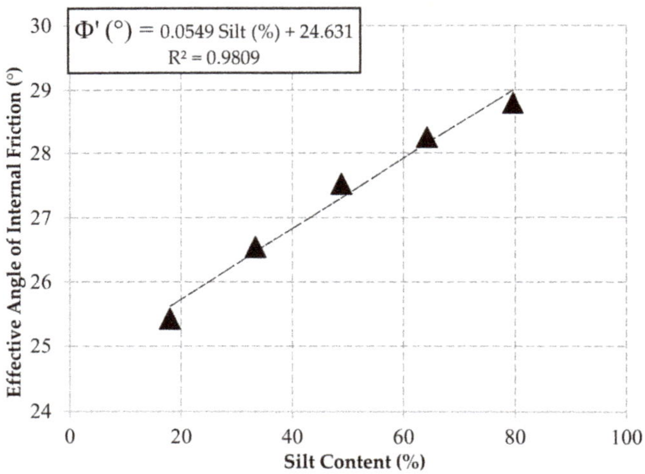

Figure 8. Correlation between effective internal friction angle and silt content.

The semi-logarithmic plot of CSLs and NCLs derived from consolidated isotropic undrained shear triaxial compression testing results is shown in Figure 9. In the undrained shearing stage, the void ratio of each specimen remains constant. The water content of each tested specimen was measured as introduced previously. Therefore, the void ratio, as well as the specific volume in CSLs and NCLs, were determined. The effective stress paths shift to the left horizontally, as shown in Figure 9, because the drainage condition was undrained during the shear phase. It can be observed that the CSLs and NCLs are parallel, as shown in Figure 9, for the S1, S2, and S3 groups' specimens. The gradients of the respective CSLs seem to be becoming non-parallel to the gradients of their corresponding NCLs with the increase in silt content, as indicated by the CSLs and NCLs of S4. For the S5 group specimens, the gradients of the CSLs are obviously not parallel to those of the NCLs, indicating that the silt particle content starts to dominate the behaviour of the soil. Therefore, it can be concluded that S1, S2, and S3 show typical clay behaviour, while S5 shows typical loose to medium-dense sand behaviour. S4 is in a transitional form.

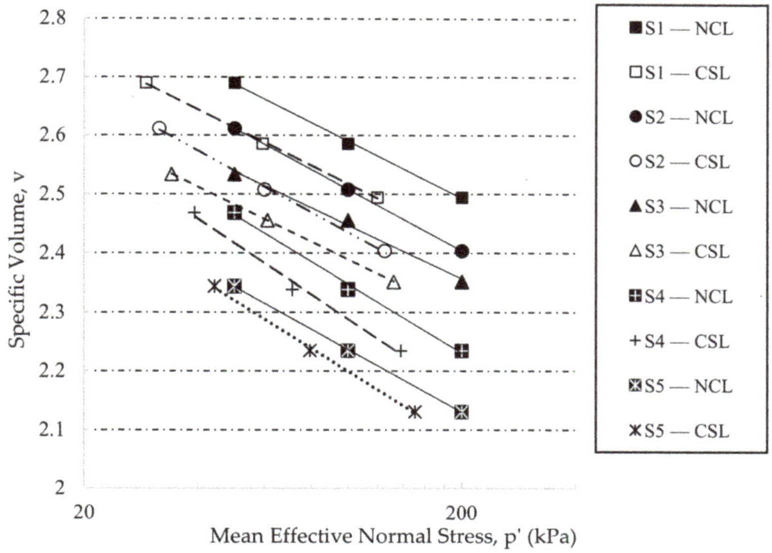

Figure 9. V-lnp relationship on a semi-logarithmic scale.

Furthermore, the gradients of the compression lines of the tested specimens were derived from the NCL and CSL results of each group of specimens, as shown in Table 5 and Figure 9. The ratios of the gradients of compression lines (λ (NCL)/λ (CSL)) of S1, S2, S3, S4, and S5 are 1.02, 0.99, 0.97, 0.90, and 0.88, respectively. This also indicates that the NCLs and the corresponding CSLs of S1, S2, and S3 are relatively more parallel than those of S4, and S5. This indicates that specimens with higher clay contents are likely to have more parallel CSLs and NCLs. The R^2 of the λ (NCL)/λ (CSL) and silt content for the tested specimens is 0.951 and the correlation equation is:

$$\lambda\ (NCL)/\lambda\ (CSL) = -0.0023\ \text{Silt content\%} + 1.07 \quad (5)$$

Table 5. Summary of CSL and NCL parameters.

Soil Groups	Silt Content (%)	λ (NCL)	λ (CSL)	λ (NCL)/λ (CSL)
S1	18.20	0.14	0.14	1.02
S2	33.41	0.15	0.15	0.99
S3	48.81	0.13	0.13	0.97
S4	64.22	0.16	0.18	0.90
S5	79.62	0.15	0.17	0.88

3.2. *Unconfined Compressive Strength of Stabilised Soils*

The unconfined compressive strength (UCS) results for all the tested specimens are summarised in Table 6. The cement additive contents adopted were 10, 15, 20, 25, and 30%. Four groups of specimens with different PoB soft soil and kaolin soil mixing ratios were considered, as shown in Table 6. Therefore, in total 20 sets of specimens were prepared, and each set includes three specimens to determine the average results.

Table 6. Summary of stabilised specimens for unconfined compressive strength tests.

Soil Group	Port of Brisbane (PoB) Soft Soil	Kaolin Soil	Cement Content (%)
S1	100%	0%	10, 15, 20, 25, 30
S2	75%	25%	10, 15, 20, 25, 30
S3	50%	50%	10, 15, 20, 25, 30
S4	25%	75%	10, 15, 20, 25, 30

For clarity, the correlation between UCS results and the additive content of cement and the content of silt are shown from the plotted Figures 10 and 11. The correlation between the UCS and the additive content for each group of stabilised specimens is plotted in Figure 10. The results of UCS for each group of specimens increased significantly as the cement additive content increased from 10% to 30%. This is consistent with the outcomes in other studies conducted for soft soils with a high moisture content [31,68,69]. However, it is interesting to find that the influence of the cement on each type of soil is different. The increase in the cement additive content has the biggest influence for the S1 group of specimens, which contain 100% Port of Brisbane soft soil. With the increase in silt content, the influence in UCS as a result of the increase in the cement additive content becomes more and more slight. For example, the addition of cement increased the strength of specimens in group S4 (25% PoB soft soil + 75% kaolin soil), but a further increase in cement content does not contribute significantly to the improvement of strength.

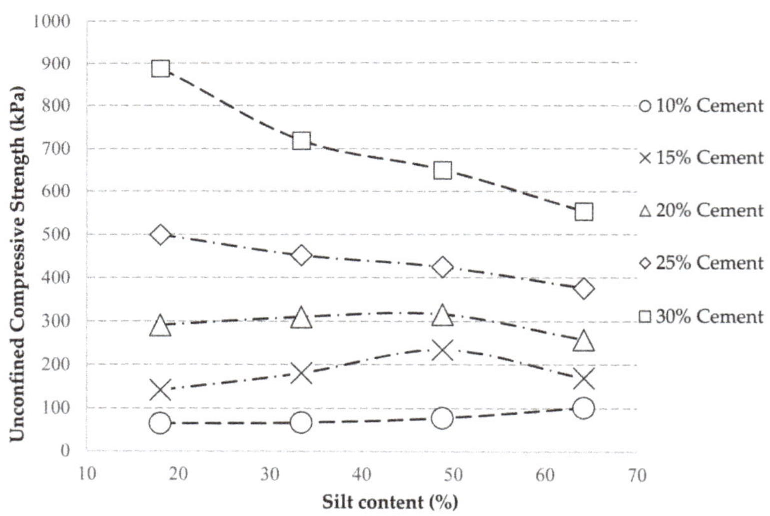

Figure 10. Correlation between Unconfined Compressive Strength (UCS) and silt content.

Based on the experimental results in this study, the correlation between silt content and the strength of specimens is related to the cement additive content. Figure 10 shows the relationship between the silt content and UCS of the specimens under varying cement additive contents. From Figure 10, it can be concluded that silt content is a key factor affecting the strength of the stabilised specimens with the same moisture content. Also, the cement additive content has significant influence in terms of the correlation between silt content and the UCS of samples.

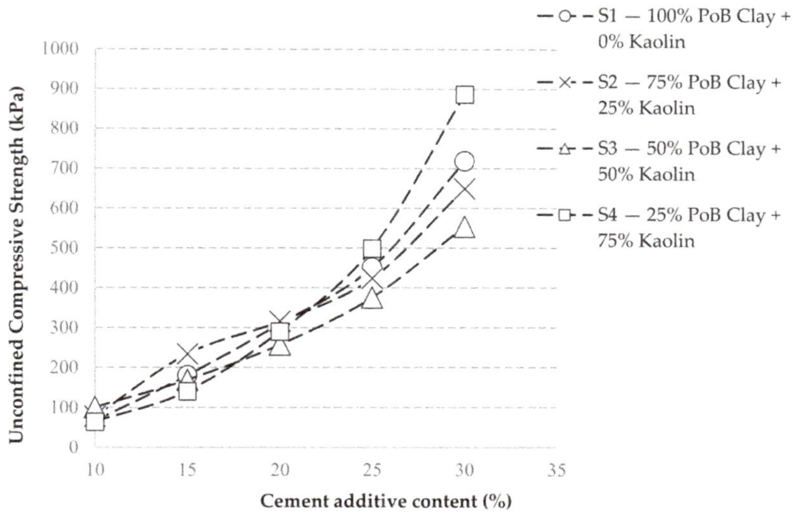

Figure 11. Correlation between Unconfined Compressive Strength (UCS) and cement content.

In particular, there is an increase in the UCS with the increase of silt content when the cement additive content is only 10%. By comparing the results of specimens in groups S1 (100% PoB soft soil), S2 (75% PoB soft soil + 25% kaolin soil), and S3 (50% PoB soft soil + 50% kaolin soil), it can be seen that for the specimens with a cement additive content of 15 or 20%, the UCS increases with the increase in silt content at first. However, then there is a decrease in the UCS of specimens with a further increase in the silt content, since the UCS of the specimens in S4 (25% PoB soft soil + 75% kaolin soil) is lower than that in S3 at 15% or 20% of cement, as shown in Figure 10. That means that, for the specimens with a cement additive content of 15 and 20%, the UCS of specimens increases with the increase in silt content at first. However, once the silt content reaches a particular percentage or a saturation point, further increases in silt content will result in a decrease in the UCS. For the specimens with a high cement content, such as 25 and 30%, there is a negative correlation between the UCS of the specimens and the cement additive content, as shown in Figure 10. The UCS of specimens in the S1 (100% PoB soft soil) group is higher than that of the S2 (75% PoB soft soil + 25% kaolin soil), S3 (50% PoB soft soil + 50% kaolin soil), and S4 (25% PoB soft soil + 75% kaolin soil) groups, with 25 or 30% cement content.

The addition of cement can aggregate the soil particles to form stronger bonds between the particles, which results in the increase of strength. When the cement additive content is high, such as 25 or 30%, there are not enough clay particles to fill the voids between the silt and cement particles, which results in the decrease in strength with the increase in silt content with the cement additive content of 25% and 30%.

Therefore, it is observed based on the results plotted in Figure 10 that there is not always a positive correlation between the UCS of stabilised specimens and the silt content. Both the silt content and cement additive content have significant influence on the UCS of stabilised specimens with high-water content. Essentially, there is a linear correlation between the UCS of specimens and silt content when the cement additive content is low. When the cement additive content is in the medium range (15, or 20%), the UCS of the specimens increases with the increase in silt content at first, but then the UCS decreases once the silt content reaches a saturation point. These trends can also be observed and verified by the SEM results of specimens. The mechanism of these observed correlations is further explained by the SEM results on the aspect of the micro-structure of the specimens.

3.3. Microstructure of Stabilised Soils

To determine the relationship between strength and microstructure development, and further clarify the findings observed from the UCS results, SEM images with 5000 magnification were analysed. Figures 12–14 show the microstructure of stabilised specimens with varying silt and cement contents. Figure 12 indicates the SEM images of four types of clay silt mixture S1 (Figure 12a), S2 (Figure 12b), S3 (Figure 12c), and S4 (Figure 12d) stabilised with 10% cement. Figures 13 and 14 present the SEM images of specimens stabilised with 20% and 30% cement, respectively.

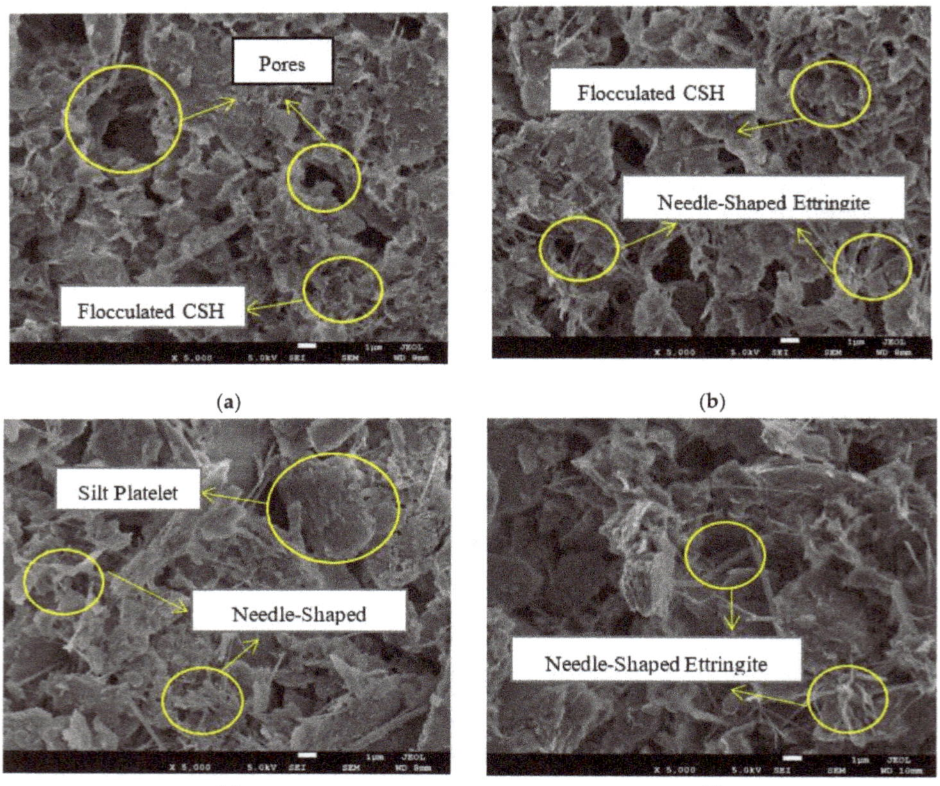

Figure 12. SEM images of specimen treated by 10% cement with different silt contents: (**a**) S1 + 10% cement, (**b**) S2 + 10% cement, (**c**) S3 + 10% cement, (**d**) S4 + 10% cement.

It can be clearly seen by comparing Figures 12a, 13a and 14a that there is a decrease in the number of large pores and an increase in small pores with the increment of cement content, which can be attributed to the production of cementitious materials formed from cement hydration and pozzolanic reaction, such as calcium silicate hydrates (CSH), calcium aluminium silicate hydrates (CASH), and ettringite. These products, which can be found in the images as reticulation (CSH, CASH), flocculation (CSH, CASH) and needle-shaped structure (Ettringite), filled pores and coated soil particles, which results in a dense aggregated structure responsible for the strength gain. The observations from SEM images are consistent with results obtained by [70]. His observations of SEM images of cement-treated marine clay in Singapore indicated that the flocculation of the clay particles results in the water to be trapped within the clay-cement cluster and, therefore, both the entrance pore diameter and particles of cluster were increased [70].

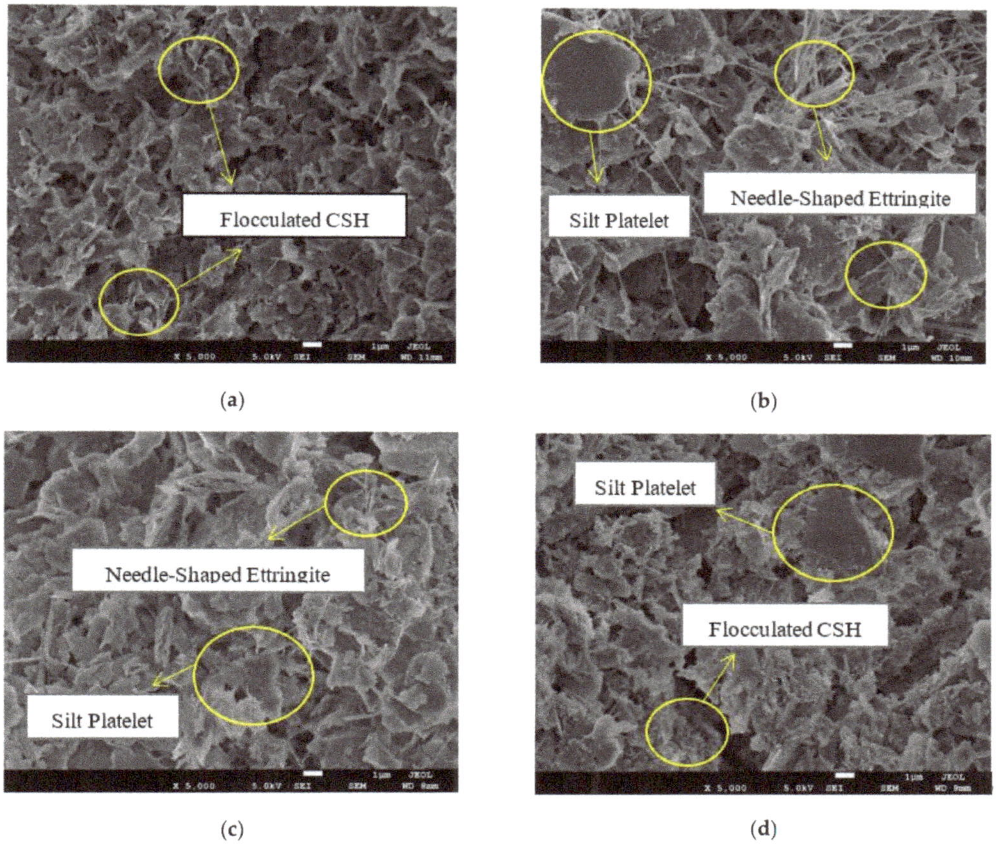

Figure 13. SEM images of specimen treated by 20% cement with different silt contents: (**a**) S1 + 20% cement, (**b**) S2 + 20% cement, (**c**) S3 + 20% cement, (**d**) S4 + 20% cement.

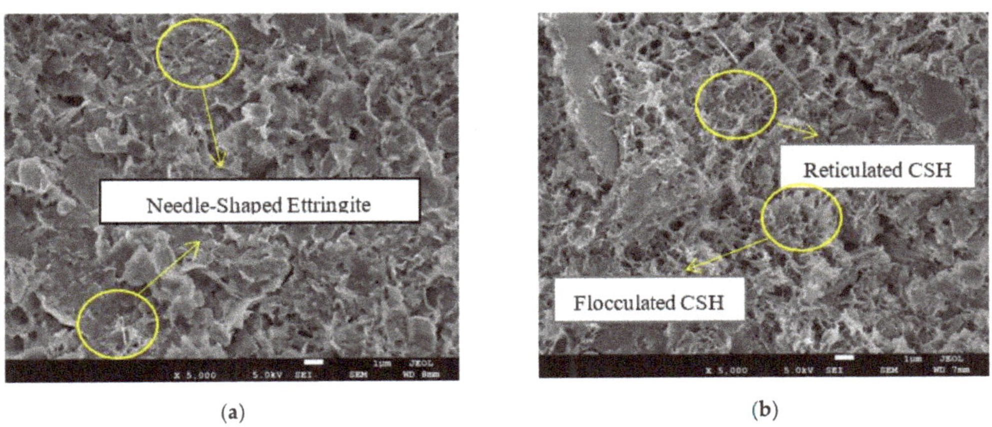

Figure 14. *Cont.*

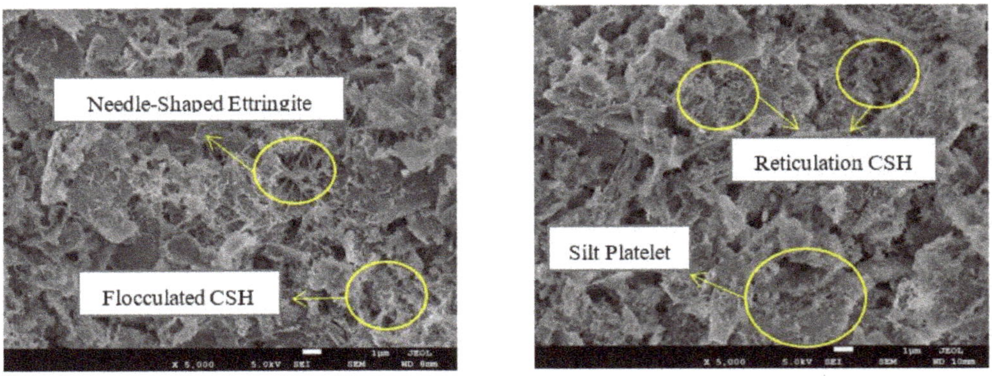

(c) (d)

Figure 14. SEM images of specimen treated by 30% cement with different silt contents: (**a**) S1 + 30% cement, (**b**) S2 + 30% cement, (**c**) S3 + 30% cement, (**d**) S4 + 30% cement.

However, when the cement content remained constant, the microstructure of the four samples with different silt content did not change consistently. At the lower additive content of cement (10%), there is an obvious growth in the number of larger silt platelets (see Figure 12a–d) with the increment of silt content which is due to the larger size of silt particles and edge-to-edge and edge-to-surface contact of silt platelets. However, the area of soil pores is slightly decreased at the same time. This can be attributed to those smaller size clay platelets which can easily fill in the large pores and aggregate together by cementitious gels to form large and dense aggregates, which improved the stability of the soil structure and increased the strength.

When the cement additive content is 20% (Figure 13a–c), the soil pores show an increasing trend, which is consistent with the trend of lower (10%) cement-stabilised soil. However, when the silt content is higher than 50% (Figure 13d), there is a significant increase in large pores, which can be attributed to the excessive number of large particles, because cement particles also have a larger size than clay particles. Thus, the strength gained from cementitious products can be partially countered by the pore size increment caused by excessive cement content, and that is the reason why this phenomenon did not perform at the 10% cement content. Thus, when the silt content exceeded the saturation point (50%), the excessive number of large particles lead to a substantial increase of large pores and the content of clay particles and formed hydration products are not sufficient to fully fill the pores, which resulted in a loose structure and lesser connections among particles. Once loaded, the soil particles and aggregates are prone to dislocation and sliding, resulting in the change of soil structure and reduction of soil strength. The aforementioned changes of microstructure are consistent with the tendency of UCS strength in Figure 10. Therefore, it can be concluded that the saturation point of silt content is approximately 50% when treated with 20% cement.

When the cement additive content is higher (30%), it can be seen from Figure 14a–d that the number of large pores increases with the increase of silt content due to the larger particle size. The large pores can lead to the decrease of the compactness of soil structure and can be easily dislocated and slide under compression. Meanwhile, it is shown that the cementitious products reduced with the increase of silt content because these cementitious materials are mainly produced by the reaction between cement and silica and alumina from clay particles. Therefore, the reduction of clay content impeded the hydration and pozzolanic reaction, resulting in the incomplete reaction of cement, which made the stabilization process imperfect. These changes in microstructure thus explains why the UCS strength continuously decreased with the increase of silt content at a high cement content (30%).

4. Conclusions

A series of triaxial compressive tests were conducted in this study for soft soils with varying silt contents and the effect of silt contents on the strength and critical state behaviours of soft soils were investigated. The empirical equations were proposed to evaluate the effect of silt content on the stress paths of reconstituted soft soils under consolidated isotropic undrained triaxial tests and also the critical state parameters. A further set of unconfined compression tests were conducted for each group of specimens stabilised by cement. As the silt content might exhibit different influence on the strength of cement- stabilised samples, a varying dosage of cement additives content were considered in this study. The experimental results indicate that silt content play a different role in soil stabilisation under different cement additive or replacement contents. The effect of cement additive content and silt content on the microstructure development of stabilised soils were also analysed by utilising the Scanning Electron Microscope (SEM) images. With the increase of cement dosage, the number of cementitious products, such as reticulated CSH and needle-shaped ettringite were notably increased, resulting in a denser structure. This can be attributed to the hydration of cement and the pozzolanic reactions.

As for the effect of silt content, since particle size plays a very important role in microstructure development, both cement and silt contents can dramatically affect the pore size distribution. When the cement additive content is lower than 10%, clay platelets can fill the pore spaces and the cementitious products can enhance the inter-cluster bond strength by aggregating clay and silt platelets together to form larger and denser aggregates responsible for the strength improvement. When the cement content is between 10% and 20%, the stabilised soil strengths increase with the increase of silt content and then decrease when silt contents are higher than 50%. This is because the strength gained from cementitious product enhancement was partially countered by the increment of pore size caused by the excessive cement and silt contents. When the cement content is higher than 20%, the strength shows a negative correlation with silt content, which can be attributed to the incomplete reaction of cement due to the reduction of clay content.

In general, any increase in the amount of cement will increase the improved soil strength (see Figure 11). However, in this study, it is very important for engineers to also realise that the increment of cement used in each cement replacement group in soils with appreciable silt percentage (see Figure 10) has a 'saturation point', after which any further increase in cement replacement content will not further increase the strength of the improved soil. The saturation points in this study were found to be (i) more than 50% silt for cement contents not exceeding 20% replacement, and (ii) regardless of silt content for cement contents exceeding 20% replacement. By knowing these important outcomes, engineers can reliably customise the soil-cement mix design for optimal strength outcomes given any type of fine-grained soils containing predominantly clay and silt particles.

It is acknowledged that there are other, more environmentally friendly binders than cement available for use, but this current manuscript only focuses on the use of cement. The authors' future manuscripts will include the use of more environmentally friendly constituents, such as fly ash and/or enhanced additives.

Author Contributions: Conceptualisation: Y.L. and D.E.L.O.; Methodology: Y.L. and Z.L., Validation: E.O. and D.E.L.O.; Formal analysis, Y.L. and Z.L.; Investigation: Y.L. and Z.L.; Resources: Y.L. and Z.L.; Data curation, Y.L. and Z.L.; Writing—original draft preparation: Y.L. and Z.L.; Writing—review and editing, E.O. and D.E.L.O.; Visualization, Y.L. and Z.L.; Supervision: E.O. and D.E.L.O.; Project administration: E.O; Funding acquisition: E.O. All authors have read and agreed to the published version of the manuscript.

Funding: This research received no external funding.

Institutional Review Board Statement: Not applicable.

Informed Consent Statement: Not applicable.

Data Availability Statement: The original experimental data are available on request from the corresponding author.

Acknowledgments: The soft soil collection in this work was supported by the Port of Brisbane Pty Ltd. The SEM images were acquired by use of the Central Analytical Research Facility hosted by the Institute for Future Environments at QUT.

Conflicts of Interest: The authors declare no conflict of interest.

References

1. Leong, H.Y.; Ong, D.E.L.; Sanjayan, J.G.; Nazari, A. A genetic programming predictive model for parametric study of factors affecting strength of geopolymers. *RSC Adv.* **2015**, *5*, 85630–85639. [CrossRef]
2. Leong, H.Y.; Ong, D.E.L.; Sanjayan, J.G.; Nazari, A. The Effect of Different Na_2O and K_2O Ratios of Alkali Activator on Compressive Strength of Fly Ash Based-geopolymer. *Constr. Build. Mater.* **2016**, *106*, 500–511. [CrossRef]
3. Leong, H.Y.; Ong, D.E.L.; Sanjayan, J.G.; Nazari, A. Strength Development of Soil-Fly Ash Geopolymer: Assessment of Soil, Fly Ash, Alkali Activators, and Water. *J. Mater. Civ. Eng.* **2018**, *30*. [CrossRef]
4. Leong, H.Y.; Ong, D.E.L.; Sanjayan, J.G.; Nazari, A.; Kueh, S.M. Effects of Significant Variables on Compressive Strength of Soil-Fly Ash Geopolymer: Variable Analytical Approach Based on Neural Networks and Genetic Programming. *J. Mater. Civ. Eng.* **2018**, *30*. [CrossRef]
5. Leong, H.Y.; Ong, D.E.L.; Sanjayan, J.G.; Nazari, A. Suitability of Sarawak and Gladstone Fly Ash to Produce Geopolymers: A Physical, Chemical, Mechanical, Mineralogical and Microstructural Analysis. *Ceram. Int.* **2016**, *42*, 9613–9620. [CrossRef]
6. Ngu, L.H.; Song, J.W.; Hashim, S.S.; Ong, D.E.L. Lab-scale atmospheric CO_2 absorption for calcium carbonate precipitation in sand Greenhouse Gases. *Sci. Technol.* **2019**, *9*, 519–528.
7. Mehdizadeh, A.; Disfani, M.M.; Evans, R.; Arulrajah, A.; Ong, D.E.L. Discussion of 'Development of an Internal Camera-Based Volume Determination System for Triaxial Testing' by S. E. Salazar, A. Barnes, and R. A. Coffman. *Geotech. Test. J.* **2016**, *39*, 165–168.
8. Mehdizadeh, A.; Disfani, M.M.; Evans, R.; Arulrajah, A.; Ong, D.E.L. Mechanical Consequences of Suffusion on Undrained Behaviour of a Gap-graded Cohesionless Soil—An Experimental Approach. *Geotech. Test. J.* **2017**, *40*, 1026–1042. [CrossRef]
9. Ong, D.E.L.; Yang, D.Q.; Phang, S.K. Comparison of finite element modelling of a deep excavation using SAGE-CRISP and PLAXIS. In Proceedings of the International Conference on Deep Excavations, Singapore, 28–30 June 2006; pp. 51–64.
10. Pham, V.N.; Ong, D.E.L.; Oh, O. Prediction of unconfined compressive strength of cement-stabilized sandy soil in Vietnam using artificial neural networks (ANNs) model. *Int. J. Geotech. Eng.* **2021**. [CrossRef]
11. Omoregie, A.I.; Senian, N.; Li, P.Y.; Hei, N.L.; Ong, D.E.L.; Ginjom, I.R.H.; Nissom, P.M. Ureolytic bacteria isolated from Sarawak limestone caves show high urease enzyme activity comparable to that of Sporosarcina pasteurii (DSM33). *Malays. J. Microbiol.* **2016**, *12*, 463–470.
12. Omoregie, A.I.; Khoshdelnezamiha, G.; Senian, N.; Ong, D.E.L.; Nissom, P.M. Experimental Optimisation of Various Cultural Conditions on Urease Activity for Isolated Sporosarcina Pasteurii Strains and Evaluation of Their Biocement Potentials. *Ecol. Eng.* **2017**, *109*, 65–75. [CrossRef]
13. Omoregie, A.I.; Ngu, L.H.; Ong, D.E.L.; Nissom, P.M. Low-cost cultivation of Sporosarcina pasteurii strain in food-grade yeast extract medium for microbially induced carbonate precipitation (MICP) application. *Biocatal. Agric. Biotechnol.* **2019**, *17*, 247–255. [CrossRef]
14. Omoregie, A.I.; Palombo, E.A.; Ong, D.E.L.; Nissom, P.M. Biocementation of sand by Sporosarcina pasteurii strain and technical-grade cementation reagents through surface percolation treatment method. *Constr. Build. Mater.* **2019**, *228*, 116828. [CrossRef]
15. Omoregie, A.I.; Ong, D.E.L.; Nissom, P.M. Assessing ureolytic bacteria with calcifying abilities isolated from limestone caves for biocalcification. *Lett. Appl. Microbiol.* **2019**, *68*, 173–181. [CrossRef]
16. Omoregie, A.I.; Palombo, E.A.; Ong, D.E.L.; Nissom, P.M. A feasible scale-up production of Sporosarcina pasteurii using custom-built stirred tank reactor for in-situ soil biocementation. *Biocatal. Agric. Biotechnol.* **2020**, *24*, 101544. [CrossRef]
17. Ong, D.E.L.; Sim, Y.S.; Leung, C.F. Performance of Field and Numerical Back-Analysis of Floating Stone Columns in Soft Clay Considering the Influence of Dilatancy. *Int. J. Geomech.* **2018**, *18*. [CrossRef]
18. Ong, D.E.L.; Leung, C.F.; Chow, Y.K. Time-dependent pile behaviour due to excavation-induced soil movement in clay. In Proceedings of the 12th Pan-American Conference on Soil Mechanics and Geotechnical Engineering, Boston, MA, USA, 22–26 June 2003; Massachusetts Institute of Technology: Cambridge, MA, USA, 2003; Volume 2, pp. 2035–2040.
19. Ong, D.E.L.; Leung, C.F.; Chow, Y.K. Piles subject to excavation-induced soil movement in clay. In Proceedings of the 13th European Conference on Soil Mechanics and Geotechnical Engineering, Prague, Czech Republic, 1 January 2003; Volume 2, pp. 777–782.
20. Chong, E.E.M.; Ong, D.E.L. Data-Driven Field Observational Method of a Contiguous Bored Pile Wall System Affected by Accidental Groundwater Drawdown. *Geosciences* **2020**, *10*, 268. [CrossRef]
21. Ong, D.E.L.; Choo, C.S. Sustainable Bored Pile Construction in Erratic Phyllite. In Proceedings of the ASEAN-Australian Engineering Congress, Kuching, Malaysia, 14–16 July 2011; pp. 30–45, ISBN 78-967-10485.

22. Ong, D.E.L.; Leung, C.F.; Chow, Y.K.; Ng, T.G. Severe Damage of a Pile Group due to Slope Failure. *J. Geotech. Geoenviron. Eng.* **2015**, *141*. [CrossRef]
23. Liu, Z.; Liu, Y.; Bolton, M.; Ong, D.E.L.; Oh, E. Effect of Cement and Bentonite Mixture on The Consolidation Behavior of Soft Estuarine Soils. *Int. J. Geomate* **2020**, *18*, 49–54. [CrossRef]
24. Sun, J.; Oh, E.; Ong, D.E.L. Influence of Degree of Saturation (DOS) on Dynamic Behavior of Unbound Granular Materials. *Geosciences* **2021**, *11*, 89. [CrossRef]
25. Peerun, M.I.; Ong, D.E.L.; Choo, C.S. Interpretation of geomaterial behavior during shearing aided by PIV technology. *J. Mater. Civ. Eng.* **2019**, *31*. [CrossRef]
26. Peerun, M.I.; Ong, D.E.L.; Choo, C.S.; Cheng, W.C. Effect of interparticle behavior on the development of soil arching in soil-structure interaction. *Tunn. Undergr. Space Technol.* **2020**, *106*, 103610. [CrossRef]
27. Cheng, W.C.; Li, G.; Ong, D.E.L.; Chen, S.L.; Ni, J.C. Modelling liner forces response to very close-proximity tunnelling in soft alluvial deposits. *Tunn. Undergr. Space Technol.* **2020**, *103*, 103455. [CrossRef]
28. Cheng, W.C.; Li, G.; Ong, D.E.L. Lubrication characteristics of pipejacking in soft alluvial deposit. *Geotech. Asp. Undergr. Constr. Soft Ground* **2021**. [CrossRef]
29. Kang, G.; Tsuchida, T.; Athapaththu, A.M.R.G. Strength mobilization of cement-treated dredged clay during the early stages of curing. *Soils Found.* **2015**, *55*, 375–392. [CrossRef]
30. Liu, Y.; Jiang, Y.; Xiao, H.; Lee, F.H. Determination of representative strength of deep cement-mixed clay from core strength data. *Geotechnique* **2017**, *67*, 350–364. [CrossRef]
31. Yao, K.; Pan, Y.; Jia, L.; Yi, J.T.; Hu, J.; Wu, C. Strength evaluation of marine clay stabilized by cementitious binder. *Mar. Georesour. Geotechnol.* **2020**, *38*, 730–743. [CrossRef]
32. Oh, E. Geotechnical and Ground Improvement Aspects of Motorway Embankments in Soft Clay. Ph.D. Thesis, Griffith University, Southeast Queensland, Australia, 2007. Available online: http://hdl.handle.net/10072/367085 (accessed on 14 May 2021).
33. Wei, X.; Yang, J. A critical state constitutive model for clean and silty sand. *Acta Geotech.* **2019**, *14*, 329–345. [CrossRef]
34. Mitchell, J.K.; Soga, K. *Fundamentals of Soil Behavior*, 3rd ed.; John Wiley & Sons: New York, NY, USA, 2005.
35. Yin, J.H. Stress-strain strength characteristics of a marine soil with different clay contents. *Geotech. Test. J.* **2002**, *25*, 459–462.
36. Wong, S.T.; Ong, D.E.L.; Robinson, R.G. Behaviour of MH silts with varying plasticity indices. *Geotech. Res.* **2017**, *4*, 118–135. [CrossRef]
37. Ekinci, A. Effect of preparation methods on strength and microstructural properties of cemented marine clay. *Constr. Build. Mater.* **2019**, *227*, 116690. [CrossRef]
38. Liu, S.Y.; Zhang, D.W.; Liu, Z.B.; Deng, Y.F. Assessment of unconfined compressive strength of cement stabilized marine clay. *Mar. Georesour. Geotechnol.* **2019**, *26*, 19–35. [CrossRef]
39. Horpibulsuk, S.; Rachan, R.; Suddeepong, A. Assessment of strength development in blended cement admixed Bangkok clay. *Constr. Build. Mater.* **2011**, *25*, 1521–1531. [CrossRef]
40. Ma, C.; Chen, L.; Chen, B. Analysis of strength development in soft clay stabilized with cement-based stabilizer. *Constr. Build. Mater.* **2014**, *71*, 354–362.
41. Kwan, P.S.; Bouazza, A.; Fletcher, P.; Ranjith, P.G.; Oh, E.Y.; Shuttlewood, K.; Bolton, M. Behaviour of cement treated Melbourne and Southeast Queensland soft clays in deep stabilization works. In Proceedings of the International Conference Deep Mixing Best Practice and Recent Advances, Stockholm, Sweden, 23–25 May 2005.
42. Uddin, K.; Balasubramaniam, A.S.; Bergado, D.T. Engineering behavior of cement-treated Bangkok soft clay. *Geotech. Eng.* **1997**, *28*, 89–119.
43. Kang, G.; Tsuchida, T.; Athapaththu, A.M.R.G. Engineering behavior of cement-treated marine dredged clay during early and later stages of curing. *Eng. Geol.* **2016**, *209*, 163–174. [CrossRef]
44. Kang, G.O.; Tsuchida, T.; Kim, Y.S. Strength and stiffness of cement-treated marine dredged clay at various curing stages. *Constr. Build. Mater.* **2017**, *132*, 71–84. [CrossRef]
45. Lee, F.H.; Lee, Y.; Chew, S.H.; Yong, K.Y. Strength and modulus of marine clay-cement mixes. *J. Geotech. Geoenviron. Eng.* **2005**, *131*, 178–186. [CrossRef]
46. Jauberthie, R.; Rendell, F.; Rangeard, D.; Molez, L. Stabilisation of estuarine silt with lime and/or cement. *Appl. Clay Sci.* **2010**, *50*, 395–400. [CrossRef]
47. Zhang, X.; Zhu, Z. Microscopic Mechanism of Cement Improving the Strength of Lime-Fly Ash-Stabilized Yellow River Alluvial Silt. *Adv. Civ. Eng.* **2020**, *2020*, 9649280.
48. Lo, S.R.; Wardani, S.P. Strength and dilatancy of a silt stabilized by a cement and fly ash mixture. *Can. Geotech. J.* **2002**, *39*, 77–89. [CrossRef]
49. Lehane, B.M. Vertically loaded shallow foundation on soft clayey silt. *Proc. Inst. Civ. Eng. Geotech. Eng.* **2003**, *156*, 17–26. [CrossRef]
50. Kelln, C.; Sharma, J.; Hughes, D.; Graham, J. Finite element analysis of an embankment on a soft estuarine deposit using an elastic–viscoplastic soil model. *Can. Geotech. J.* **2009**, *46*, 357–368. [CrossRef]
51. Ferreira, P.M.V.; Bica, A.V.D. Problems in identifying the effects of structure and critical state in a soil with a transitional behaviour. *Géotechnique* **2006**, *56*, 445–454. [CrossRef]

52. Nocilla, A.; Coop, M.R.; Colleselli, F. The mechanics of an Italian silt: An example of 'transitional' behaviour. *Géotechnique* **2006**, *56*, 261–271. [CrossRef]
53. Horpibulsuk, S.; Miura, N.; Nagaraj, T.S. Clay–water/cement ratio identity for cement admixed soft clays. *J. Geotech. Geoenviron. Eng.* **2005**, *131*, 187–192. [CrossRef]
54. Zhang, D.; Fan, L.; Liu, S.; Deng, Y. Experimental Investigation of Unconfined Compression Strength and Stiffness of Cement Treated Salt-Rich Clay. *Mar. Georesour. Geotechnol.* **2013**, *31*, 360–374.
55. Yun, J.M.; Song, Y.S.; Lee, J.H.; Kim, T.H. Strength Characteristics of the Cement-Stabilized Surface Layer in Dredged and Reclaimed Marine Clay, Korea. *Mar. Georesour. Geotechnol.* **2006**, *24*, 29–45. [CrossRef]
56. Liu, L.; Zhou, A.; Deng, Y.; Cui, Y.; Yu, Z.; Yu, C. Strength performance of cement/slag-based stabilized soft clays. *Constr. Build. Mater.* **2019**, *211*, 909–918. [CrossRef]
57. Terzaghi, K. *Erdbaumechanik Auf Bodenphysikalischer Grundlage*; Tokiwa Shoin: Chiyoda, Tokyo, 1925.
58. Zhang, X.; Zhu, Z. Study on the Relationship between Microstructure and Strength of Stabilized/Solidified Silt. *Int. J. Struct. Civ. Eng. Res.* **2018**, *7*, 65–71. [CrossRef]
59. Wijeyakulasuriya, V.; Hobbs, G.; Brandon, A. Some experiences with performance monitoring of embankments on soft clays. In Proceedings of the 8th Australia New Zealand Conference on Geomechanics, Hobart, Australia, 1 January 1999; Institution of Engineers Australia: Barton, Australia, 1999; pp. 783–788.
60. Yin, J.; Miao, Y.H. An oedometer-based method for preparing reconstituted clay samples. *Appl. Mech. Mater.* **2015**, *719*, 193–196. [CrossRef]
61. Allman, M.A.; Atkinson, J.H. Mechanical properties of reconstituted Bothkennar soil. *Géotechnique* **1992**, *42*, 289–301. [CrossRef]
62. Burland, J.B. On the compressibility and shear strength of natural clays. *Géotechnique* **1990**, *40*, 329–378. [CrossRef]
63. Karstunen, M.; Koskinen, M. Plastic anisotropy of soft reconstituted clays. *Can. Geotech. J.* **2008**, *45*, 314–328. [CrossRef]
64. Hyodo, M.; Tanimizu, H.; Yasufuku, N.; Murata, H. Undrained cyclic and monotonic triaxial behaviour of saturated loose sand. *Soils Found.* **1994**, *34*, 19–32. [CrossRef]
65. Head, K.H. *Manual of Soil Laboratory Testing*; Effective stress tests; John Wiley & Sons: Singapore, 1980; Volume 3.
66. ASTM. *D2166/D2166M—16 Standard Test Method for Unconfined Compressive Strength of Cohesive Soil*; ASTM: Conshohocken, PA, USA, 2000.
67. Wang, S.; Luna, R. Monotonic behavior of Mississippi River Valley silt in triaxial compression. *J. Geotech. Geoenviron. Eng.* **2012**, *138*, 516–525. [CrossRef]
68. Horpibulsuk, S.; Rachan, R.; Chinkulkijniwat, A.; Raksachon, Y.; Suddeepong, A. Analysis of strength development in cement-stabilized silty clay from microstructural considerations. *Constr. Build. Mater.* **2010**, *24*, 2011–2021. [CrossRef]
69. Bolton, M. Soil Improvement Using Optimised Cementitous Materilas Design. Ph.D. Thesis, Griffith University, Southeast Queensland, Australia, 2014. Available online: http://hdl.handle.net/10072/365243 (accessed on 12 May 2021).
70. Kamruzzaman, A.H.M.; Chew, S.H.; Lee, F.H. Microstructure of cement-treated Singapore marine clay. *Proc. Inst. Civ. Eng. Ground Improv.* **2006**, *10*, 113–123. [CrossRef]

geosciences

Article

Influence of Degree of Saturation (DOS) on Dynamic Behavior of Unbound Granular Materials

Junyu Sun *, Erwin Oh and Dominic Ek-Leong Ong

School of Engineering and Built Environment, Griffith University, Southport, QLD 4222, Australia; y.oh@griffith.edu.au (E.O.); d.ong@griffith.edu.au (D.E.-L.O.)
* Correspondence: junyu.sun@griffithuni.edu.au; Tel.: +86-153-4083-4006

Abstract: The extensive application of natural unbound granular materials (UGMs) motivates studies into the mechanical properties of alternatives such as processed crushed rocks employed commonly as base or subbase layers. The rutting and settlement generated in base and subbase layers is widely restricted in many specifications and standards. In this research, the dynamic behavior including the resilient modulus (M_r) and the plastic strain (ε_p) of the crushed rocks collected from Queensland in Australia will be tested by a series of repeated load triaxial test (RLT) tests to investigate the behavior of UGMs under the fluctuation of the degree of saturation (DOS) (59%–100%). In particular, the RLT specimens were prepared in the laboratory through proper gradation under optimum moisture content (OMC) and 100% standard proctor maximum dry unit weight. Results from the RLT tests showed that UGM specimens soaked at higher DOS generated lower resilient modulus and weaker resistance to heavy traffic volumes with significant accumulation of plastic strain. The M_r and ε_p of the tested aggregates under different cyclic deviator stresses of 425 kPa and 625 kPa approximately linearly decreased and approximately linearly increased as the DOS increased with a certain number of cycles up to 50,000, respectively.

Keywords: unbound granular materials (UGMs); degree of saturation (DOS); repeated load triaxial test (RLT); resilient modulus; plastic strain

Citation: Sun, J.; Oh, E.; Ong, D.E.-L. Influence of Degree of Saturation (DOS) on Dynamic Behavior of Unbound Granular Materials. *Geosciences* **2021**, *11*, 89. https://doi.org/10.3390/geosciences11020089

Academic Editors: Eric J. R. Parteli and Jesus Martinez-Frias

Received: 31 December 2020
Accepted: 11 February 2021
Published: 15 February 2021

1. Introduction

Unbound granular pavement materials (UGMs), which mainly comprise rocks, gravels, and manufactured crushed rock, are generally heavily compacted in pavement structural systems as base and subbase materials to distribute the traffic loads from the surface layer above to the subgrade layer [1,2]. Natural crushed rock is the most widely applied product in pavement materials due to the mature craft of breaking the mining rocks down to reach the proper size by crushers [3,4].

Flexible pavement is a complex structure constructed with the main purpose of supporting traffic loading. Typically, pavements classified into rigid and flexible pavements are complicated structures that are constituted of several layers such as structural surface layer (asphalt surface concrete surface), base course, subbase course, and subgrade soil [5]. For the differences between flexible and rigid pavement components, in rigid pavements with the concrete surface layer, the base layer is applied to level and structurally strengthen week subgrade [6]. Moreover, for flexible pavements with an asphalt surface layer, the base and subbase layers are structural components that should provide enough strength to decrease the stress to levels that can be supported by the subgrade [6]. In particular, the function of the subbase is to support the above base layer adequately and distribute the transferred load to the subgrade layer with the facilitated drainage path [5]. In the meantime, a separation between the surface layer and subgrade is provided by the subbase layer to avoid the subgrade fines being pumped to the surface layer through the joints during the traffic loading [6].

Between November and April each year, tropical cyclones with large amounts of concentrated rainfall assault Australia and induce flooding and waterlogging to infrastructure and road closure [7,8]. The Queensland Tropical Cyclone Warning Center is impacted by five tropical cyclones on average, with 207 detected cyclones affecting the east coast since 1885 [9]. In some tropical areas such as the southeast Queensland region, a significant strength loss and even premature failure of pavements exposed to extreme moisture infiltration and increase in base saturation may be induced with sudden rainfall [7]. Sultana et al. [10] reported that most areas in eastern Australia encountered a considerable wet weather period from August to December of 2010. During the final week of December 2010 to January 2011, widespread flooding and waterlogging of roads were recorded with three-quarters of the Queensland state declared a disaster zone [10]. Roads encountered asset damage when Queensland experienced extreme flooding in January 2011 with rainfalls between 600 and 100 mm recorded in the southeast Queensland area during December 2010 and January 2011 [10]. Moreover, in March 2017, a recent tropical cyclone, Debbie, which was the strongest tropical cyclone in the Queensland region since 2015, brought 747 mm rainfall within two days and induced significant damage in pavements due to flooding and waterlogging [7,9]. Due to the potential disruption of service and extra maintenance cost, this issue and related vulnerability evaluations are remarkable in construction design and construction policy [2,6].

The significant function of pavements as part of the transportation system is to support a high amount of passenger travel without extreme maintenance requirements. Under the seasonal cycles of wet and dry periods, the formation of severe pavement cracks is potentially caused by significant stress concentration in the base, subbase, and subgrade soil [10,11]. According to the research of Sultana et al. [12] using the falling weight deflectometer (FWD) and surface condition data, a comparison of structural and functional performance was developed of the before and after flood data on flooded and non-flooded sections of the tested pavements. The results of this research showed that loss of structural strength with an accelerated deterioration rate of the flood-affected pavements was encountered. After a period of traffic loading imposed on the saturated pavement, a loss in the subgrade CBR (California bearing ratio) up to 67% and structural number up to 50% was detected by the long-term monitoring of pavements before and after the flooding event [10].

Ahmed et al. [13] modified the standard pavement maintenance policy and developed a deformation prediction model applying topographical surveying and moisture sensor monitoring concerning the rainfall at any time of the year in Texas, USA. Pavement engineers need to study the changes in the dynamic responses of materials in fluctuating moisture conditions to predict the long-term deformation and corresponding deterioration for pavement design and management. However, Sultana et al. [11,12] pointed out that most of the available pavement deterioration models considered the pavements under a normal climatic condition such as average rainfall and design traffic. Based on the gathered historical data and continually monitored data, Sultana et al. [11,12] developed new deterioration models including rutting, roughness, and structural strength to describe the rapid deterioration generated in the partially or fully saturated pavement sections subject to extreme weather events such as flooding and extreme rainfall in the southeast Queensland region. The research was conducted to investigate whether the pavement response or deterioration prediction after extreme weather could provide engineering knowledge for strategic plans such as the timing of the duration of the road closures and re-openings after a waterlogging event [11,14].

However, after the flood water receded and dried out, Sultana et al. [10] also found that further reduction of the structural strength and unexpected rehabilitation effectiveness could be caused by underlying problems with ongoing strength loss of the base, subbase, and subgrade layers after the post-flooding rehabilitation works. Therefore, the deterioration pattern of the mechanical properties of the base and subbase materials under the partially or fully saturated situations should be further investigated to provide evidence for the rehabilitation strategy.

1.1. Moisture-Damage Mechanisms in Unbound Granular Pavement Materials (UGMs)

Two main environmental factors are regarded to affect pavement performance and pavement structure design: temperature and rainfall. As mentioned by the standards including American Association of State Highway and Transportation Officials (AASHTO) [15] and Austroads [16], the design of roads is according to the moisture and temperature reflecting the historical climate of the location. The variations in the rainfall, temperature, and evaporation situations can change the moisture balance in the pavement foundations and further affect the reduction of the structural strength of the roads [17]. Since investigations into the mechanism of road distress related to moisture variation and the influence of rainfall in the prediction of pavement remaining life are important for the management of both the maintenance and rehabilitation of the pavements. Moreover, as for the description by AASHTO [15], variations in moisture content over time in the unbound base and subbase layers are also related to water infiltration through the cracks of the top asphalt concrete layers during raining, melting of ice, capillary action, and water table changes seasonally [18]. Consequently, induced higher moisture influences the resilient modulus of the unbound granular base and subbase layers, which further leads to premature failure in roads and decreases service life [18].

Road foundations are generally constructed above the water table since the pavement layers are under an unsaturated condition [19]. As the base and subbase layers are compacted close to the maximum dry density as determined in a standard proctor test, considerable variations in the degree of saturation (DOS) can be achieved by small fluctuations in moisture content [19]. Meanwhile, the load-bearing mechanism of the UGMs mainly depends on the adequate contact stresses and interparticle friction between aggregates under draining conditions. The extra moisture infiltration could undermine the bonding in the aggregate skeleton and consequently induce the losses in strength and stiffness in the UGMS layers under repeated traffic loading [5]. Dense graded base and subbase layer materials show a high dry density that coincides with defective permeability and weak draining conditions because of the low void ratio [5]. When water fills all the voids during a relative non-draining condition, this scenario results in the neutralization of the particle contact pressures and yields a decrease in material strength because of the cumulative excess pore water pressure in the voids of the aggregates, especially with the association of heavy traffic load [20]. According to the regulations of Austroads [21], the low resistance to both resilient and permanent deformations under dynamic load is also caused by the increased pore water pressure at a high DOS. On the other hand, the permanent deformation in the base and subbase layers is generated slowly when the DOS is low because the main function of the water in voids is acting as a lubricating agent in the skeleton to shift the particles into the most effective packing [21]. Consequently, the pavement deterioration is accelerated by the diminutions of effective stress, bearing capacity, and deformation resistance of base or subbase layers [20]. Moreover, poor load spread competence of the base and subbase layers can be triggered by the loss of stiffness due to the softening effect of extreme water content [19]. The induced stress concentration may cause an excessive permanent deformation in the beneath layers such as the lower subbase or subgrade [19].

To eliminate the influence of excess moisture from pavement layers on their mechanical response as far as possible, some preventive techniques can be used to restrict the water entrance such as setting barriers and enhancing drainage systems [22].

1.2. Optimum Moisture Content and the Degree of Saturation (DOS)

Water content or moisture content is the parameter that indicates the quantity of water contained in the void of porous materials such as soils, rock, ceramics, and wood on a gravimetric or volumetric basis. This property is applied in a wide range of scientific and technical fields. The quality of moisture, defined as the optimum moisture content (OMC) within the UGMs, corresponds to the maximum dry density (MDD) that the UGMs, which the materials can achieve when a standard compaction process is applied [21]. To control the moisture content during the construction of the pavement base and subbase layers, the raw material is prepared below the OMC to achieve the OMC by adjustments such as in situ spraying of additional moisture and assorted moisture maintenance measures [15]. If the moisture content of UGMs is over the OMC due to rainfall and inapposite on site operation with wave shape after compaction, the partially compacted layers should be harrowed, and wait until the OMC is dried back by solar heating, air drying, and evaporation [21]. A well restricted UGM compaction with optimum density can ensure a designed strength and resistance ability to settlement and volume change [21].

In soil mechanics and petroleum engineering, the term degree of saturation, S, is applied to indicate the saturation condition. To determine the DOS for moisture management during compaction, the moisture content measured by further drying and compacted dry density needs to be determined [23]. The DOS can be calculated by Equation (1) as follows:

$$S = \frac{w}{\frac{\rho_w}{\rho_d} \times \frac{1}{\rho_{st}}} \quad (1)$$

where
S = degree of saturation (%);
w = moisture content of sample (%);
ρ_w = water density (t/m^3)taken as 1.000 t/m^3;
ρ_d = compacted dry density of sample (t/m^3) and
ρ_{st} = apparent particle density (t/m^3).

1.3. *Traffic Loading Response and Repeated Load Triaxial Test*

Initially, the flexible pavement design program was empirical and tended to determine the thickness of the considered material layers that could provide sufficient strength to support upper structural loading and protect the weaker subgrade soils below. The adopted failure criterion of the design was the subgrade shear failure mode and the experience from previous pavement construction projects. In the early 1960s, the trend of flexible pavement design shifted from empirical methods to a combined mechanistic–empirical method [24,25], which is more reasonable because an empirical model is supplemented with the theory of mechanics and the observed performance of the pavement structure in the design and analysis procedures [5]. For the details of the adopted mechanistic approach, the pavement response parameters including stresses, strains, and deformations were taken into account to estimate the influences of traffic loading and environmental factors.

As far as the main failure criteria used in the mechanistic–empirical (M–E) method, two main criteria of flexible pavement design are introduced here. The first failure criterion is the restricted vertical compressive strain on the surface of the subgrade soil layer to eliminate permanent deformation, which is claimed and recommended by [26]. After this, the second criterion is the restricted horizontal tensile strain at the bottom of the asphalt concrete layer and cemented base layer to avoid the potential of fatigue cracking, which was developed by [27]. However, the base and subbase layers are designed as ideal materials without any deterioration according to the most recent flexible pavement structural design guide in Australia [28]. UGMs are sensitive to permanent deformation (PD) and cause deterioration such as rutting under long-term transport loads [29]. Meanwhile, Austroads [28] reported that no appropriate understanding is available to describe the PD and resilient modulus of UGMs in flexible pavements in Australia. Therefore, the traffic loading response of UGMs is essential for the performance and reliability of the pavement structure.

As a universal technique to simulate the dynamic response of pavement materials in the laboratory environment, the repeated load triaxial test (RLT) is commonly suggested to evaluate the pavement material behavior under specific traffic loading magnitude and frequency, which are stipulated by standards such as Standards Australia and Austroads [1,2]. Compared with other testing techniques such as the California bearing ratio (CBR), unconfined compression strength, and triaxial shear strength, RLT could observe the dynamic response of materials under a relatively low repeated loading level under a simulated stress state [30]. Lekarp et al. [31,32] and Castelli [33] presented the dynamic behavior of UGMs and subgrade soils under the repeated wheel loads of RLT test equipment. Under the repeated wheel loads, the UGMs showed a complex nonlinear behavior (time-dependent elastoplastic response) that has been studied by many researchers through experiments [5,19]. A theoretical assumption of the response of a characteristic pavement element under cyclic loading separated the strain response into a resilient strain (recoverable deformation) and plastic strain (permanent deformation) components according to whether the component of strain can recover during the unloading process, as shown in Figure 1. Under repeated loading conditions, the strain response is generally assumed where the resilient strain is mainly elastic and the plastic strain is defined by the accumulation of the plastic strain under cyclic loading [34]. With the increase in the number of cycles, the plastic strain induced by each load application decreases, and the plastic component of strain tends to accumulate continuously [31,34]. The accumulation characteristics of plastic strain are commonly logarithmic with an increasing number of cycles [35]. Therefore, it leads to the potential rutting phenomenon, which affects the pavement structure and traffic conditions, as discussed above [31,32]. The residual performance of the pavement materials is the key parameter that is useful to define the residual performance capacity of the pavement under the imposed long-term traffic loads [33]. Under a large number of cycles, the tested specimen would present degradation phenomena with a sudden decrease in its mechanical characteristics and large strain level [33].

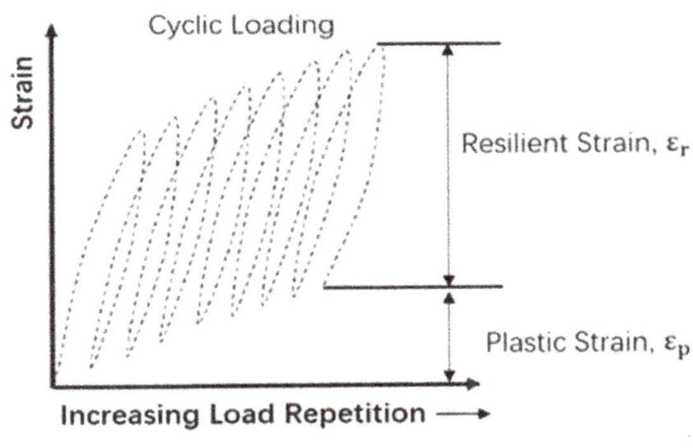

Figure 1. Characteristic stress–strain behavior [36].

As suggested by the AASHTO and Austroads design guide [37,38], the plastic strain (ε_p) the resilient modulus (M_r), notably the M_r of the base and subbase layers, has been extensively applied as a structural input parameter for pavement engineering through M–E design methods. As shown in Figure 1, the plastic strain (ε_p) can be calculated as a difference between the total strain (ε_T) in one cycle to the resilient strain (ε_r) in one cycle. The cyclic deviator stress (σ_d) can be calculated as a difference between dynamic axial compressive stress (σ_1) to radial confining stress (σ_3). As presented in Equation (2), the parameter of resilient modulus, M_r, is defined as the ratio of the magnitude of the cyclic deviator stress, σ_d to the axial resilient strain, and ε_r in the RLT test [39]. The values of M_r and ε_p are usually tested by applying the repeated load triaxial test (RLT), which is the most direct test to simulate the field traffic loads [37,38].

$$M_r = \frac{\sigma_d}{\varepsilon_r} = \frac{\sigma_1 - \sigma_3}{\varepsilon_r} \quad (2)$$

where
M_r = resilient modulus (MPa);
σ_d = cyclic deviator stress (kPa);
σ_1 = dynamic axial compressive stress (kPa);
σ_3 = radial confining stress (kPa); and
ε_r = axial resilient strain.

Some of the traditional design methods treat the resilient modulus as a constant parameter (linear stress–strain behavior), while other researchers believé it to be a function of the dynamic stress conditions and materials related factors [37,38]. Theoretically, M_r is stress-dependent, consequently, it means that M_r is associated with tire loads, load duration, frequency of loading, and load sequence subjected by the granular base and subbase materials [35,37]. Typically, a greater degradation is observed when a higher frequency of loading is applied under long-term cyclic loading [35]. The material affecting factors of UGMs include material density, particle gradation, aggregate type, and DOS [19].

1.4. Effect of DOS on the Dynamic Response of UGMs

According to previous research by experiment investigation, the resilient modulus and plastic strain accumulation of UGMs are significantly affected by the DOS. The resilient modulus (M_r) of pavement materials has been investigated as a sensitive function of the post-construction moisture content. Commonly, the value of M_r and the stiffness of pavement materials reduce when the higher moisture content is achieved [40–42]. Based on this, Heydinger [43] reported that the moisture content of the tested material was the basic variable to predict the seasonal variation of the resilient modulus of pavement materials. The M_r of all the loading sequences decreased with the increase in moisture content, while the sample compacted and soaked before the test showed critical lower values of M_r [5]. However, in practice, Uzan [44] stated that the pavement materials represented an increase in moisture content up to 30% higher than the plastic limit or the equilibrium moisture content of the soils during the initial five years of pavement service life. It can be summarized from previous studies that there is no agreement about the effect of DOS on the M_r of UGMs under repeated loads.

As mentioned by Theyse [45], who conducted a series of RLT tests in the University of California's pavement research center, a significant effect of DOS variation on the plastic strain accumulation of the UGMs, and the variation range can be up to 30% to 100%. Guo et al. [46] reported that there was a positive relationship between the plastic strain and the confining pressure using the specimens prepared under OMC because the existence of the particle skeleton was more critical under low confining pressure. Using the UGM specimens with moisture contents below OMC, Arulrajah et al. [47–49] and Soliman [50] pointed out that the resilient modulus and the resistance of the plastic strain of UGMs under repeated loads declined with the increase in DOS. Zhalehjoo et al. [51] and Cerni et al. [52] also illustrated that the resilient modulus of UGMs declined with the increase in DOS under varied bulk stresses. Austroads [53,54] describes that the growth of traffic loading and the prediction of extreme weather should be the main requirement of the quality and strength characteristics of UGMs. However, only very few studies have been conducted to investigate the effect of moisture condition changing over OMC on the plastic strain under repeated traffic loads using base or subbase UGMs with various DOS.

Moreover, the research related to the ratcheting effect on the plastic strain of partially saturated UGMs is also limited. The gradual accumulation of plastic strain in UGMs subjected to repeated loading has been described as ratcheting in soil mechanics [55]. Garcia-Rojo et al. [56] and Sun et al. [57] pointed out that the triaxial specimens under OMC were initially unstable with plastic strain accumulated linearly with the increase in the number of cycles under large deviator stress and large load frequency. The influence of varied moisture content on the plastic strain of partially saturated UGMs needs to be further investigated under a large number of cycles.

1.5. Identification of Problems and Objectives

It can be seen from the discussion above that there is no agreement about the effect of DOS and in particular, the extra moisture than OMC on the resilient modulus and plastic strain of UGMs under repeated traffic loads. Therefore, this research focused on the extra moisture effect on the dynamic behavior of Class 3 subbase crushed rock under repeated loading. The specimens compacted on OMC were further treated and tested by the RLT test under different DOS to simulate the dynamic response of the subbase layer with moisture infiltration. The main objective of this research was to investigate the response and performance reduction of UGMs with extra water infiltration after being compacted (aggregates are standardly compacted to maximum dry density) as a method to design a porous UGM layer under water-induced damage. Consequently, the deterioration pattern of partially saturated UGMs observed in this research may be used for regional pavement management such as the temporary traffic control after flooding in Australia.

2. Experimental Investigation

To conduct a series of RLT tests, crushed rock subbase materials were collected from a pavement construction material stockyard in Australia. After that, the grading requirement and material classification were determined by sieve analysis. The compaction test was conducted to identify the maximum dry density (MDD) and corresponding optimum moisture content (OMC) for the specimen preparation. Based on this, the samples were compacted under MDD and OMC and were further soaked and dried to achieve the gradual increased setting of DOS. Then, repeated load triaxial (RLT) tests were operated to measure the dynamic response of the specimens under different stress levels. The DOS of the tested RLT specimens extracted from the triaxial cell after loading can be evaluated using Equation (1).

2.1. Material and Specimen Preparation

The particle size distribution of the tested UGMs met the grading requirements for Class 3 at 40 mm used as a subbase material according to the Austroads design guide [21]. Limestone rock is the origin of the crushed Class 3, which was collected from a pavement construction material stockyard in Queensland, Australia, and delivered in plastic boxes. It can be clearly distinguished in Figure 2 that the particle size distribution of the tested specimens is a dense-graded material and fits inside the Austroads specification limitation envelope of Class 3: upper and lower limits.

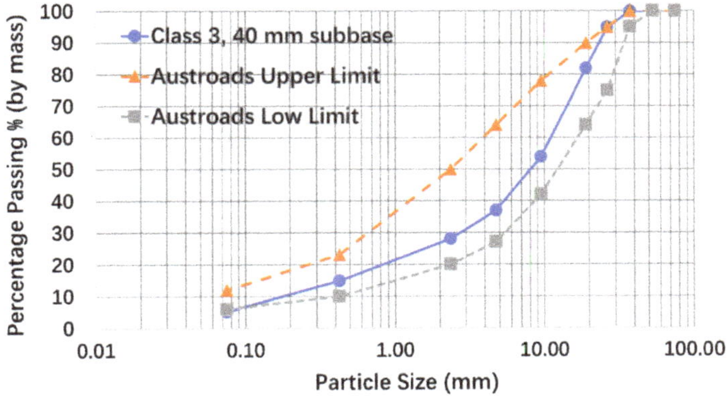

Figure 2. Particle size distribution of the tested aggregates.

In the laboratory, the compaction test was conducted to determine the sample at the maximum dry density and compact the sample at optimum moisture content by six split cylindrical molds according to OMC and MDD. Figure 3 presents the test results of the compaction test of the used aggregates. The dry density of the specimens increased correspondingly with the increase in aggregate moisture content until the OMC at 13.9% was achieved with the MDD at 1.80 ton/m^3

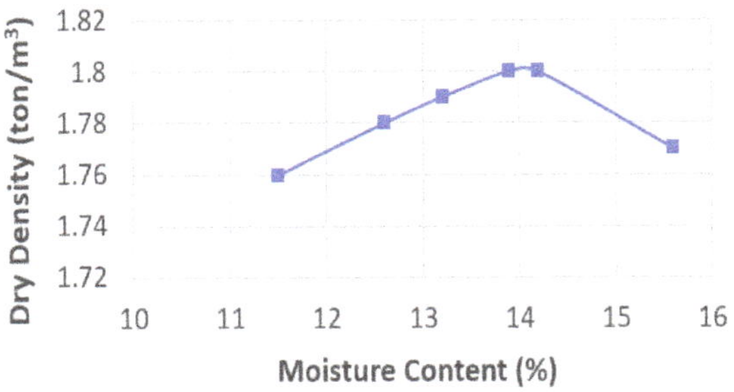

Figure 3. Dry density–moisture content relationship.

Furthermore, each specimen of the RLT tests was compacted into three layers using a standard compaction effect with 25 blows per layer. According to the specifications in AS 1289.6.8.1 (1995) [1] and Austroads, AG-PT/T053 [2], the maximum particle size of the triaxial specimens was restricted to not exceeding 19 mm for the RLT equipment. The aggregate material passing the 19 mm sieve was compacted to the 100% standard proctor maximum dry unit weight utilizing dynamic compaction at different degrees of saturation (DOS) levels. The triaxial specimen size was trimmed to a 200 mm high and 100 mm diameter. To eliminate the membrane penetration, double membranes (0.3 mm thickness for each) were applied for the RLT samples with a bottom fine porous stone (80 kPa). The triaxial cell contained a bronze bushing without an O-ring. The specimens of this project were prepared with a latex membrane in an AS modified split mold placed directly on the bottom plate of the triaxial cell. The specimen preparation procedure was used in accordance with Standards Australia [1] and Austroads [2].

2.2. Testing Equipment

The UTM-5P (Universal Testing Machine-5 kN, Pneumatic), coupled with recent developments in pneumatic control valves and digital control technology, is one smart equipment of Universal Testing Machines on the RLT test using cylindrical specimens. A range of transducers, sample loading jigs, actuator motor, feed-back controlled pneumatic, application computer control system, and internal and external displacement transducer of this system can be applied for different standards, general-purpose, and user-programmable tests. A constant confining stress was adopted for the sample using water in the triaxial cell. Furthermore, the axial deformation of the specimen was measured by three linear variable displacement transducers (LVDT) assembled internally and externally to the cell. The internal LVDTs had an accuracy of ±0.005 mm, while the external LVDTs had an accuracy of ±0.0015 mm. The RLT testing equipment can test the samples compacted at different DOS and corresponding densities under varied axial and radial stresses, which were developed to reflect the different elements in the road base or subbase in terms of different stress levels.

2.3. Testing Sequences

A rectangular waveform of 0.33 Hz was applied to impose the deviator stress on samples under undrained conditions. To simulate the condition of the road pavement affected by flooding in Australia [7–12], the partially saturated UGM specimens were tested under a water-logging situation. Due to the stiffness loss in the saturated upper layers such as the base layer when an extreme water-logging event occurred, a condition about the induced stress concentration in lower pavement layers should be considered.

Therefore, high confinement and high deviator stresses need to be selected for the partially saturated UGM specimens, which were prepared in the laboratory to represent the upper subbase layer or the lower base layer with the potential extreme stress condition due to the weakening of upper layers. The suggested range of contact stress (p) was from 550 to 750 kPa in AS 1289.6.8.1 [1]. Meanwhile, the ratios of σ_1/p and σ_3/σ_1 of the plastic strain test were recommended as 0.94–1.00 and 0.167–0.300 in AS 1289.6.8.1 [1], respectively. In this case, the ratios of σ_1/p and σ_3/σ_1 were determined as 1.00 and 0.167 with a constant cell pressure of 125 kPa. According to Standards Australia AS 1289.6.8.1 [1], the resilient modulus and the plastic strain tests of the constant DOS were conducted under a constant cell pressure of 125 kPa and the pulsating deviator stresses ($\sigma_1 - \sigma_3$ of 425 kPa and 625 kPa, which were selected to simulate the light and heavy traffic loading conditions with the σ_1 equal to 550 kPa and 750 kPa for this project, respectively. The plastic strain test recording was corrected at 100 cycles to eliminate compliance effects.

Repeated deviator stress, static confining stress, targeted DOS, treating type. and actual tested DOS of the RLT testing program is presented in Table 1. The experimental stop conditions of each RLT test were that the number of cycles of 100,000 was involved or the plastic strain of one test achieved 8%, which was pre-set as the failure situation [1,2].

Table 1. Repeated loading triaxial testing program of unbound aggregates.

σ_3 (kPa)	σ_d (kPa)	Target DOS (%)	Treat Type	Actual DOS (%)
125	425	60%–100%	OMC	59%
			Soaking	71% 80% 95% 100%
125	625	60%–75%	OMC	59%
			Soaking	69% 74%

Note: σ_3: confining stress. σ_d: deviator stress. DOS: the degree of saturation.

3. Results and Discussion

3.1. RLT Test Results with Different Applied Deviator Stresses

The test result related to the resilient modulus of tested aggregates under an applied deviator stress of 425 kPa is illustrated in Figure 4a. The specimen with 100% of the DOS was prepared to investigate the dynamic behavior of the granular material under the saturated condition. For all the prepared water contents from 59% to 100%, the resilient modulus of the specimens showed a stable trend with the increase in the number of cycles. Moreover, for all number of cycles, the corresponding resilient modulus decreased with the increase in the DOS of specimens. Particularly for the DOS of 100%, the sample presented a sudden failure after the number of cycles of 5000.

The test result of the plastic strain of the tested aggregates under an applied vertical stress of 550 kPa is presented in Figure 4b. For the extreme moisture condition with a DOS of 100%, compared to other degrees of saturation, the highest plastic strain was recorded as approximately 8.7% before the sudden failure at 50,000 cycles. Meanwhile, the tested aggregates with a DOS of 59% had the lowest plastic strain percentage of approximately 0.8% after 100,000 cycles. Furthermore, the plastic strain ε_p at DOS of 71%, 80%, and 95% were approximately 1.1%, 1.9%, and 7.9% with the number of cycles up to 100,000, respectively.

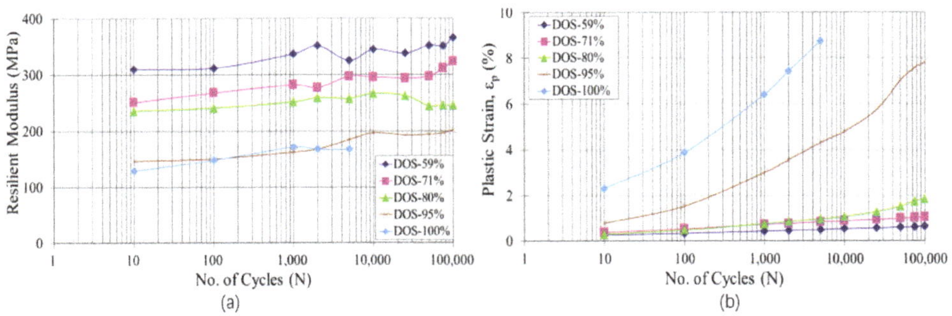

Figure 4. The relationship between dynamic response and the number of cycles with an applied deviator stress of 425 kPa: (**a**) resilient modulus; (**b**) plastic strain.

Generally, it was observed that the specimen with high moisture content generated a large plastic strain with a higher increase rate when the number of cycles increased. Moreover, if the DOS of the granular samples was lower than a specific value (80%), there was no significant plastic strain observed under the applied deviator stress of 425 kPa.

Due to the increase in the applied vertical stress, relatively lower degrees of saturation (69% to 74%) of the samples were prepared to avoid the failure induced by the extreme deformation after a larger number of cycles. The result of the resilient modulus of the tested aggregates under the applied deviator stress of 625 kPa is illustrated in Figure 5a. From this figure, the range of the resilient modulus under all degrees of saturation increased with the rise in the applied vertical load. Compared to the trend of resilient modulus with the increase in the number of cycles obtained from the samples under low-stress level, the results under the applied deviator stress of 625 kPa presented a relatively obvious reduction trend after 10,000 number of cycles. Similarly, it was observed that the value of the resilient modulus of the tested aggregates reduced with the increase in the DOS for all number of cycles.

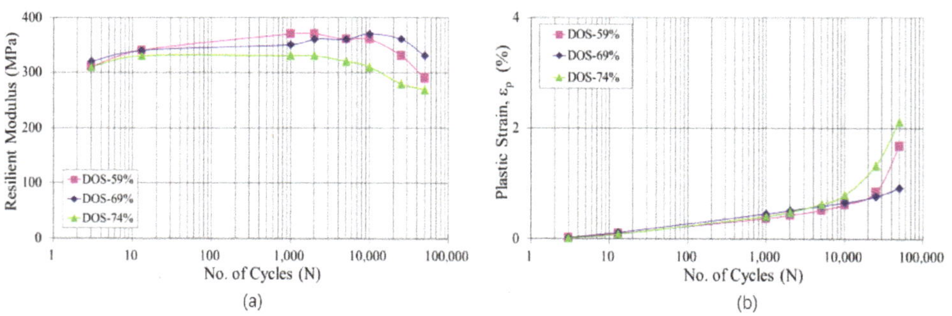

Figure 5. The relationship between the dynamic response and the number of cycles with an applied deviator stress of 625 kPa: (**a**) resilient modulus; (**b**) plastic strain.

The test result related to the observed accumulation trend of plastic strain with the increase in the number of cycles is stated in Figure 5b. The plastic strain ε_p at a DOS of 59%, 69%, and 74% were approximately 0.9%, 1.8%, and 2.1%, respectively, with the number of cycles of 50,000. Compared to the result related to the low-stress level, the plastic strains for the tested aggregates at a deviator stress of 625 kPa seemed to be higher, depending on the number of cycles. This means that this comparison was not observable with a low number of load cycles, whereas this comparison tended to be significant with the increase in the number of cycles.

3.2. The Comparison between Different Deviator Stresses at Similar DOS

To analyze the effect of the DOS on the resilient modulus and the plastic strain accumulation, Figure 6a–d provides more detailed comparisons of the results corresponding to vertical repeated deviator stresses of 425 kPa and 625 kPa under similar DOS. As shown in Figure 6a, under a similar DOS of around 70%, the sample imposed a deviator stress of 425 kPa accumulated a larger plastic strain than the specimen under the deviator stress of 625 kPa before 50,000 cycles. However, with the increase in the number of cycles, the rate of the plastic strain accumulation increased significantly for the sample under a high-stress level.

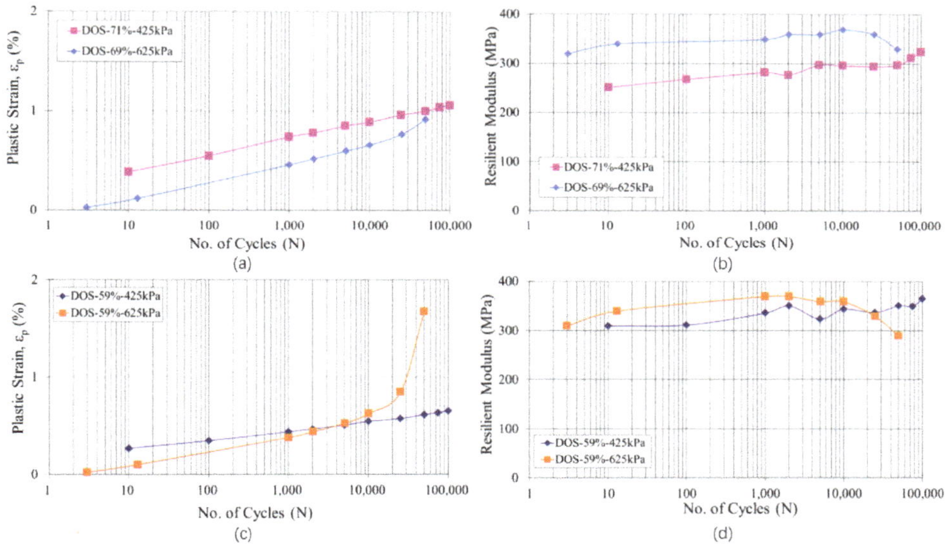

Figure 6. The comparison of resilient modulus and plastic strain versus the number of cycles at similar DOS: (**a**) plastic strain of DOS at around 70%; (**b**) resilient modulus of DOS at around 70%; (**c**) plastic strain at DOS around 59%; and (**d**) resilient modulus at DOS around 59%.

As for the effect of the applied vertical stress on the resilient modulus, Figure 6b illustrates the comparison of the variation of resilient modulus tested under different applied axial deviator stresses. With the enhancement of the applied vertical stress or the deviator stress, the resilient modulus of the aggregates increased from 270 MPa to 350 MPa at approximately 100 cycles. This means that the effect of the deviator stress on the resilient modulus is significant. The effect of the deviator stress on the resilient modulus for unbound granular materials is supported in the literature [5,6].

Similarly, the RLT test results of the samples with the DOS at 59% are compared in Figure 6c,d for the discussion related to the influence of the vertical stress on the accumulation of plastic strain and the resilient modulus. From Figure 6c, the values of the accumulated plastic strains of the samples applied axial deviator stresses of 425 kPa and 625 kPa gradually increased to approximately 0.6% before10,000 cycles. After that, the plastic strain induced by the 425 kPa vertical loading gradually accumulated to around 0.7% at 100,000 cycles. However, the plastic strain of the sample imposed a high-level vertical loading, indicating a significant increase trend to approximately 1.6% after 50,000 cycles. This was consistent with a result under around 70% DOS. Therefore, the influence of applied vertical loading on the accumulation of plastic strain is critical after a larger number of cycles.

As far as the influence of the applied vertical loading on the resilient modulus under 59% DOS, Figure 6d presents a similar result to that under 70% DOS. With the enhancement of the applied vertical stress, the resilient modulus increased from 310 MPa to 360 MPa approximately. Similar to the result under around 70% DOS, under the vertical deviator stress of 625 kPa, there was a sudden decrease in the resilient modulus of the test after 10,000 cycles. This means that a larger number of cycles tended to induce a sudden degradation of the resilient modulus of the granular materials imposed by the large vertical stress. Due to extreme water infiltration, the subbase materials with high DOS would show weaker resistance to the degradation of the resilient modulus and plastic strain, especially under a higher loading level and larger loading repetitions. Therefore, a limitation of traffic volume or road closure should be adopted when the roads suffer waterlogging or flooding during extreme weather events.

3.3. Resilient Modulus and Plastic Strain with Certain Number of Cycles under Varied DOS

The RLT tests under different degrees of saturation were applied to determine the deformation and the resilient modulus curves for a range of stress conditions and moisture conditions to investigate the rutting of the pavement. As described in Figure 7a, the resilient modulus of the tested aggregates applied a cyclic deviator stress of 425 kPa approximately linearly decreased as the DOS increased from 59% to 100% with a certain number of cycles up to 50,000. The resilient modulus of the tested aggregates as road subbase materials was impacted by the DOS.

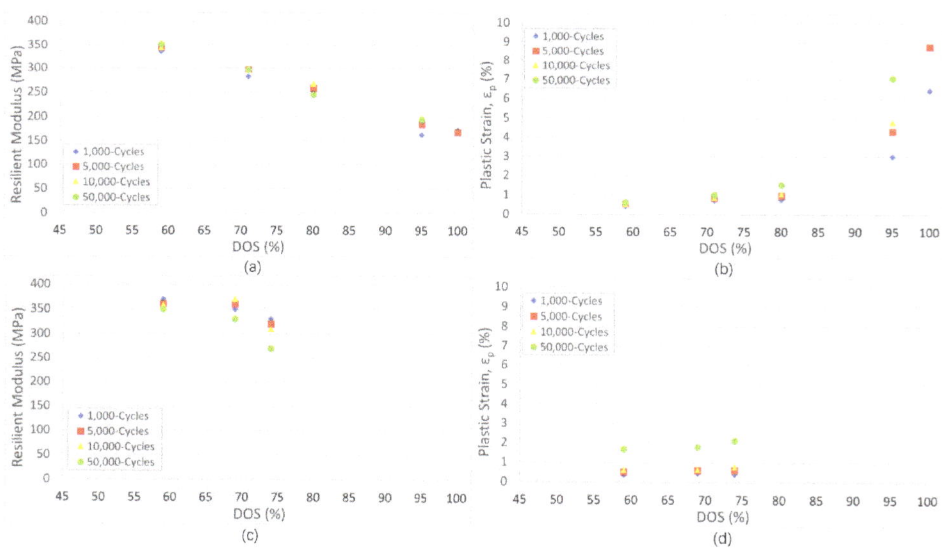

Figure 7. The relationship between the repeated load triaxial test (RLT) test results and DOS at a different number of cycles: (**a**) resilient modulus with applied deviator stress of 425 kPa; (**b**) plastic deformation with applied deviator stress of 425 kPa; (**c**) resilient modulus with applied deviator stress of 625 kPa; and (**d**) plastic deformation with applied deviator stress of 625 kPa.

As presented in Figure 7b, for a certain number of cycles, the accumulated plastic strain of the tested aggregates adopting a cyclic deviator stress of 425 kPa increased approximately linearly as the DOS increased from 59% to 80%. After that, the plastic strain increases significantly as the sample tends to be fully saturated. Moreover, with a high number of load cycles, the plastic strain exhibited was observable, especially for the DOS of 95% and 100%.

Similarly, as illustrated in Figure 7c, for a certain number of cycles up to 50,000, the resilient modulus of the tested aggregates adopting a vertical deviator stress of 625 kPa also approximately linearly reduced with the increase in DOS from 69% to 74%. As indicated in Figure 7d, with a certain number of cycles, an approximately positively linear relationship was observed between the plastic strain and the DOS of the specimens imposed a vertical deviator stress of 625 kPa. The relatively linear response illustrates the onset of moisture sensitivity with the increase of plastic strain.

4. Conclusions

Based on the RLT tests conducted on the specimens with various DOS and different deviator stress levels, the conclusions that were obtained from the repeated load triaxial tests are as follows:

(i) The resilient modulus of tested UGMs that had both 425 and 625 kPa deviator stresses applied illustrated a decreasing trend in the increase in the DOS. After that, the accumulated plastic strain of the tested aggregates of both the 425 and 625 kPa deviator stresses applied presented an increasing trend with the increase in the DOS. This type of behavior was reported previously by Arulrajah et al. [47,48] and Soliman and Shalaby [50]. With regard to the sensitivity to the moisture of the tested UGMs, Arulrajah et al. [47,48] claimed that there were higher limits of plastic strain and lower limits of resilient modulus, especially at higher DOS. This phenomenon was also observed in this research. The resilient modulus declined and the plastic strain at certain loading cycles increased with the rise in DOS at an increasing rate. When the DOS was over 75%, the plastic strain of the tested subbase UGMs showed a particularly higher sensitivity to moisture.

(ii) With the increase in the applied deviator stress, the resilient modulus of the tested aggregates under around 60% and 70% degrees of saturation presented an increasing tendency. The increasing trend of resilient modulus with a larger deviator stress of UGMs was studied by Craciun [19] and Arulrajah et al. [49].

(iii) With the increase in the applied vertical cyclic stress, the accumulated plastic strain of the aggregates under 60% and 70% degrees of saturation increased significantly after a large number of cycles of 20,000.

(iv) The resilient modulus of the specimens, which were applied a larger vertical cyclic load, reduced more distinctly after the number of cycles increased to 10,000.

(v) The resilient modulus of the tested aggregates applied different cyclic deviator stresses of 425 kPa and 625 kPa approximately linearly decreased as the DOS increased with a certain number of cycles up to 50,000. For a certain number of cycles, the accumulated plastic strain of the tested aggregates increased approximately linearly as the DOS increased from 59% to 80%. After that, the plastic strain increased significantly as the sample tended to be fully saturated. At a higher DOS level, the resistance to the plastic strain of tested UGMs was found to decline beyond the accepted limit. The results of the repeated load triaxial tests could indicate that the tested subbase UGMs at a DOS of around 75% is a feasible subbase material when the pavements encounter extreme moisture infiltration. The results of the study by Atherulrajah et al. [48,49] for UGMs showed a similar picture. They found that moisture contents in the range of 65–90% were the optimum moisture content of pavement subbase UGMs.

This paper suggests that Class 3 subbase UGMs could provide relatively sufficient resilient modulus and plastic strain resistance to low traffic volumes before the DOS increase to a critical level, which was around 75% in this research.

5. Further Research

The recommendations for further investigations are included in this research, focusing on the extension of this study to carry out tests with more series of applied stresses of the granular material for the development of empirical equations. Moreover, the recording of the pore pressure for saturated samples should be performed to study the relationship between the accumulation of pore pressure and the changing of the resilient modulus and the plastic strain.

Author Contributions: Data curation, J.S.; Formal analysis, J.S.; Funding acquisition, E.O.; Investigation, J.S.; Methodology, J.S.; Project administration, E.O. and D.E.-L.O.; Resources, E.O. and D.E.-L.O.; Software, J.S.; Supervision, E.O.; Validation, J.S.; Visualization, J.S.; Writing—original draft, J.S.; Writing—review & editing, E.O. and D.E.-L.O. All authors have read and agreed to the published version of the manuscript.

Funding: This research received no external funding.

Conflicts of Interest: The authors declare no conflict of interest.

References

1. Australian Standard, Soil Strength and Consolidation Tests: Determination of the Resilient Modulus and Permanent Deformation of Granular Unbound Pavement Materials, 1289.6.8.1. *Standards Australia*, Sydney, Australia. 1995. Available online: https://trid.trb.org/view/1203019 (accessed on 14 February 2021).
2. Austroads, Austroads Repeated Load Triaxial Test Method: Determination of Permanent Deformation and Resilient Modulus Characteristics of Unbound Granular Materials Under Drained Conditions, AG-PT/T053. Austroads Publication. 2007. Available online: https://austroads.com.au/publications/pavement/agpt-t053-07 (accessed on 14 February 2021).
3. Huang, Y. *Pavement Analysis and Design*; Pearson Prentice Hall: Upper Saddle River, NJ, USA, 2004.
4. USGS. *Mineral Yearbook*; United State Geological Services: New York, NY, USA, 2000.
5. Nguyen, B. Resilient Behavior of Fine-Grained and Granular Materials for the Design of Flexible Pavements. Ph.D. thesis, RMIT University, Melbourne, Australia, 2016.
6. Werkmeister, S.; Dawson, A.; Wellner, F. Pavement design model for unbound granular materials. *J. Transp. Eng.* **2004**, *130*, 665–674. [CrossRef]
7. Cogan, J.; Gratchev, I.; Wang, G. Rainfall-induced shallow landslides caused by ex-Tropical Cyclone Debbie, 31st March 2017. *Landslides* **2018**, *15*, 1215–1221. [CrossRef]
8. Podlaha, A.; Bowen, S.; Darbinyan, C.; Lörinc, M. Global Catastrophe Recap-April 2017 (PDF). Aon Benfield Analytics. 2017. Available online: http://thoughtleadership.aonbenfield.com/Documents/201707-if-1h-global-recap.pdf (accessed on 14 February 2021).
9. Bureau of Meteorology. Tropical Cyclones in Queensland. Available online: http://www.bom.gov.au/cyclone/about/eastern.shtml#history (accessed on 20 November 2017).
10. Sultana, M.; Chai, G.; Martin, T.; Chowdhury, S. A study on the flood affected flexible pavements in Australia. In Proceedings of the 9th International Conference on Road and Airfield Pavement Technology, Dalian, China, 9–13 August 2015; pp. 9–13.
11. Sultana, M.; Chowdhury, S.; Chai, G.; Martin, T. Modelling rapid deterioration of flooded pavements. *Road Transp. Res. A J. Aust. N. Z. Res. Pract.* **2016**, *25*, 3–14.
12. Sultana, M.; Chai, G.; Chowdhury, S.; Martin, T.; Anissimov, Y.; Rahman, A. Rutting and Roughness of Flood-Affected Pavements: Literature Review and Deterioration Models. *J. Infrastruct. Sys.* **2018**, *24*, 04018006. [CrossRef]
13. Ahmed, A.; Hossain, M.; Pandey, P.; Sapkota, A.; Thian, B. Deformation Modeling of Flexible Pavement in Expansive Subgrade in Texas. *Geosciences* **2019**, *9*, 446. [CrossRef]
14. Hedayati, M. Rainfall Induced Distress in Low Volume Pavements. Ph.D. Thesis, The University of Texas at Arlington, Arlington, TX, USA, 2014.
15. AASHTO. Guide for Design of Pavement Structures. AASHTO. 1993. Available online: https://habib00ugm.files.wordpress.com/2010/05/aashto1993.pdf (accessed on 14 February 2021).
16. Austroads, Predicting Structural Deterioration of Pavements at a Network Level–Interim Models, Sydney, NSW, Australia. 2010. Available online: https://austroads.com.au/publications/asset-management/ap-t159-10 (accessed on 14 February 2021).
17. Papagiannakis, A.T.; Masad, E.A. *Pavement Design and Material*; John Wiley & Sons: Hoboken, NJ, USA, 2007.
18. Haider, S.; Masud, M. Effect of moisture infiltration on flexible pavement performance using the AASHTOWare Pavement-ME. Advances in Materials and Pavement Prediction. In Proceedings of the International Conference on Advances in Materials and Pavement Performance Prediction (AM3P 2018), Doha, Qatar, 16–18 April 2018.
19. Craciun, O. Matric Suction Response of Unbound Granular Base Materials Subject to Cyclic Loading. Ph.D. Thesis, University of New South Wales, Sydney, Australia, 2009.
20. Rahman, M.; Erlingsson, S. Influence of moisture on Resilient Deformation behaviour of Unbound Granular Materials. In Proceedings of the 12th International Conference on Asphalt Pavements, ISAP 2014, Raleigh, NC, USA, 1–5 June 2014.
21. Austroads, Guide to Pavement Technology Part 4A: Granular Base and Subbase Materials, AGPT04A-08. Austroads Publication. 2008. Available online: https://austroads.com.au/publications/pavement/agpt04a (accessed on 14 February 2021).
22. Park, S.; Lytton, R.; Benson, F. Characterizing pavement subgrades in Texas. *Env. Eng. Geosci.* **1999**, *3*, 353–357.

23. Transport and Main Roads, QLD, Test Method Q146, Degree of Saturation of Soils and Crushed Rock, Materials Testing Manual, Edition 4, Amendment 4. 2017. Available online: https://www.ipweaq.com/MRTS05%20Mod%20-%2031%20October%202017%20final%20version%20style.pdf (accessed on 14 February 2021).
24. Brown, S. Soil mechanics in pavement engineering. *Géotechnique* **1996**, *46*, 383–426. [CrossRef]
25. Mamlouk, M.; Zaniewski, J.; He, W. Analysis and design optimization of flexible pavement. *J. Transp. Eng.* **2000**, *126*, 161–167. [CrossRef]
26. Kerkhoven, R.; Dormon, G. Some Considerations on the California Bearing Ratio Method for the Design of Flexible Pavements. Shell Petroleum Company. 1953. Available online: https://www.worldcat.org/title/some-considerations-on-the-california-bearing-ratio-method-for-the-design-of-flexible-pavements/oclc/475100213?referer=di&ht=edition (accessed on 14 February 2021).
27. Saal, R.; Pell, P. Fatigue of bituminous road mixes. *Colloid Polym. Sci.* **1960**, *171*, 61–71. [CrossRef]
28. Austroads. Guide to Pavement Technology Part 2: Pavement Structural Design. Austroads Publication. 2008. Available online: https://austroads.com.au/publications/pavement/agpt02 (accessed on 14 February 2021).
29. Berthelot, C.; Podborochynski, D.; Marjerison, B. Saskatchewan field case study of triaxial frequency sweep characterization to predict failure of a granular base across increasing fines content and traffic speed applications. *J. Transp. Eng.* **2009**, *135*, 907–914.
30. Sorsa, A.; Senadheera, S.; Birru, Y. Engineering Characterization of Subgrade Soils of Jimma Town, Ethiopia, for Roadway Design. *Geosciences* **2020**, *10*, 94.
31. Lekarp, F.; Isacsson, U.; Dawson, A. State of the art. I: Resilient response of unbound aggregates. *J. Transp. Eng.* **2000**, *126*, 66–75. [CrossRef]
32. Lekarp, F.; Isacsson, U.; Dawson, A. State of the art. II: Permanent strain response of unbound aggregates. *J. Trans. Eng.* **2000**, *126*, 76–83. [CrossRef]
33. Castelli, F.; Cavallaro, A.; Grasso, S.; Lentini, V. Undrained Cyclic Laboratory Behavior of Sandy Soils. *Geosciences* **2019**, *9*, 512.
34. Huang, H.; Luo, J.; Moaveni, M. *Advanced Analytical Tool for Flexible Pavement Design and Evaluation. Airfield and Highway Pavements 2019: Design, Construction, Condition Evaluation, and Management of Pavements*; American Society of Civil Engineers: Reston, VA, USA, 2019; pp. 61–71.
35. Kucharczyk, K.; Głuchowski, A.; Miturski, M.; Sas, W. Influence of Load Frequency on Cohesive Soil Respond. *Geosciences* **2018**, *8*, 468. [CrossRef]
36. Alnedawi, A.; Nepal, K.; Al-Ameri, R. Mechanistic behavior of open and dense graded unbound granular materials under traffic loads. *Int. J. Geomate.* **2018**, *14*, 124–129. [CrossRef]
37. AASHTO. Guide T307: Standard Test Method for Determining the Resilient Modulus of Soil and Aggregate Materials. AASHTO. 2007. Available online: https://global.ihs.com/doc_detail.cfm?document_name=AASHTO%20T%20307&item_s_key=00489204 (accessed on 14 February 2021).
38. Austroads. *Pavement Design-A Guide to the Structural Design of Road Pavements*; Austroads Publication: Sydney, Australia, 2012.
39. Seed, H.; Chan, C.; Lee, C. Resilience characteristics of subgrade soils and their relation to fatigue failures in asphalt pavements. In Proceedings of the International Conference on the Structural Design of Asphalt Pavements, Michigan, MI, USA, 20–24 August 1962.
40. Rada, G.; Witczak, M. Comprehensive evaluation of laboratory resilient moduli results for granular material. *Transp. Resear. Rec.* **1981**, *810*, 23–33.
41. Wang, M.; Kong, L.; Zhao, C. Dynamic characteristics of lime-treated expansive soil under cyclic loading. *J. Rock Mech. Geotech. Eng.* **2012**, *4*, 352–359.
42. Naji, K. Resilient modulus-moisture content relationships for pavement engineering applications. *Int. J. Pavement Eng.* **2018**, *19*, 651–660. [CrossRef]
43. Heydinger, A. Evaluation of seasonal effects on subgrade soils. *Transp. Resear. Rec.* **2003**, *1821*, 47–55. [CrossRef]
44. Uzan, J. Characterization of granular material. *Transp. Resear. Rec.* **1985**, *1022*, 52–59.
45. Theyse, H. *Stiffness, Strength, and Performance of Unbound Aggregate Material: Application of South African HVS and Laboratory Results to California Flexible Pavements*; Report produced under the auspices of the California Partnered Pavement Research Program for the California Department of Transportation; University of California: Berkeley, CA, USA, 2002.
46. Guo, L.; Wang, J.; Cai, Y. Undrained deformation behavior of saturated soft clay under long-term cyclic loading. *Soil Dyn. Earthq. Eng.* **2013**, *50*, 28–37. [CrossRef]
47. Arulrajah, A.; Piratheepan, J.; Aatheesan, T.; Bo, M. Geotechnical properties of recycled crushed brick in pavement applications. *J. Mater. Civ. Eng.* **2011**, *23*, 1444–1452.
48. Arulrajah, A.; Piratheepan, J.; Ali, M.; Bo, M. Geotechnical properties of recycled concrete aggregate in pavement sub-base applications. *Geotech. Test. J.* **2012**, *35*, 743–751. [CrossRef]
49. Arulrajah, A.; Piratheepan, J.; Disfani, M.; Bo, M. Resilient moduli response of recycled construction and demolition materials in pavement subbase applications. *J. Mater. Civ. Eng.* **2013**, *25*, 1920–1928. [CrossRef]
50. Soliman, H.; Shalaby, A. Permanent deformation behavior of unbound granular base materials with varying moisture and fines content. *Transport. Geotech.* **2015**, *4*, 1–12. [CrossRef]
51. Zhalehjoo, N.; Tolooiyan, A.; Mackay, R.; Bodin, D. The effect of instrumentation on the determination of the resilient modulus of unbound granular materials using advanced repeated load triaxial testing. *Transport. Geotech.* **2018**, *14*, 190–201. [CrossRef]

52. Cerni, G.; Corradini, A.; Pasquini, E.; Cardone, F. Resilient behaviour of unbound granular materials through repeated load triaxial test: Influence of the conditioning stress. *Road Mater. Pavement Des.* **2015**, *16*, 70–88.
53. Austroads. Guide to Pavement Technology Part 2: Pavement Structural Design, AGPT02-17. Austroads Publication. 2017. Available online: https://trid.trb.org/view/1498339 (accessed on 14 February 2021).
54. Austroads. Long-Term Pavement Performance Study-Final Report, AP_T342-19. Austroads Publication. 2019. Available online: https://austroads.com.au/publications/asset-management/ap-t342-19 (accessed on 14 February 2021).
55. Alonso-Marroquin, F.; Herrmann, H.J. Ratcheting of granular materials. *Phys. Rev. Lett.* **2004**, *92*, 054301. [CrossRef]
56. García-Rojo, R.; Herrmann, H.J. Shakedown of unbound granular material. *Granul. Matt.* **2005**, *7*, 109–118. [CrossRef]
57. Sun, Q.D.; Indraratna, B.; Nimbalkar, S. Effect of cyclic loading frequency on the permanent deformation and degradation of railway ballast. *Géotechnique* **2014**, *64*, 746–751. [CrossRef]

geosciences

MDPI

Article

Elastic Settlement Analysis of Rigid Rectangular Footings on Sands and Clays

Lysandros Pantelidis [1,*] and Elias Gravanis [1,2]

[1] Department of Civil Engineering and Geomatics, Cyprus University of Technology, Limassol 3036, Cyprus; elias.gravanis@cut.ac.cy
[2] Eratosthenes Centre of Excellence, Cyprus University of Technology, P.O. Box 50329, Limassol 3603, Cyprus
* Correspondence: lysandros.pantelidis@cut.ac.cy; Tel.: +357-2500-2271

Received: 16 November 2020; Accepted: 30 November 2020; Published: 4 December 2020

Abstract: In this paper an elastic settlement analysis method for rigid rectangular footings applicable to both clays and sands is proposed. The proposed method is based on the concept of equivalent shape, where any rectangular footing is suitably replaced by a footing of elliptical shape; the conditions of equal area and equal perimeter are satisfied simultaneously. The case of clay is differentiated from the case of sand using different contact pressure distribution, whilst, additionally, for the sands, the modulus of elasticity increases linearly with depth. The method can conveniently be calibrated against any set of settlement data obtained analytically, experimentally, or numerically; in this respect the authors used values which have been derived analytically from third parties. Among the most interesting findings is that sands produce "settlement x soil modulus/applied pressure" values approximately 10% greater than the respective ones corresponding to clays. Moreover, for large Poisson's ratio (v) values, the settlement of rigid footings is closer to the settlement corresponding to the corner of the respective flexible footings. As v decreases, the derived settlement of the rigid footing approaches the settlement value corresponding to the characteristic point of the respective flexible footing. Finally, corrections for the net applied pressure, footing rigidity, and non-elastic response of soil under loading are also proposed.

Keywords: elastic settlement; shallow foundations; rigid footings; contact pressure; equivalent shape

1. Introduction

The problem of settlement of shallow foundations is among the more important ones in classical soil mechanics. During the last several decades a great number of approaches has been proposed in the literature for the title problem, an exact solution of which is still not available. In this respect, solutions have been provided by various authors [1–11]. Numerical results for perfectly smooth, uniformly loaded rectangular rafts, of any rigidity resting on a homogeneous elastic layer which is underlain by a rough rigid base, were presented in graphical form by Fraser and Wardle [12]. Semi-empirical approaches combining the theory of elasticity with experimental and/or numerical results also exist in the literature [13–17]. Schiffman and Aggarwala [18] and Mayne and Poulos [15] used the concept of equivalent ellipse and circle respectively (equivalency as for the footprint area).

In this paper an elastic settlement analysis method for rigid rectangular footings applicable to both clays and sands is proposed. The proposed method is based on the concept of equivalent shape, where any rectangular $B \times L$ footing is suitably replaced by a footing of elliptical shape. The suitability of the equivalent shape concept is deeply investigated aiming at the production of analytical expressions for the elastic settlement analysis of rigid rectangular footings on sands and clays. As shown, this concept is only valid when the equivalent shape satisfies both the condition of equal area and equal perimeter length at the same time. The case of clay is differentiated from the case of sand using different contact pressure distribution, whilst the modulus of elasticity for the case of sands additionally increases linearly with depth.

2. Settlement of Rigid Elliptical Footing

Boussinesq [19] solved the problem of stresses produced at any point in a homogenous, elastic, isotropic, and semi-infinite medium as the result of a point load applied on the surface. Following the notation of Figure 1, the increase in normal stresses caused by the point load P is

$$\Delta\sigma_x = \frac{P}{2\pi}\left(\frac{3x^2z}{L^5} - (1-2v)\left(\frac{x^2-y^2}{Lr^2(L+z)} + \frac{y^2z}{L^3r^2}\right)\right), \tag{1}$$

$$\Delta\sigma_y = \frac{P}{2\pi}\left(\frac{3y^2z}{L^5} - (1-2v)\left(\frac{y^2-x^2}{Lr^2(L+z)} + \frac{x^2z}{L^3r^2}\right)\right), \tag{2}$$

$$\Delta\sigma_z = \frac{3P}{2\pi}\frac{z^3}{L^5}, \tag{3}$$

where $L = \sqrt{x^2+y^2+z^2}$, $r = \sqrt{x^2+y^2}$ and v is the Poisson's ratio (see [20–22]).

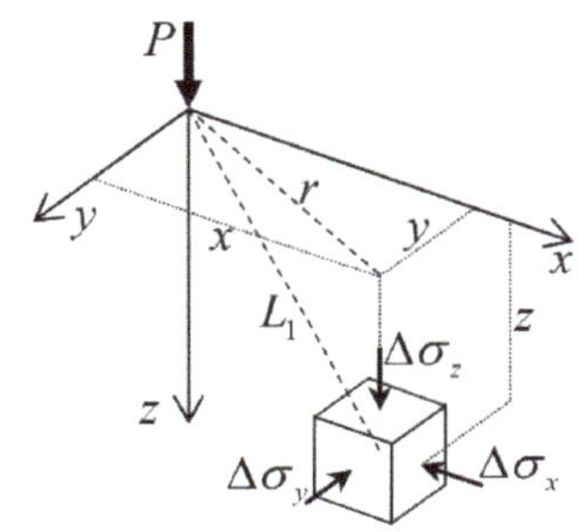

Figure 1. Stresses in an elastic medium caused by a point load acting on the surface of a semi-infinite mass [23].

The above equations along with the equation effectively representing the contact pressure distribution leads to the elastic settlement of footing following a standard procedure (see [19,24]). In this respect, by using the following contact pressure distribution (see [18]):

$$p = \frac{P_{ell.}}{\pi ab}\frac{1}{2\sqrt{1-\frac{x^2}{a^2}-\frac{y^2}{b^2}}}. \tag{4}$$

The settlement of a rigid elliptical footing can be obtained in a way similar to the case of Boussinesq's [19] rigid circular footing (see also [24]); where $P_{ell.}$ is the total load on the elliptical rigid footing (the origin of the axes lies on the centroid of the elliptical disc), whilst a and b are the two radii defining the ellipse.

The infinitesimal point load P at any point (x,y) inside the ellipse is, thus, given by

$$P = pdxdy = \frac{q}{2\sqrt{1-\frac{x^2}{a^2}-\frac{y^2}{b^2}}}dxdy, \tag{5}$$

where

$$q = \frac{P_{ell.}}{\pi ab}, \tag{6}$$

is the load per unit area of the footing.

Hence, the increase in the normal stress in the i-axis due to loading over the elliptical surface area can be found by integrating the respective stress increase $\Delta\sigma_i$ ($i = x, y$ or z) (see Equations (1)–(3)) over this area using Equation (5). That is:

$$\Delta\sigma_{i,Ell.} = \iint_{ellipse} \Delta\sigma_i(x, y, z). \tag{7}$$

Note that the differentials dx and dy are included in P (see Equation (5)). The integration can conveniently be done by changing to polar variables r and θ, setting $x = ar\cos\theta$ and $y = br\sin\theta$, where the ranges of the radial and angular variables, respectively, are $0 \leq r \leq 1$ and $0 \leq \theta \leq 2\pi$. For the area element in the polar variables then it stands that $dxdy = abrdrd\theta$.

The unit vertical strain is, then, calculated from the constitutive relationship of Hooke's Law:

$$\varepsilon_z = \frac{1}{E}\left[\Delta\sigma_{z,Ell.} - v\left(\Delta\sigma_{x,Ell.} + \Delta\sigma_{y,Ell.}\right)\right] \tag{8}$$

and finally, the footing settlement, ρ_r, corresponding to a layer of thickness H is

$$\rho_r = \int_{z=0}^{H} \varepsilon_z dz = \frac{qb}{E}\beta, \tag{9}$$

where

$$\beta = \frac{1}{b}\int_{z=0}^{H} I_z dz = \int_{\zeta=0}^{H/b} \frac{(1+v)(1-2v) + 2(1-v^2)(1+k^2)\zeta^2 + (1+v)(3-2v)k^2\zeta^4}{2(1+\zeta^2)^{3/2}(1+k^2\zeta^2)^{3/2}} d\zeta, \tag{10}$$

with $\zeta = z/b$ and $k = b/a$; I_z is the strain influence factor. The integration as for z can easily be implemented numerically in spreadsheet so that the effect of varying soil modulus with depth to be taken into account in a way similar to Schmertmann's method [14].

In the case of homogenous semi-infinite mass, the settlement becomes:

$$\rho_r = \frac{qb}{E}kK(e)\left(1 - v^2\right), \tag{11}$$

where $e = \sqrt{1-k^2}$ is the eccentricity of the ellipse and $K(x)$ is the complete elliptic integral of the first kind [25]. Due to the fact that in the circular case $K(0) = \pi/2$ (because $e = 0$ and also, $k = 1$), Equation (11) goes over to the well-known expression of Boussinesq [19,24]:

$$\rho_r = \frac{\pi qa}{2E}(1 - v^2). \tag{12}$$

3. Elastic Settlement of the Equivalent $B \times L$ Rectangular Footing

From the theory of elasticity for flexible footings over semi-infinite media, it is well known that the settlement at the center of a uniformly loaded square footing having edge B is approximately equal to the respective settlement of a circular footing of equal area (radius of footing equal to $B/\sqrt{\pi}$) [26]; the percentage difference in settlement between these two special cases is only 0.53%.

The concept of shape equivalency has been used in the past by Mayne and Poulos [15], where any $B \times L$ rectangular footing is reduced to a circular footing of equal area. The radius of Mayne and Poulos' [15] equivalent circle is

$$a_{eq} = \sqrt{BL/\pi}. \tag{13}$$

Several years earlier, Schiffman and Aggarwala [18] suggested that the two radii defining the equivalent elliptical footing of a $B \times L$ rectangular footing be

$$\left.\begin{array}{l} a = L/\sqrt{\pi} \\ b = B/\sqrt{\pi} \end{array}\right\} \text{with } a \geq b. \tag{14}$$

Schiffman and Aggarwala [18] considered the problem of an elastic half-space loaded on the surface by a rigid elliptical footing. They developed general formulae for the stresses and displacements within and on the surface of the solid. It is noted that Equation (14) derives from the condition of equal area and from $B/L = b/a = k$; the condition of equal perimeter is not satisfied.

Sovinc [8] worked analytically on the problem of settlement of smooth, rectangular footing resting on the surface of a finite layer with $\nu = 0.5$, providing a "calibration platform" for the proposed method. He expressed the settlement of footing with respect of the long footing dimension, L, as follows:

$$\rho_r = \frac{qL}{E}\beta. \tag{15}$$

Unfortunately, Sovinc published β vs. H/L curves and not a closed-form expression for β; for readers' convenience, these curves can be found in the freely available book of Poulos and Davis [27] (www.usucger.org/PandD/complete_book.pdf). Sovinc's [8] values were compared by the authors against the respective values given by Butterfield and Banerjee ([10]; also in Poulos and Davis [27]) providing, in turn, the necessary validation of the "calibration platform" mentioned above. The results of these studies are in full agreement.

From the β vs. H/L comparison chart of Figure 2 it is clear that Schiffman and Aggarwala's [18] equivalent shape (see Equation (14)) is not a reliable choice.

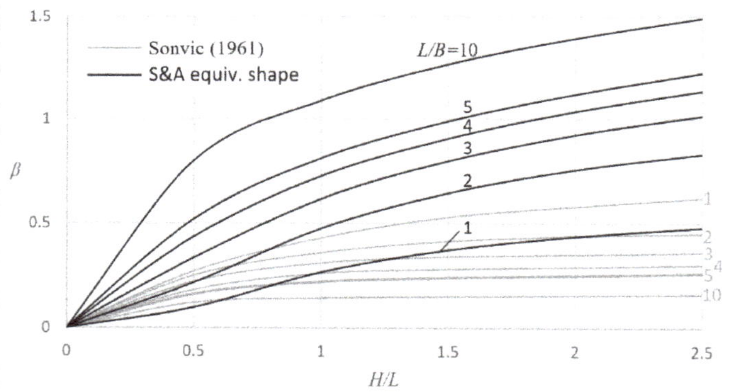

Figure 2. Comparison chart between Sovinc's [8] β factor for rigid rectangular footing ($\nu = 0.5$) and the respective one based on Schiffman and Aggarwala ([18]; denoted as S&A on the figure) equivalent ellipse.

The authors also considered the case of equivalent ellipse satisfying both the condition of area and perimeter. The area of the ellipse is exact and given by the expression πab. The perimeter is exactly given by $2aE(e)$, where $E(x)$ is the complete elliptic integral of second kind [25]. This is a transcendental expression which can be calculated explicitly through infinite series. Instead, one may use Ramanujan's approximate formula giving values very close to the exact perimeter values:

$$\Pi \approx \pi\left(3(a+b) - \sqrt{(3a+b)(a+3b)}\right). \tag{16}$$

The two radii of the equivalent ellipse are then:

$$\left.\begin{array}{l} a = \frac{1}{2\pi}\left(B+L+\frac{1}{3}\left(\eta - \sqrt{6}\sqrt{(B+L)(\eta+2(B+L))-5\pi BL}\right)\right) \\ b = \frac{BL}{a\pi} \end{array}\right\}, \tag{17}$$

with $\eta = \sqrt{3(B+L)^2 + 2\pi BL}$.

The new equivalent shape gave β vs. H/L curves close to those given by Sovinc indicating the need for calibration with respect of the footing shape. The new β_R factor referring to the equivalent rectangular footing is a function of the aspect ratio of footing, L/B. The calibration against Sovinc's [8] chart gave:

$$\beta_R = I_s \cdot \beta(\zeta_R), \tag{18}$$

where

$$I_s = 2\left(1+\log\frac{L}{B}\right)\frac{B_o}{L}, \tag{19}$$

and B_o = 1 m (unit width). The notation $\beta(\zeta_R)$ denotes the use of Equation (10) but replacing ζ with

$$\zeta_R = \left(0.85\lambda + \frac{z}{b}\right)^{\frac{\lambda}{0.85}},\ \lambda = \min\left\{\left(1+\log\frac{L}{B}\right)^{1/3} - 0.05,\ 1.1\right\}. \tag{20}$$

The expression for the settlement, thus, becomes

$$\rho_r = \frac{qL}{E}\beta_R, \tag{21}$$

or, in the more familiar form,

$$\rho_r = \frac{qB}{E}I_H I_{s1}\left(1-\nu^2\right), \tag{22}$$

with

$$I_H = \frac{B_o}{B}\int_{\zeta=0}^{H/b}\frac{1+2(1+k^2)\zeta_R^2(1-v)-2v+k^2\zeta_R^4(3-2v)}{2\left((1+\zeta_R^2)(1+k^2\zeta_R^2)\right)^{3/2}(1-v)}d\zeta, \tag{23}$$

and

$$I_{s1} = 2\left(1+\log\frac{L}{B}\right). \tag{24}$$

The integral in Equation (23) requires numerical evaluation, which can be easily done using one of the numerous proprietary and non-proprietary mathematical programs available.

As shown is Figure 3, the new β_R factor compares very well with Sovinc's β factor. Although, the proposed method was calibrated manually against Sovinc's [8] dimensionless values, any set of relevant data could be used instead, while the procedure could be facilitated by the use of a statistical analysis program. The use of elastic settlement values derived from 3D finite element analysis is subject matter of work by the authors.

It is interesting that, Equations (19) and (20) contain the term $(1+\log(L/B))$, best known from the influence depth of footings proposed by Terzaghi et al. [28], i.e., $z_I = 2B(1+\log(L/B))$

The effect of footing shape on the modulus of elasticity has not yet been considered in the analysis. In this respect, Terzaghi et al. [28] suggested that

$$\frac{E}{E_{axis.}} = \left(1+0.4\log\frac{L}{B}\right), \tag{25}$$

with $1 \leq L/B \leq 10$, where, $E_{axis.}$ the triaxial modulus of elasticity of soil. For $1 \leq L/B \leq 10$, the $E/E_{axis.}$ ratio ranges between 1 and 1.4 for the axisymmetric and plane strain condition, respectively. Very recently, Pantelidis [29,30] showed that the correct relationship is

$$\frac{E}{E_{\text{axis.}}} = \left(1 + \log\frac{L}{B}\right). \tag{26}$$

Thus, including the effect of footing shape on the modulus of soil, the footing settlement will be

$$\rho_r = \frac{2qB}{E_{\text{axis.}}} I_H\left(1 - \nu^2\right) \text{ or } \frac{2qB_o}{E_{\text{axis.}}}\beta(\zeta_R). \tag{27}$$

The triaxial modulus of soil, $E_{\text{axis.}}$, can be obtained by the triaxial compression test or continuous probing tests.

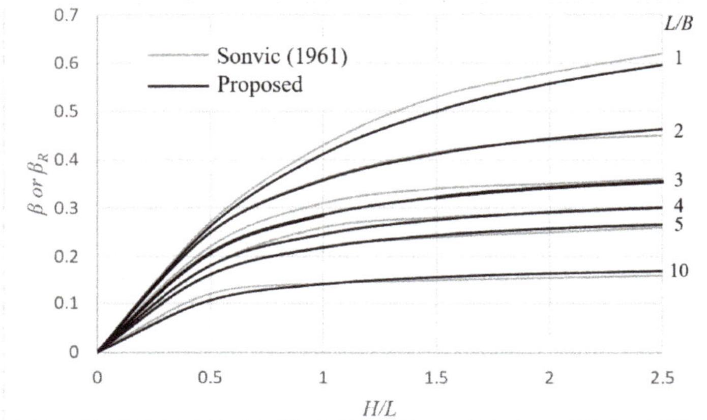

Figure 3. Comparison chart between Sovinc's [8] β factor for rigid rectangular footing and the proposed β_R; all curves refer to $\nu = 0.5$.

4. Elastic Settlement Analysis of Rigid Footings on Sands

The above formulations were based on the contact pressure distribution of Equation (4). This contact pressure of parabolic form, where the minimum and the maximum pressure value is found at the center and the edges of footing respectively (Figure 4a), is, however, more suitable for clays. For smooth rigid footings on clean sand, the minimum and maximum contact pressure are observed at the edges and the center of the footing respectively, as shown in Figure 4b [28,31].

Referring to an elliptical rigid footing over sand, the contact pressure distribution can be effectively represented by a half spheroid attached under the footing. The infinitesimal point load P in Equations (1)–(3) can, then, be expressed as follows:

$$P = \frac{3}{2}q\sqrt{1 - \frac{x^2}{a^2} - \frac{y^2}{b^2}}dxdy. \tag{28}$$

In addition, as known sands are compacted under their own weight, where their modulus of elasticity usually appears to increase linearly with depth obeying Gibson's law [32]:

$$E(z) = E_o + k_E z, \tag{29}$$

where E_o = the soil modulus directly beneath the foundation base ($z = 0$), k_E = the rate of increase of modulus with depth (units of E per unit depth) and z = the depth having modulus $E(z)$. This has also been proved experimentally by the first author [29] performing a series of Dynamic Probing Light (DPL) tests on clean quarry sand. Indeed, this phenomenon occurs immediately after deposition and it is due to the increase of the density caused by the overburden mass in combination with the increased confinement at greater depths. In a similar manner, any surcharge on the surface of the sand material will have an analogous impact. In this respect, it was suggested [29] that

$$E(z) = E_o + k_E \frac{\sigma_z}{\gamma} = E_o + k_E\left(z + \frac{q}{\gamma} I_\sigma\right) = E_o\left[1 + \frac{k_E}{E_o}\left(z + \frac{q}{\gamma} I_\sigma\right)\right], \tag{30}$$

where γ is the unit weight of soil, $\sigma_z = \gamma z + qI_\sigma$ and I_σ is the well-known stress influence factor; I_σ values for various footing shapes can be found in any soil mechanics book (e.g., [26]). Therefore, the settlement is given as follows:

$$\rho_r = \frac{qL}{E_o}\beta_R = \frac{qL}{E_o} I_s \beta(\zeta_R). \tag{31}$$

The factor $\beta(\zeta_R)$ is calculated using the following Equation (recall Equation (10))

$$\beta = \frac{1}{b}\int_{z=0}^{H} \frac{I_z(z)}{1 + k_E(z + qI_\sigma/\gamma)/E_o} dz = \int_{\zeta=0}^{H/b} \frac{I_z(\zeta)}{1 + k_E b(\zeta + qI_\sigma/(b\gamma))/E_o} d\zeta, \tag{32}$$

with

$$I_z(\zeta) = (\nu + 1)\left[\frac{3}{2\sqrt{(1+\zeta^2)(1+k^2\zeta^2)}} - \nu F_4\right], \tag{33}$$

and

$$F_4 = 3 - \frac{3k^2\zeta}{2\pi}\int_0^{2\pi} \frac{\cot^{-1}\left(\frac{k\zeta}{\sqrt{1-(1-k^2)\sin^2\theta}}\right)}{\left(1-(1-k^2)\sin^2\theta\right)^{3/2}} d\theta, \tag{34}$$

whilst ζ must again be replaced by ζ_R (recall Equation (20)) as indicated in Equation (31). It is reminded that I_s is given by Equation (19). The integral of Equation (34) has been transformed to polar variables with $x = ar\cos\theta$ and $y = br\sin\theta$ for the sake of convenience. Unfortunately, the integral in Equation (34) requires numerical evaluation. For avoiding such a procedure, the F_4 term is given in chart form in Figure 5.

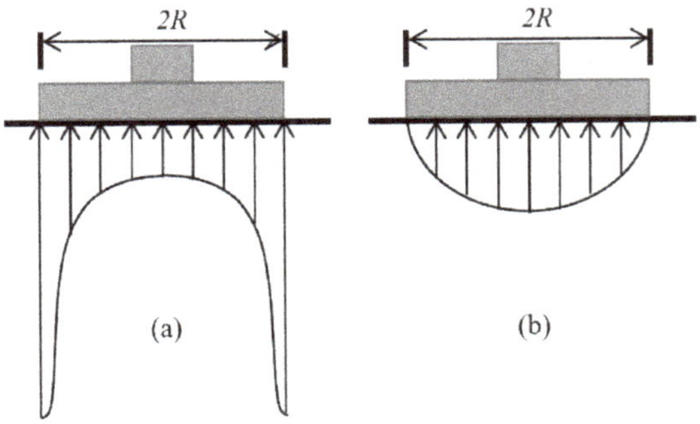

Figure 4. Assumed contact pressure distribution of smooth rigid footing on (**a**) cohesive soil and (**b**) cohesionless soil.

It is finally mentioned that, for the special case of $H \to \infty$ and $k_E = 0$ the factor β can be obtained fully analytically (see Appendix A).

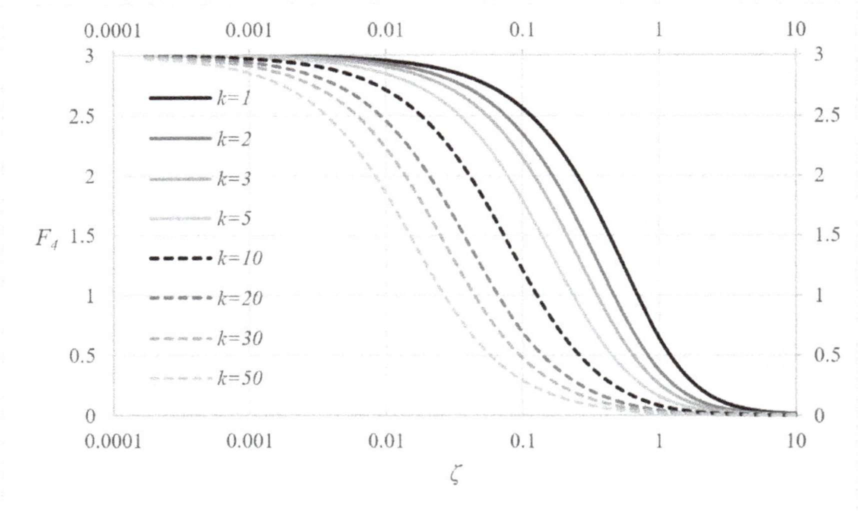

Figure 5. Chart giving the F_4 factor with respect to ζ.

5. Other Factors Affecting Elastic Settlement of Shallow Foundations

The modulus of elasticity and in turn, the elastic settlement of shallow foundations may seriously be affected by various factors, such as the shape of the footing [29,30,33], the footing load (phenomenon occurring in sands; [29]), the embedment depth of footing [15,34–38], and a possible water table rise into the influence depth of footing [39–48]. Creep in sands may also affect the settlement of shallow foundations (see [13,28]). The rigidity of footing, the net applied pressure and the non-elastic response of the ground are also factors affecting settlement of shallow foundations for which corrections are suggested below.

5.1. Correction for Footing Rigidity

According to ACI Committee 336, DIN 4018 and IS 2950-1 [49–51], whether a foundation behaves as a rigid or a flexible structure depends on the relative stiffness of the structure and the foundation soil. The relative stiffness factor, K_r, is given as follows:

$$K_r = \frac{E_b I}{E L^3 B}, \tag{35}$$

where $E_b I$ is the flexural rigidity of the superstructure and foundation per unit length at right angles to B and E_b is the modulus of concrete. An approximate value of $E_b I$ per unit width of building can be determined by summing the rigidity of foundation, $E_b I_F$, the rigidity of beams, $E_b I_b$, and the rigidity of shearwalls, $E_b t_w h_w^3/12$:

$$E_b I = E_b\left(I_F + \sum I_b + t_w h_w^3/12\right), \tag{36}$$

where t_w and h_w are the thickness and height of the shearwalls respectively (e.g., see [49–52]. Ignoring the effect of superstructure, the relative stiffness factor simplifies to

$$K_r = \frac{E_b}{12E}\left(\frac{d}{L}\right)^3, \tag{37}$$

because $I = Bd^3/12$, where, d is the thickness of foundation.

Alternatively, the expression including the Poisson's ratio values can be used (e.g., see [53]):

$$K_r = \frac{E_b\left(1-\nu^2\right)}{12E\left(1-\nu_b^2\right)}\left(\frac{d}{L}\right)^3. \tag{38}$$

Generally, the foundation is considered to be rigid when $K_r > 0.5$ and flexible when $K_r < 0.5$ (e.g., [51]). Based on the finite element analysis carried out by Brown [54] (see also [15]), however, a footing can be regarded as flexible if $K_r < 0.05$, rigid if $K_r > 5$ and of intermediate rigidity if $0.05 \le K_r \le 5$. Indeed, between the limits defining the intermediate condition, a linear relationship exists between K_r and settlement [15]. In this respect, the following linear interpolation for calculating the settlement of footings (at the center) of intermediate rigidity is suggested:

$$\rho_{\text{int}} = \rho_r I_F, \tag{39}$$

where

$$I_F = 1 + \frac{5-K_r}{4.95}\left(\frac{\rho_{f,center}}{\rho_r} - 1\right). \tag{40}$$

$\rho_{f,center}$ is calculated using Equation (A7) (see Appendix B).

5.2. Correction for the Net Applied Pressure

When the structure lies at the bottom of excavation of depth D_f, as usually this is the case, the settlement should be better calculated according to the following equation:

$$\rho_r = \frac{\left(\gamma D_f\right)\beta L}{M_{R1}} I_E + \frac{\left(q-\gamma D_f\right)\beta L}{E} I_E = \left(\frac{q}{E} - \gamma D_f\left(\frac{1}{E} - \frac{1}{M_{R1}}\right)\right)\beta L I_E, \tag{41}$$

where M_{R1} the first resilient modulus of soil (it refers to the first reloading). An empirical relationship connecting E with the overconsolidation ratio (OCR) of soils could be used for the estimation of M_{R1}, e.g., Duncan and Buchignani's [55] E_u/c_u - PI - OCR chart for undrained clays (where E_u and c_u are the undrained modulus and the undrained cohesion respectively, and PI the plasticity index of soil; see also [56]) or Bowles's [42] $\sqrt{OCR}$ multiplier. According to Mayne and Kulhawy [57], the square-root of OCR is applicable for both sands and clays. In addition, for saturated overconsolidated clays subjected to unloading, Mesri et al. [58] observed that, in every case, the excess negative pore pressures is dissipated faster than predicted from the Terzaghi's one dimensional consolidation theory; thus, Equation (41) could be used in every case.

The overconsolidation, here, is due to γD_f, which is excavated for placing the construction at depth D_f. Alternatively, adopting the square root rule of Bowles [42], the first resilient modulus will be

$$M_{R1} = E\sqrt{(q+\gamma D_f)/q}, \tag{42}$$

and thus, Equation (41) simplifies to

$$\rho_r = \frac{\Delta q \beta L}{E} I_E, \tag{43}$$

with

$$\Delta q = q - \gamma D_f\left(1 - \sqrt{\frac{q}{q+\gamma D_f}}\right). \tag{44}$$

By ignoring the term in brackets, that is, if Δq is equal to $q - \gamma D_f$, the calculated settlement is on the unfavorable side.

5.3. Non-Elastic Response of Soil under Loading

In reality, the soil medium response to loading is not elastic, even at low working stresses. Based on parametric finite element analysis, D'Appolonia et al. [59] proposed a correction factor to the settlement of perfectly flexible footings on saturated clays ($\nu = 0.5$). An alternative and more general procedure is proposed below.

First, a load-settlement response is assumed, relying on the initial gradient of this relationship and the ultimate bearing capacity of the soil medium, q_u, which are both known; a third (calibration) parameter is also necessary (this is discussed later). In a q-ρ diagram, the initial gradient of the curve is $\tan\delta = E/\left(\beta L I_f\right)$ (with I_f being a factor for taking into account e.g., the influence of the embedment depth, a possible water table rise etc.), whilst q_u is calculated based on a well-established theory (in this respect, the formulations given in EN 1997-1 [60] could be used). In the case of stratified soil mediums, the equivalent modulus of soils concept shall be used (see [61]). Mathematically, the load-settlement response can be approximated by an equation of the form:

$$q = a_1 \rho_{PL}^{a_2} + \tan\delta \cdot \rho_{PL}. \tag{45}$$

where ρ_{PL} is the plastic settlement of footing.

Equation (45) presents maximum at

$$\rho_{PL,q_u} = \left(-\frac{\tan\delta}{a_1 a_2}\right)^{\frac{1}{a_2-1}}. \tag{46}$$

Relating the plastic settlement at q_u with the respective elastic one, i.e.,

$$\rho_{PL,q_u} = n_{PL}\rho_{EL,q_u} = n_{PL} q_u / \tan\delta, \tag{47}$$

from Equations (46) and (47)

$$a_1 = -\frac{n_{PL} q_u}{a_2}\left(\frac{\tan\delta}{n_{PL} q_u}\right)^{a_2}, \tag{48}$$

where n_{PL} is the calibration parameter denoting how many times the plastic settlement at q_u is greater than the respective elastic one (see Figure 6); n_{PL} takes values greater than or equal to 2.

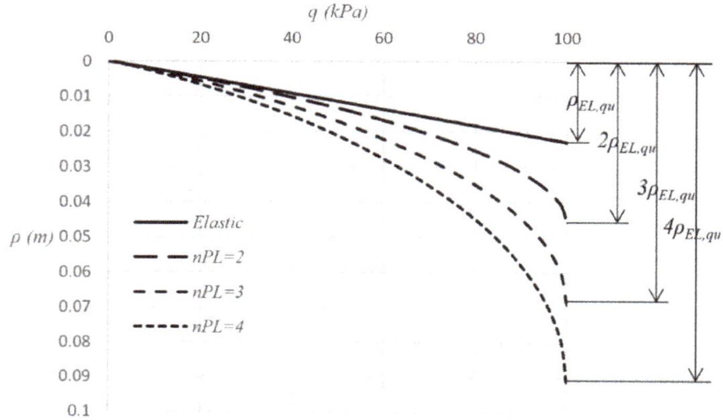

Figure 6. Proposed approach for non-elastic response of soil; n_{PL} can be any positive real number greater than 2.

At the ultimate loading, from Equations (45), (46), and (48):

$$q_u = \left\{ a_1 \rho_{PL,q_u}^{a_2} + \tan\delta \cdot \rho_{PL,q_u} \right\} = q_u n_{PL} \frac{a_2 - 1}{a_2}, \tag{49}$$

and solving the latter as for a_2:

$$a_2 = \frac{n_{PL}}{n_{PL} - 1}. \tag{50}$$

The parameter n_{PL} can be estimated based on (a) local experience, i.e., known measured settlements of neighbor structures, (b) plate bearing test data given that the soil and soil conditions are relatively homogenous with depth, or (c) 2D plastic finite element analysis by interpolating between the load-settlement response of the respective axisymmetric and plane strain problem (circular and strip footing both having radius and width respectively equal to B). The calibration should refer to similar loading conditions, i.e., drained or undrained.

The plastic settlement, ρ_{PL}, corresponding to q is, finally, obtained from Equation (45). The latter has two solutions, from which the smaller one is adopted. Depending on the n_{PL} value used, numerical approximation may be required.

An example q -ρ chart referring to a 1 m × 2 m rigid footing lying on the surface of a semi-infinite homogenous clay soil with E = 5000 kPa and ν = 0.1 (drained conditions) is given in Figure 6. In this respect, four q -ρ relationships were drawn, i.e., for perfectly elastic medium and for plastic medium with n_{PL} = 2, 3 and 4. The ultimate bearing capacity value (q_u) is equal to 100 kPa. Moreover, the correction regarding the effect of footing shape on the modulus of soil (recall Equation (26)) was ignored in this example.

6. Comparison Examples

6.1. Comparison with Existing Methods and Approaches

The proposed method is compared with Mayne and Poulos' [15] method, as well as with the "characteristic point" approach [62,63]. Third factors affecting settlement, such as the embedment depth, soil creep, are ignored, whilst the mass is considered to be semi-infinite and homogenous.

Since Mayne and Poulos' method deals with footings of any rigidity, both the cases of perfectly flexible and perfectly rigid footing will be presented. The settlement values obtained from the theory of elasticity for both the center and the corner of footing, considering that the latter is perfectly flexible are presented as boundary values (the elastic settlement of the respective rigid footing should be an intermediate value). The average settlement of perfectly flexible footing is also given. All settlement expressions used are given in the Appendix B.

Regarding the so-called "characteristic point", at this point the settlement of a flexible loaded area is considered to be the same as the uniform settlement of a rigid area, both having the same area and shape and bearing the same total load. In an ideally elastic soil, the characteristic point at a $B \times L$ rectangular footing is located at distance $0.37B$ and $0.37L$ from the center [62–65]. What is not widely known, however, is that the above distances refer to the simple case of $\nu = 0$ [62,63,66].

$\rho E/q$ vs. L/B curves are given for ν = 0.1 and 0.5; see Figure 7. These values were selected because, based on accurate measurements using local strain devices mounted midlevel on soil specimens and measured internally to the triaxial cell, the appropriate value of ν to use in elastic continuum solutions for drained loading is $0.1 < \nu < 0.2$ for all soil types [15,67–69]. For undrained conditions involving short-term loading of clays, it remains appropriate to use the value from isotropic elastic theory of $\nu = 0.5$.

From Figure 7 it is inferred that, Mayne and Poulos' [15] method largely overestimates settlement (see also [70]); indeed, it gives settlement values for the rigid footing greater than the values corresponding to the center of the respective flexible footing (upper boundary). On the other hand, the $\rho E/q$ vs. L/B curves derived from the proposed method lie between the two boundary curves derived from the theory of elasticity, whilst moreover, they are parallel to these boundaries. In addition, the proposed $\rho E/q - L/B$ curve for $\nu = 0.1$ compares well with the respective curve corresponding to the characteristic point. The latter can also be considered indicative of the success of the calibration of the proposed method. Furthermore, as shown in Figure 7, the average settlement curve overestimates settlement for small ν values and underestimates settlement for great ν values.

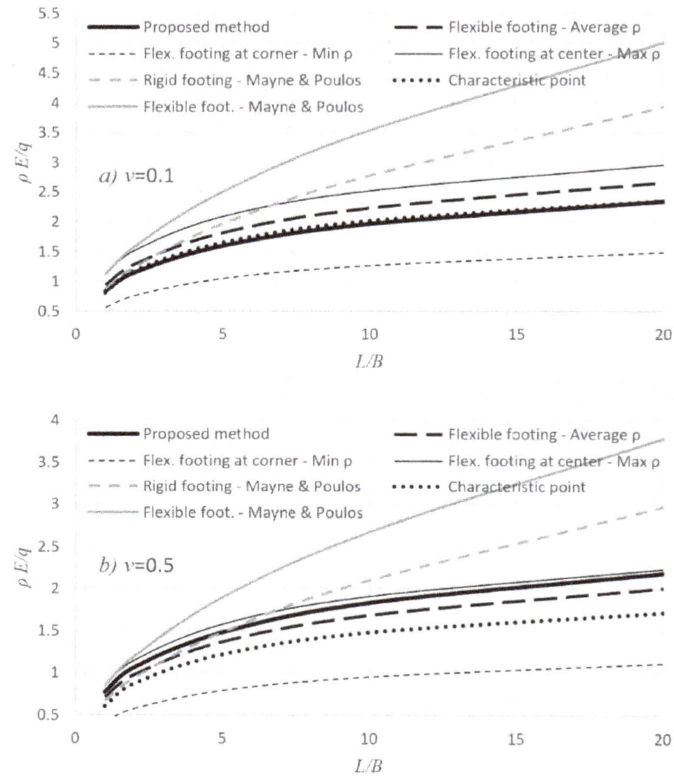

Figure 7. Comparison example: (**a**) $v = 0.1$ and (**b**) $v = 0.5$.

In addition, from Figure 8 it is inferred that for the same ν value, the $\rho E/q$ term for the rigid footing on sand is greater than the respective one for the same footing on clay. The goal of Figure 8 is to compare the effect of the different contact pressure distributions, assuming the same E value, which is also constant with depth. In reality, E may vary with depth; this is usually observed in sands, where E increases with depth (Gibson profile [32]), even immediately after deposition [29]. It is also very interesting that, as shown in Figure 9, when the effect of footing shape on the modulus of elasticity of soil is considered in the analysis, the L/B ratio of footing has, practically, minor effect on the derived settlement. In this respect, revisiting existing experimental and theoretical studies, Pantelidis [29,30] draw the conclusion that the elastic settlement appears to be independent of the aspect ratio of footing; it seems that the adverse effect of the second dimension of footing, which is responsible for the iso-stress bulbs to dive deeper into the ground, is compensated by the increase in E due to the increased confinement induced by the longer (and thus, greater) footprint of the footing. In this respect, he suggested that, the elastic settlement of a $B \times L$ footing be calculated considering a circular footing with diameter equal to B (or better, $2B/\sqrt{\pi}$) and the bearing pressure q of the original problem. A problem, however, arises when soil medium is stratified. This can be solved by replacing the original stratified medium with an equivalent homogenous one having the same elastic constant values throughout its body. A reliable procedure for that is given by Pantelidis [33,61].

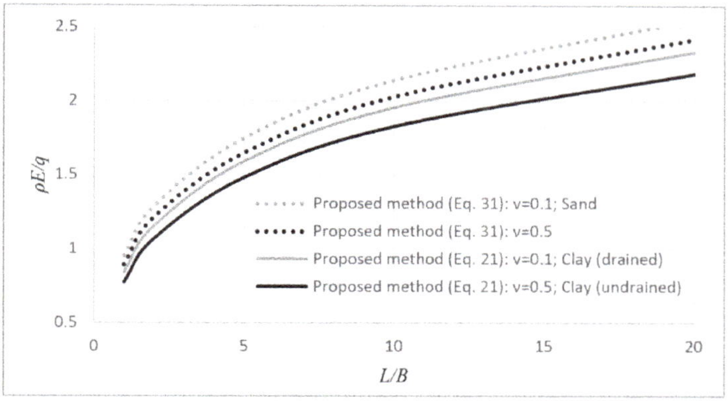

Figure 8. Effect of the contact pressure distribution (simulating rigid footing on sands and clays) on $\rho E/q$.

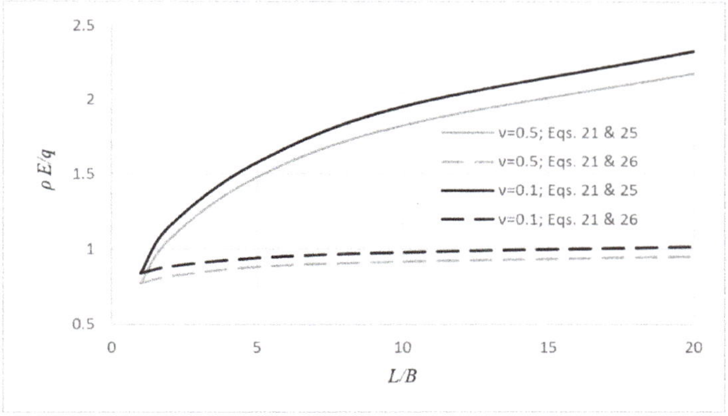

Figure 9. Effect of footing shape on the modulus of soil and, in turn, on $\rho E/q$.

6.2. Comparison against 3D Finite Element Elastic Settlement Analysis

Fraser and Wardle [12] gave a complete elastic settlement example using 3D numerical analysis. They considered a square raft foundation with edge 10 m and thickness d = 0.5 m, lying on a soil medium of thickness 40 m. The raft is subjected to a uniform load of 0.1 MPa. The elastic constants of raft are E_b = 15,000 MPa and v_b = 0.2, whilst the elastic constants of soil are E = 83.2 MPa and v = 0.3.

Following the proposed approach, the settlement of the raft considering that it is perfectly rigid is ρ_r = 8.728 mm, whilst following Harr [71], the settlement at the center of the respective flexible footing is $\rho_{f,center}$ = 10.796 mm. Based on Equation (38), K_r = 1.78, thus, the footing is of intermediate rigidity ($0.05 \le K_r \le 5$) with I_F = 1.154 (Equation (40)). Therefore, the corrected settlement value is ρ_{int} = 10.07 mm (Equation (39)), which compares very well with the 10.7 mm given by Fraser and Wardle. Applying Mayne and Poulos' method, the settlement of the same footing is 8.71 mm (Mayne and Poulos' factors: I_E = 1, I_F = 0.792 and I_G = 0.89).

7. Summary and Conclusions

In this paper an elastic settlement analysis method for rigid rectangular footings applicable to both clays and sands has been proposed. The case of clay is differentiated from the case of sand using different contact pressure distribution, whilst, moreover, the modulus of elasticity for the case of sands increases linearly with depth. The proposed method is based on the concept of equivalent shape, where, any $B \times L$ footing is suitably replaced by a footing of elliptical shape. The analysis showed that an equivalent ellipse returns satisfactory results only if the conditions of equal area and equal perimeter length are satisfied simultaneously. The equivalent area condition alone, as it has been shown, gives elastic settlement values that greatly differ from the respective values referring to the mother rectangular footing.

Herein, the derived equations were calibrated manually against Sovinc's [8] dimensionless values. However, any set of relevant data could be used, while the procedure could be facilitated by the use of a statistical analysis program. The use of elastic settlement values derived from 3D finite element analysis is subject matter of work by the authors.

Among the most interesting findings is that sands produce "settlement x soil modulus/applied pressure" values approximately 10% greater compared to the respective ones corresponding to clays. Moreover, for large Poisson's ratio (v) values, the settlement of a rigid footing is closer to the settlement corresponding to the center of the respective flexible footing. On the other hand, as v decreases, the derived settlement of the rigid footing approaches the settlement value corresponding to the characteristic point of the respective flexible footing; it is also noted that, the derived settlement value for small v values is smaller than the average settlement of the footing considering that the latter is perfectly flexible. Last but not least is that, considering the effect of footing shape on the modulus of soil, the settlement of rigid footings is rather insensitive to the L/B ratio. In this respect, it seems that the adverse effect of the second dimension of footing (the length L), which is responsible for the iso-stress bulbs to dive deeper into the ground, is compensated by the increase in the modulus of soil due to the increased confinement induced by the longer (and thus, greater) footprint of the footing. Thus, it is suggested that, the elastic settlement of a $B \times L$ footing be calculated considering a circular footing with diameter equal to $2B/\sqrt{\pi}$ and the bearing pressure q of the original problem.

Finally, corrections for the net applied pressure, footing rigidity and non-elastic response of soil under loading are also offered. Regarding the net applied pressure, it has been shown that when the $q - \gamma D_f$ bearing pressure is used, the calculated settlement is on the unfavorable side; instead a new, yet simple expression is suggested, allowing for the first resilient modulus of the soil to be taken into account for structure loading equal to γD_f.

Author Contributions: Conceptualization, L.P.; methodology, L.P. and E.G.; software, L.P. and E.G.; validation, L.P. and E.G.; formal analysis, L.P. and E.G.; writing—original draft preparation, L.P. and E.G.; writing—review and editing, L.P. and E.G.; visualization, L.P. and E.G.; supervision, L.P. All authors have read and agreed to the published version of the manuscript.

Funding: This research received no external funding.

Conflicts of Interest: The authors declare no conflict of interest.

Appendix A

In spite of the difficulty to evaluate analytically F_4 that would provide the β factor for arbitrary H, the β factor for $H \to \infty$ can be obtained fully analytically by first integrating with respect to ζ and then with respect to θ:

$$\beta = \frac{3}{4}(1 - v^2)\Big(K(e) + \frac{1}{k}K(ie/k)\Big), \tag{A1}$$

where $K(x)$ is the complete elliptic integral of the first kind, and i is the imaginary unit. Hence,

$$\rho_r = \frac{3b}{4E}(1 - v^2)\Big(K(e) + \frac{1}{k}K(ie/k)\Big). \tag{A2}$$

For the case of circular footing ($k = 1$) over sand, Equation (33) simplifies to

$$I_z = 3(\nu+1)\left(\frac{1}{2(1+\zeta^2)} - \nu\left(1-\zeta\cot^{-1}\zeta\right)\right), \tag{A3}$$

and by integrating the latter from $z = 0$ to H, we get the factor β

$$\beta = \frac{3}{4}(\nu+1)\left(\nu\left(\nu\frac{H}{b}-2\right)\left(\frac{H}{b}\right)+2\left(1-\nu-\nu\left(\frac{H}{b}\right)^2\right)\tan^{-1}\left(\frac{H}{b}\right)\right). \tag{A4}$$

For $H \to \infty$ Equation (A4) gives

$$\beta = \frac{3\pi}{4}(1-\nu^2), \tag{A5}$$

and, thus, the settlement expression becomes

$$\rho_r = \frac{3\pi qa}{4E}\left(1-\nu^2\right) = 1.5\left(\frac{\pi qa}{2E}\left(1-\nu^2\right)\right), \tag{A6}$$

meaning that the term $\rho_r E/q$ for a circular rigid footing on sands is by 50% greater than the respective one for clays.

Appendix B

Settlement under the corner and center of a $B \times L$ (flexible) rectangular footing [71]

$$\rho(z) = \frac{a_{cc}qB'}{2E_s}\left(1-\nu^2\right)\left(A_{cc} - \frac{1-2\nu}{1-\nu}B_{cc}\right), \tag{A7}$$

$$A_{cc} = \frac{1}{\pi}\left(\ln\frac{\sqrt{1+m^2+n^2}+m}{\sqrt{1+m^2+n^2}-m} + m\ln\frac{\sqrt{1+m^2+n^2}+1}{\sqrt{1+m^2+n^2}-1}\right), \tag{A8}$$

$$B_{cc} = \frac{n}{\pi}\tan^{-1}\frac{m}{n\sqrt{1+m^2+n^2}}, \tag{A9}$$

with $a_{cc} = 1$, $B' = B$, $m = L/B$ and $n = z/B$ for the corner of the footing, whilst $a_{cc} = 4$, $B' = B/2$, $m = L/B$ and $n = 2z/B$ for the center; $z = 0$ (and thus, $n = 0$) for the settlement on the surface of a homogenous, semi-infinite mass. For the settlement of footing underlain by a layer of finite thickness H, the principle of superposition stands, i.e., $\rho = \rho(0) - \rho(H)$.

Average settlement of a $B \times L$ (flexible) rectangular footing on semi-infinite homogenous mass [72]

$$\rho(z) = \frac{q\sqrt{BL}}{E_s}\left(1-\nu^2\right)A_{av}, \tag{A10}$$

$$A_{av} = \frac{1}{\pi\sqrt{m}}\left(\ln\frac{\sqrt{1+m^2}+m}{\sqrt{1+m^2}-m} + m\ln\frac{\sqrt{1+m^2}+1}{\sqrt{1+m^2}-1} - \frac{2}{3}\frac{\left(1+m^2\right)^{\frac{3}{2}}-\left(1+m^3\right)}{m}\right), \tag{A11}$$

with m as defined above.

Mayne and Poulos [15] expressed the settlement as follows:

$$\rho(z) = \frac{q}{E_s}\sqrt{\frac{4BL}{\pi}}\left(1-\nu^2\right)I_G I_F I_E, \tag{A12}$$

where $I_F = \pi/4$ and 1 for perfectly rigid and perfectly flexible footing respectively, $I_E = 1$ for foundation on the surface and $I_G = 1$ for homogenous, semi-infinite mass. For more about the I-factors, please see the original publication [15].

References

1. Noble, B. The numerical solution of the singular integral equation for the charge distribution on a flat rectangular plate. In Proceedings of the PICC Symposium on Differential and Integral Equations, Rome, Italy, 20–24 September 1960; Birkhäuser: Basel, Switzerland; Stuttgart, Germany, 1960; pp. 530–543.
2. Borodachev, N.M.; Galin, L.A. Contact problem for a stamp with narrow rectangular base: PMM vol. 38, n≗ 1, 1974, pp. 125–130. *J. Appl. Math. Mech.* **1974**, *38*, 108–113. [CrossRef]
3. Gorbunov-Possadov, M.; Serebrjanyi, R. Design of Structures on Elastic Foundations. In *Proceedings of the Fifth International Conference on Soil Mechanics and Foundation Engineering, 17–22 July, 1961*; Dunod: Paris, France, 1961; pp. 643–648.
4. Borodachev, N.M. Contact problem for a stamp with a rectangular base: PMM vol. 40, n≗ 3, 1976, pp. 554–560. *J. Appl. Math. Mech.* **1976**, *40*, 505–512. [CrossRef]
5. Mullan, S.J.; Sinclair, G.B.; Brothers, P.W. Stresses for an elastic half-space uniformly indented by a rigid rectangular footing. *Int. J. Numer. Anal. Methods Geomech.* **1980**, *4*, 277–284. [CrossRef]
6. Brothers, P.W.; Sinclair, G.B.; Segedin, C.M. Uniform indentation of the elastic half-space by a rigid rectangular punch. *Int. J. Solids Struct.* **1977**, *13*, 1059–1072. [CrossRef]
7. Panek, C.; Kalker, J.J. A solution for the narrow rectangular punch. *J. Elast.* **1977**, *7*, 213–218. [CrossRef]
8. Sovinc, I. Displacements and inclinations of rigid footings resting on a limited elastic layer on uniform thickness. In *Proceedings of the 7th International Conference on Soil Mechanics and Foundation Engineering*; Sociedad Mexicana de Mecanica: Mexico City, Mexico, 1961; pp. 385–389.
9. Dempsey, J.P.; Li, H. A rigid rectangular footing on an elastic layer. *Géotechnique* **1989**, *39*, 147–152. [CrossRef]
10. Butterfield, R.; Banerjee, P.K. A rigid disc embedded in an elastic half space. *Geotech. Eng.* **1971**, *2*, 35–52.
11. Fabrikant, V.I. Flat punch of arbitrary shape on an elastic half-space. *Int. J. Eng. Sci.* **1986**, *24*, 1731–1740. [CrossRef]
12. Fraser, R.A.; Wardle, L.J. Numerical analysis of rectangular rafts on layered foundations. *Géotechnique* **1976**, *26*, 613–630. [CrossRef]
13. Schmertmann, J.H. Static cone to compute static settlement over sand. *J. Soil Mech. Found. Div.* **1970**, *96*, 1011–1043.
14. Schmertmann, J.H.; Hartman, J.P.; Brown, P.R. Improved strain influence factor diagrams. *J. Geotech. Geoenviron. Eng.* **1978**, *104*, 1131–1135.
15. Mayne, P.W.; Poulos, H.G. Approximate displacement influence factors for elastic shallow foundations. *J. Geotech. Geoenviron. Eng.* **1999**, *125*, 453–460. [CrossRef]
16. Foye, K.C.; Basu, P.; Prezzi, M. Immediate Settlement of Shallow Foundations Bearing on Clay. *Int. J. Geomech.* **2008**, *8*, 300–310. [CrossRef]
17. Lee, J.; Salgado, R. Estimation of footing settlement in sand. *Int. J. Geomech.* **2002**, *2*, 1–28. [CrossRef]
18. Schiffman, R.L.; Aggarwala, B.D. Stresses and displacements produced in a semi-infinite elastic solid by a rigid elliptical footing. In Proceedings of the 5th International Conference on Soil Mechanics and Foundation Engineering, Paris, France, 17–22 July 1961; Dunod: Paris, France, 1961; Volume 1, pp. 795–801.
19. Boussinesq, J. *Application des Potentiels à L'étude de L'équilibre et du Mouvement des Solides Élastiques: Principalement au Calcul des Déformations et des Pressions que Produisent, dans ces Solides, des Efforts Quelconques Exercés sur une Petite Partie de Leur Surface*; Gauthier-Villars: Paris, France, 1885; Volume 4.
20. Das, B.M. *Fundamentals of Geotechnical Engineering*, 3rd ed.; Thomson-Engineering: Mobile, AL, USA, 2007; ISBN 10:0-495-29572-8.
21. Hudson, J.A.; Harrison, J.P. *Engineering Rock Mechanics: An Introduction to the Principles*; Elsevier: Amsterdam, The Netherlands, 2000; ISBN 0080530966.
22. Sadd, M.H. *Elasticity: Theory, Applications, and Numerics*; Academic Press: Cambridge, MA, USA, 2009; ISBN 0080922414.
23. Pantelidis, L. Strain Influence Factor Charts for Settlement Evaluation of Spread Foundations based on the Stress–Strain Method. *Appl. Sci.* **2020**, *10*, 3822. [CrossRef]
24. Kézdi, Á.; Réththe, L. *Soil Mechanics of Earthworks, Foundations and Highway Engineering, Handbook of Soil Mechanics*; Elsevier: New York, NY, USA, 1988; Volume 3, ISBN 0-444-98929-3.
25. Abramowitz, M.; Stegun, I.A. *Handbook of Mathematical Functions: With Formulas, Graphs, and Mathematical Tables*; Courier Corporation: North Chelmsford, MA, USA, 1965; Volume 55, ISBN 0486612724.

26. Das, B.M. *Shallow Foundations: Bearing Capacity and Settlement*, 3rd ed.; CRC Press: Boca Raton, FL, USA, 2017; ISBN 9781315163871.
27. Poulos, H.G.; Davis, E.H. *Elastic Solutions for Soil and Rock Mechanics*; John Wiley: New York, NY, USA, 1991; ISBN 0471695653.
28. Terzaghi, K.; Peck, R.B.; Mesri, G. *Soil Mechanics in Engineering Practice*; John Wiley: New York, NY, USA, 1996; ISBN 0471086584.
29. Pantelidis, L. Elastic Settlement Analysis for Various Footing Cases Based on Strain Influence Areas. *Geotech. Geol. Eng.* **2020**, *38*, 4201–4225. [CrossRef]
30. Pantelidis, L. The effect of footing shape on the elastic modulus of soil. In Proceedings of the 2nd Conference of the Arabian Journal of Geosciences (CAJG), Sousse, Tunisia, 25–28 November 2019; Springer Nature: Sousse, Tunisia, 2019.
31. Barnes, G. *Soil Mechanics: Principles and Practice*; Palgrave Macmillan: London, UK, 2016; ISBN 1137512210.
32. Gibson, R.E. Some Results Concerning Displacements and Stresses in a Non-Homogeneous Elastic Half-space. *Géotechnique* **1967**, *17*, 58–67. [CrossRef]
33. Pantelidis, L. On the modulus of subgrade reaction for shallow foundations on homogenous or stratified mediums. In Proceedings of the 3rd International Structural Engineering and Construction Conference (EURO-MED-SEC-03), Limassol, Cyprus, 3–8 August 2020; Vacanas, Y., Danezis, C., Yazdani, S., Singh, A., Eds.; ISEC Press: Limassol, Cyprus, 2020.
34. Fox, L. The mean elastic settlement of a uniformly loaded area at a depth below the ground surface. In Proceedings of the Second International Conference on Soil Mechanics and Foundation Engineering, Rotterdam, The Netherlands, 21–30 June 1948; Volume 1, p. 129.
35. Janbu, N.; Bjerrum, L.; Kjaernsli, B. Veiledning ved lo/sniffing av fundamenterings oppgaver. In *Norwegian with English Summary; Soil Mechanics Applied to Some Engineering Problems*; Norwegian Geotechnical Institute, Publication: Oslo, Norway, 1956.
36. Burland, J.B. Discussion of session A. In *Proc. of the Conf. on In Situ Investigations in Soils and Rocks*; British Geotechnical Society: London, UK, 1970; pp. 61–62.
37. Christian, J.; Carrier, D. Janbu, Bjerrum and Kjaernsli's chart reinterpreted. *Can. Geotech. J.* **1978**, *15*, 123–128. [CrossRef]
38. Díaz, E.; Tomás, R. Revisiting the effect of foundation embedment on elastic settlement: A new approach. *Comput. Geotech.* **2014**, *62*, 283–292. [CrossRef]
39. Teng, W.C. *Foundation Design*; Prendice-Hall Inc.: New York, NY, USA, 1962.
40. Alpan, I. Estimating the settlements of foundations on sands. *Civ. Eng Public Work. Rev. UK* **1964**, *59*, 1415–1418.
41. Bazaraa, A.R. *Use of the Standard Penetration Test for Estimating Settlements of Shallow Foundations on Sand*; University of Illinois: Champaign, IL, USA, 1967.
42. Bowles, L.E. *Foundation Analysis and Design*; McGraw-Hill: New York, NY, USA, 1996; ISBN 0079122477.
43. Peck, R.B.; Hanson, W.E.; Thornburn, T.H. *Foundation Engineering*; Wiley: New York, NY, USA, 1974; Volume 10.
44. Terzaghi, K.; Peck, R.B. *Soil Mechanics in Engineering Practice*; John and Wiley and Sons: Hoboken, NJ, USA, 1967; ISBN 0471852732.
45. Agarwal, K.B.; Rana, M.K. Effect of ground water on settlement of footings in sand. In Proceedings of the 9th European Conference on Soil Mechanics and Foundation Engineering, Dublin, Ireland, 31 August–3 September 1987; Balkema: Rotterdam, The Netherlands, 1987; pp. 751–754.
46. NAVFAC. *Soil Mechanics Design Manual (NAVFAC DM 7.1)*; Naval Facilities Engineering Command: Alexandria, VA, USA, 1982.
47. Shahriar, M.A.N.; Sivakugan, N.; Das, B.M. Settlement correction for future water table rise in granular soils: A numerical modelling approach. *Int. J. Geotech. Eng.* **2013**, *7*, 214–217. [CrossRef]
48. Shahriar, M.A.; Sivakugan, N.; Das, B.M.; Urquhart, A.; Tapiolas, M. Water Table Correction Factors for Settlements of Shallow Foundations in Granular Soils. *Int. J. Geomech.* **2015**, *15*, 06014015. [CrossRef]
49. Ulrich, E.J.; Shukla, S.N.; Baker, C.N., Jr.; Ball, S.C.; Bowles, J.E.; Colaco, J.P.; Davisson, T.; Focht, J.A., Jr.; Gaynor, M.; Gnaedinger, J.P.; et al. Suggested analysis and design procedures for combined footings and mats. *J. Am. Concr. Inst.* **1988**, *86*, 304–324.

50. *DIN 4018 Berechnung der Sohldruckverteilung unter Flächengründungen Einschl*; Deutsche Normen: Berlin, Germany, 1974.
51. *IS 2950-1 Code of Practice for Design and Construction of Raft Foundations, Part 1: Design IS 1904:1986 Code for Practice, Design and Construction of Foundations in Soil*; Bureau of Indian Standards: New Delhi, India, 1981.
52. Varghese, P.C. *Foundation Engineering*; PHI Learning Pvt. Ltd.: Delhi, India, 2005; ISBN 8120326520.
53. Milovic, D. *Stresses and Displacements for Shallow Foundations*; Elsevier: Amsterdam, The Netherlands, 1992; ISBN 0-444-88349-5.
54. Brown, P.T. Numerical analyses of uniformly loaded circular rafts on deep elastic foundations. *Geotechnique* **1969**, *19*, 399–404. [CrossRef]
55. Duncan, J.M.; Buchignani, A.L. *An Engineering Manual for Settlement Studies*; Department of Civil Engineering, University of California: Berkley, CA, USA, 1976.
56. US Army Corps of Engineers. *Engineering and Design-Settlement Analysis, EM 1110-1904*; U.S. Army Corps of Engineers: Washington, DC, USA, 1990.
57. Mayne, P.W.; Kulhawy, F.H. K0-OCR relationships in soil. *J. Geotech. Eng.* **1982**, *108*, 851–872.
58. Mesri, G.; Ullrich, C.R.; Choi, Y.K. The rate of swelling of overconsolidated clays subjected to unloading. *Géotechnique* **1978**, *28*, 281–307. [CrossRef]
59. D'Appolonia, D.J.; Poulos, H.G.; Ladd, C.C. Initial settlement of of structures on clay. *J. Soil Mech. Found. Div. ASCE* **1971**, *97*, 1359–1377.
60. EN 1997-1. *Eurocode 7 Geotechnical Design—Part 1: General Rules*; European Committee for Standardization (CEN): Brussels, Belgium, 2004.
61. Pantelidis, L. The equivalent modulus of elasticity of layered soil mediums for designing shallow foundations with the Winkler spring hypothesis: A critical review. *Eng. Struct.* **2019**, *201*, 109452. [CrossRef]
62. Kany, M. *Berechnung von Flächengründungen*, 2nd ed.; Ernst u. Sohn: Berlin, Germany, 1974.
63. Grasshoff, H. Setzungsberechnungen starrer Fundamente mit Hilfe des kennzeichnenden Punktes. *Bauingenieur* **1955**, *30*, 53–54.
64. Kaniraj, S.R. *Design Aids in Soil Mechanics and Foundation Engineering*; Tata McGraw-Hill: New York, NY, USA, 1988; ISBN 0074517147.
65. Fellenius, B.H. *Basics of Foundation Design*; BC BiTech Publishers Limited: Richmond, BC, Canada, 2006.
66. Kempfert, H.-G.; Gebreselassie, B. *Excavations and Foundations in Soft Soils*; Springer Science & Business Media: Berlin, Germany, 2006; ISBN 3540328955.
67. Tatsuoka, F.; Teachavorasinskun, S.; Dong, J.; Kohata, Y.; Sato, T. Importance of Measuring Local Strains in Cyclic Triaxial Tests on Granular Materials. In *Dynamic Geotechnical Testing II*; ASTM International: West Conshohocken, PA, USA, 1994; pp. 288–302.
68. Jamiolkowski, M.; Lancellotta, R.; LoPresti, D.C.F. Remarks on the stiffness at small strains of six Italian clays. In Proceedings of the Int. Symp. on Pre-Failure Deformation of Geomaterials, Sapporo, Japan, 12–14 September 1994; Balkema: Rotterdam, The Netherlands, 1995; pp. 817–836.
69. Pincus, H.; Lo Presti, D.; Pallara, O.; Puci, I. A Modified Commercial Triaxial Testing System for Small Strain Measurements: Preliminary Results on Pisa Clay. *Geotech. Test. J.* **1995**, *18*, 15. [CrossRef]
70. Cho, H.; Kim, N.; Park, H.; Kim, D. Settlement Prediction of Footings Using VS. *Appl. Sci.* **2017**, *7*, 1105. [CrossRef]
71. Harr, M.E. *Foundations of Theoretical Soil Mechanics*; McGraw-Hill Inc.: New York, NY, USA, 1966; ISBN 10:0070267413.
72. Schleicher, F. Schleicher Zur Theorie der Baugrundes. *Bauingenieur* **1926**, *48*, 931–935.

Publisher's Note: MDPI stays neutral with regard to jurisdictional claims in published maps and institutional affiliations.

geosciences

MDPI

Article

Validation of Analytical Solutions for Predicting Drilled Pile Behaviour under Bi-Directional Static Load Tests

Runshen Wang [1,2,*], Dominic E. L. Ong [1,2], Jialin Zhou [3,4], Siwei Liu [5] and Erwin Oh [4]

1 School of Engineering and Built Environment, Griffith University, Nathan, QLD 4111, Australia; d.ong@griffith.edu.au
2 Cities Research Institute, Griffith University, Southport, QLD 4215, Australia
3 Blade Pile Group, 12 Junction Road, Burleigh Heads, QLD 4220, Australia; jack@pierandpile.com.au
4 School of Engineering and Built Environment, Griffith University, Gold Coast Campus, Southport, QLD 4222, Australia; y.oh@griffith.edu.au
5 Shanghai Municipal Engineering Design Institute (Group) Co., Ltd., 901 Zhongshan North 2nd Road, Yangpu District, Shanghai 200031, China; liusiwei@smedi.com
* Correspondence: jason.wang2@griffithuni.edu.au

Abstract: A bi-directional static load test (BDSLT) is one of the most effective methods for accurately estimating pile bearing capacity, in which the test pile is divided into two portions by activating the single-loading device welded along the pile shaft. BDSLT, thus, eliminates the safety concerns and space limitations imposed by the reaction system, as compared to conventional static load tests (kentledge). Based on this study's project requirements, two loading devices (supercells) were welded along the pile shaft to provide sufficient bearing capacity under the BDSLT, and an equivalent method was applied to interpret the measured load–settlement response. Since the sacrificial loading device welded along the pile shaft cannot be re-used, BDSLTs lead to increased construction costs; however, their capacity for rapid set-up in a limited space and reliable application for long piles are benefits that easily justify their use. Therefore, researchers must understand how BDSLTs perform, especially regarding double-loading devices. As informed by site investigation, this paper validates the conventional analytical solutions regarding test piles in preliminary designs, including Alpha and Beta and semi-empirical methods. In terms of a soil stiffness reduction model, modified closed-form analytical solutions based on Randolph's analytical method were applied to predict the load–settlement response.

Keywords: load–settlement response; bi-directional static load test; supercell; analytical solution; Randolph method; the modified closed-form analytical solutions

Citation: Wang, R.; Ong, D.E.L.; Zhou, J.; Liu, S.; Oh, E. Validation of Analytical Solutions for Predicting Drilled Pile Behaviour under Bi-Directional Static Load Tests. *Geosciences* **2022**, *12*, 284. https://doi.org/10.3390/geosciences12080284

Academic Editors: Hongyuan Liu and Jesus Martinez-Frias

Received: 14 May 2022
Accepted: 19 July 2022
Published: 22 July 2022

1. Introduction

Construction loads are transferred from superstructures to the ground through soil–pile interactions, which are influenced by the study of interface soil–pile and load–settlement interdependency that are subsequently typical of deep foundations, pavement–subgrade behaviour and tunnelling. The present study uses Randolph's (1979) method via analytical solutions to back-analyse the pile load–settlement behaviour to demonstrate this concept [1].

The proposed viaduct at Jining Avenue plays an essential role in the urban high-speed traffic networks located on the G105 and G327 national highways in Jining. It comprises eight parallel ramps, with a length of 7.25 km, widths between 25.5–33.5 m and an estimated road area of 240,400 m^2. The present study focuses on a section of the proposed project, from the Western Circumferential Expressway to Ningan Avenue (as shown in Figure 1).

Considering the complexity of the subsurface conditions and the safety of the adjacent structure, this viaduct project selected cast in-situ pile foundations for load transfer from the shallower, looser sand stratum to the deeper, denser sand stratum. Meanwhile, the dynamic lateral loads derived from the vehicles and the significant vertical dead loads required a large pile bearing capacity to support the superstructure. Therefore, this study included cast in-situ bored piles that were constructed with 45–65 m length and diameters of 1 m and 1.5 m. Consequently, polymer slurry was used to stabilise the bored cavity during the wet construction technique, thus affecting the soil properties and pile load–settlement response.

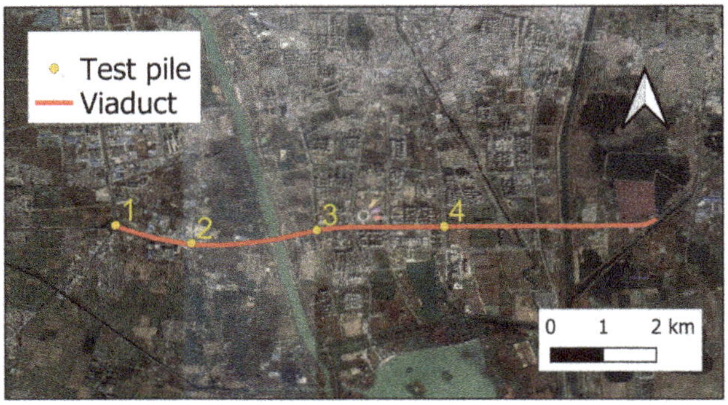

Figure 1. Test pile location (in scale).

As field test results are influenced by various factors, reliable field pile tests, such as conventional static load tests (compression-proof load test) and bi-directional static load tests (BDSLTs), are used to further verify pile performance. Conventional static load tests (SLTs) have been frequently used to validate the ultimate pile capacities of bored piles; however, using the kentledge system for conventional SLTs might entail safety risks, such as the kentledge platform toppling. In the present study, the conventional SLT's hydraulic system is required to provide a large reaction force. An extensive kentledge reaction system must thus be installed, such as the concrete blocks and supporting steel reaction beams. Due to space limitations at the project site, a limited workspace was available for erecting the bulky kentledge system. Therefore, the BDSLT was deemed a viable alternative to the kentledge load test.

Comprising Osterberg cells, or equivalent 'supercells', the BDSLT was originally developed by Osterberg [2], and it has been widely adopted as a proof test due to its proven accuracy and reliability [3–5]. It involves welding the load cell at a specific location on the pile reinforcement cage, thus providing steady vertical load increments in a bi-directional fashion to the in-situ pile. Such bi-directional load application effectively 'splits' the pile into an upper and lower portion with respect to the supercell location [6]. The embedded cell is then loaded to generate the load–displacement relationships, which are then used to interpret the mobilisation of pile resistance and soil–structure behaviour.

Setting up a BDSLT requires a small footprint, thus reducing installation difficulty and guaranteeing construction safety. It can also be used to measure end bearing capacity or shaft capacity independently and accurately, thereby contributing to a clear understanding of pile load–settlement responses so that engineers can conduct optimum pile designs; however, as the supercells cannot be retrieved after testing, due to being sacrificial, the BDSLT might be relatively costly and thus not ideal for small-scale projects [7]. To solve this issue and improve understanding of the BDSLT, Baca (2020, 2021) conducted several studies concerning numerical simulation and laboratory scale tests in medium-density sand [6,8]. Baca (2017) also validated the practicability of numerical simulation and laboratory scale tests with field tests [9].

Typically, the BDSLT only uses a single supercell installed in a single pile; however, in cases in which the piles are extensively long, a single supercell might not be able to fully mobilise pile end bearing and shaft capacities. Therefore, this study focused on two supercells to provide sufficient reaction forces to mobilise the ultimate capacity of the 60 m-long bored pile being proof tested. In the reaction system's preliminary design, the supercell location was determined by the estimated improved bearing capacity via post-grouting technology.

This article used analytical solutions based on reliable soil characterisation to present a case study on the design methodology used to interpret the BDSLT field test results. Four BDSLT tests were conducted with different pile lengths and diameters. Further, conventional analytical solutions, such as the Alpha and Beta methods, were assessed in comparison to the results derived using the enhanced, modified closed-form analytical solutions [7] to more accurately determine the load–settlement responses derived from the BDSLTs.

2. Geotechnical Conditions

As aligned with the local Chinese standard (Code for Geotechnical Investigation) and site investigations, the soil layers primarily comprised (1) silty sand that was well graded, with standard penetration test (SPT) N values ranging from 39 to 143 (increased with larger overburden pressure), being selected as the bearing stratum around the pile end; (2) silt that is dense; and (3) plastic silty clay. Based on laboratory and in-situ test results, the detailed soil parameters adjacent to the tested piles are outlined in Table 1. The friction angle and cohesion strength were determined using a direct shear test with a high shearing rate, and the cone tip and shaft friction resistance were measured using a cone penetration test (see Figure 2 for test results).

Table 1. Parameters of three typical soils.

Soil Type	Unit Weight	Friction Angle	Cohesive Strength	Cone Tip Resistance	Shaft Friction Resistance	Unconfined Compressive Strength	Compression Modulus	Poisson's Ratio
	γ	ϕ	c	q_c	f_s	q_u	E	ν
Unit	kN/m^3	degrees	kPa	MPa	kPa	kPa	MPa	
Silty sand	19.0	28	0	23.8	386.3	N/A	26.55	0.3
Silt	19.6	27.53	8.65	8.0	160.1	39	6.38	0.3
Silty clay	19.3	15.20	43.80	8.4	261.0	73	6.91	0.3

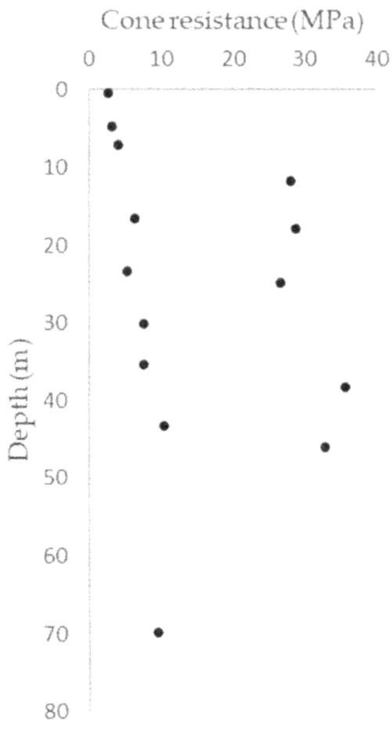

(a)

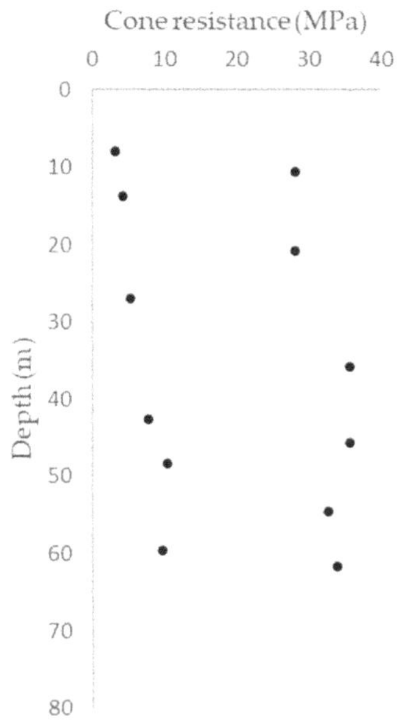

(b)

Figure 2. *Cont.*

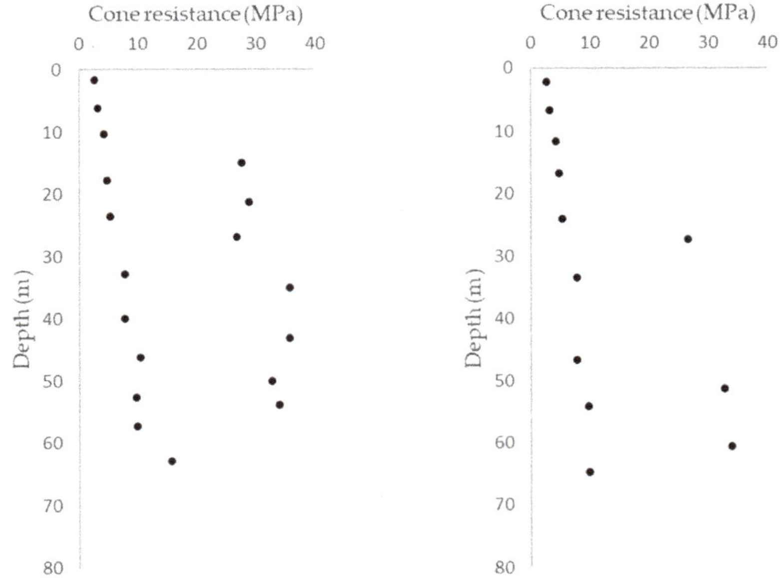

(c) (d)

Figure 2. Cone penetration test results for (**a**) Test Pile 1, (**b**) Test Pile 2, (**c**) Test Pile 3 and (**d**) Test Pile 4.

Further, Figure 3 illustrates the subsurface conditions of four test piles based on boreholes BH1, BH2, BH3 and BH4. Meanwhile, double supercells (upper and lower supercells) were welded at the specific locations along the pile shaft and separated the test pile into Parts A, B and C.

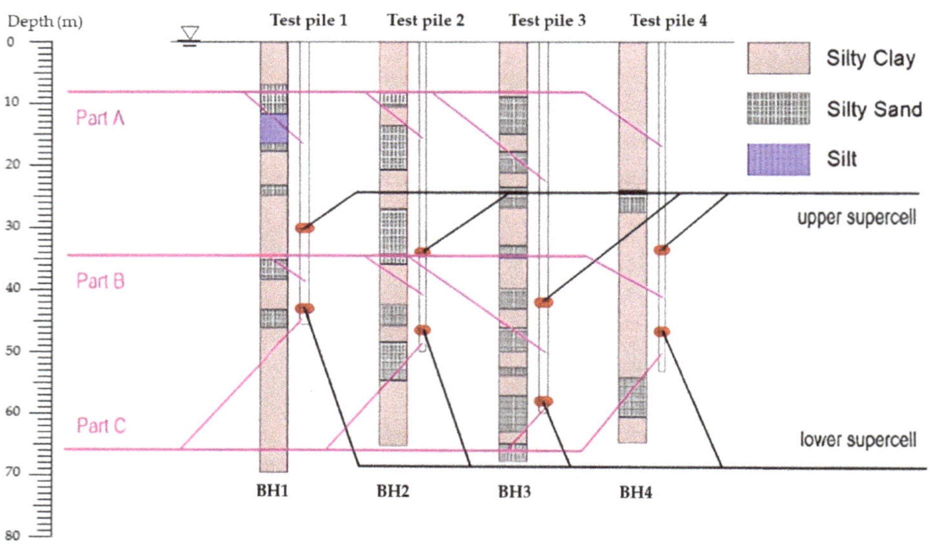

Figure 3. Subsurface conditions along test piles.

3. Bi-Directional Static Load Test

3.1. Test Pile Description

Test Piles 1 and 3 were conducted with a 1.5-m diameter and 45.6 m and 60 m lengths, respectively. Test Piles 2 and 4 were conducted with a 1-m diameter and 50 m and 53 m lengths, respectively. To determine the loading device's location, shaft and end bearing capacities (see Table 2) were estimated with a semi-empirical method in terms of Local Code JGJ 94 [10].

Table 2. Dimensions and estimated bearing capacity of test piles.

Test Pile Code	Related Borehole	Pile Length (m)	Pile Diameter (m)	End Bearing Capacity (kN)	Shaft Bearing Capacity (kN)
1	BH1	45.6	1.5	3004.15	17,301.63
2	BH2	50	1.0	1335.18	12,543.75
3	BH3	60	1.5	3004.15	21,618.18
4	BH4	53	1.0	1335.18	15,580.41

3.2. Loading Device Selection

Ougan Technology manufactures three types of loading devices: donut-shaped, solid and multiple supercells. As shown in Figure 4a, the donut-shaped supercell is traditionally applicable to small piles; in Figure 4b, the solid supercell is mostly used at the pile tip, in which the aim is to determine the end bearing; and in Figure 4c, multiple supercells are selected in the research as its large bearing capacity.

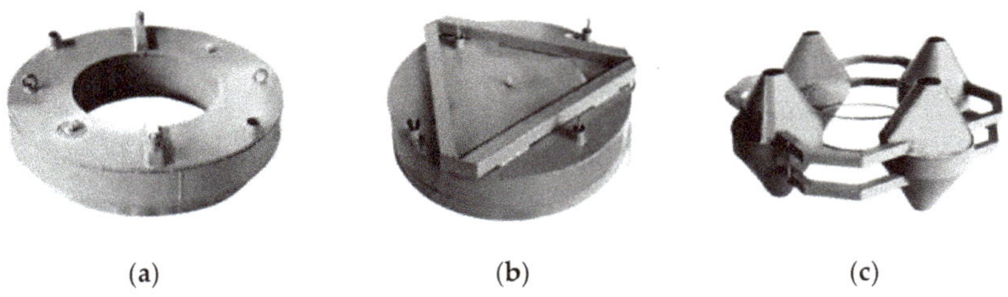

Figure 4. Supercell products: (**a**) donut-shaped supercell, (**b**) solid supercell and (**c**) multiple supercells.

As the single-loading device could not provide enough testing capacity for the BDSLT, this case study included two supercells that were welded onto the specific locations to separate the test pile into Parts A, B and C. Test procedure indicated that the estimated bearing capacity of Parts A, B and C should satisfy certain relationships to establish a stable reaction system for the BDSLT.

During the lower supercell's loading, the ultimate bearing capacity of Part C was conducted, and Parts A and B worked as a reaction system for Part C. Therefore, the estimated total bearing capacity of Parts A and B should be larger than that of Part C ($Q_A + Q_B > Q_C$). To generate enough reaction force on Part B, the estimated bearing capacity of Part A should be larger than that of Part B ($Q_A > Q_B$), and to establish a proper reaction system for Part A, the grouting technology was applied to the pile tip. Base grouting significantly reinforced Part C and fully mobilised its bearing capacity. Therefore, the estimated total bearing capacity of Parts B and C should now be larger than that of post-grouted Part A ($Q_B + Q_C + X > Q_A$, in which X represents the improvement in pile base capacities being grouted (the end bearing capacities with full mobilisation). The length and estimated bearing capacity of each part were determined, as shown in Table 3.

Table 3. Test pile length and estimated bearing capacity.

Test Pile Code	Part A		Part B		Part C	
	Length (m)	Bearing Capacity (kN)	Length (m)	Bearing Capacity (kN)	Length (m)	Bearing Capacity (kN)
1	30.1	11,851.09	13	5157.24	2.5	3865.20
2	34.0	8563.23	12.5	3090.32	3.5	2225.38
3	42.0	15,735.61	16	5617.92	2.0	3268.80
4	33.5	9090.51	13	3593.98	6.5	2895.92

3.3. Displacement Measurement Device

After constructing the test piles, the automated data acquisition system and multiple supercells were connected via a 32 mm telltale casing and 16 mm rod extensometer (see Figure 5). Two sets of displacement transducers, one for the top and one for the bottom, were installed on the supercell for settlement measurement to record the upward and downward displacement, respectively. Additionally, four Linear Variable Differential Transformers (LVDTs) were mounted above the pile head on the reference beam to determine the pile head's upside displacement (see Figure 5), and a high-pressure pump was installed near the ground surface to apply sufficient pressure to the supercell via incompressible fluid passing through the telltale casing. An electronic manometer further measured the pressure inside the high-pressure pump, contributing to the supercell's load measurement; this equipment is further detailed below:

- Automatic data acquisition system—The digital screen and data logger controlled and collected the field test data via wired transmitters and receivers.
- Telltale casing and rod extensometer—The telltale casing and rod extensometer diameters were 32 mm and 18 mm, respectively, and were embedded in the test piles.
- LVDT—The measurement error had a maximum of 0.1% FS, and the resolution had a minimum of 0.01 mm. The upward and downward displacements at the supercell were measured by a group of displacement sensors, in which each group of displacement sensors comprised more than two LVDTs and was symmetrically arranged.
- Multiple supercells—The loading device should be selected correctly according to pile type, testing requirements and foundation pile construction technology.
- High-pressure pump—The measurement range was 0–60 MPa, and the precision was 0.4 MPa.

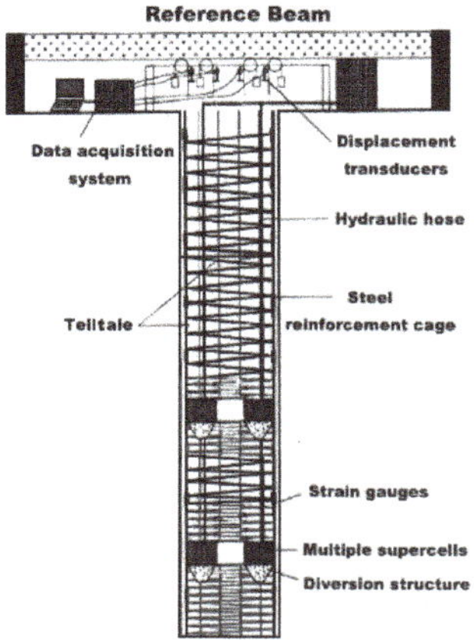

(a)

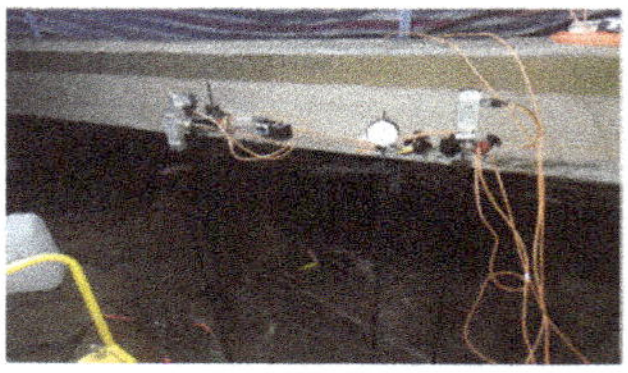

(b)

Figure 5. (**a**) Automatic data acquisition system (**b**) displacement transducers on the reference beam.

3.4. Test Setup

As described in the methodology, the cone-shaped multiple supercells (see Figure 6a) were assembled on the ground surface. High-strength concrete was cast into the supercells and tamped with a concrete vibrating spear. Supercells and 1.2 m-long funnels were then welded onto the reinforced steel cage, which was manufactured with a ϕ20 steel rod with a 10 cm space. Further, the extensometer rod and hydraulic hose were tied to the steel cage, and the reinforced steel cage was then lifted into the borehole, whose eccentricity was less than five degrees (see Figure 6b). Lastly, the concrete was cast into the borehole, and the reference beam was constructed above the pile and tied with four LVDTs (see Figure 5).

(a) (b)

Figure 6. Installation of (**a**) supercell and (**b**) reinforcement cage.

3.5. Test Procedure

After 28 days, the BDSLT with double supercells was processed by independently loading and unloading the supercells; this BDSLT procedure was divided into five loading stages (see Figure 7 and Table 4): Stage 1 involved loading the lower supercell to obtain Part C's bearing capacity, and the gap between Part B and Part C being generated; Stage 2 involved loading the upper supercell to measure Part B's bearing capacity; Stage 3 involved loading the lower supercell at least 28 days after grouting at the pile end to measure Part C's bearing capacity after grouting; Stage 4 included the upper supercell being loaded to remove the gap between Parts B and C; and Stage 5 involved the upper supercell's continued loading to measure the bearing capacity of Part A after grouting.

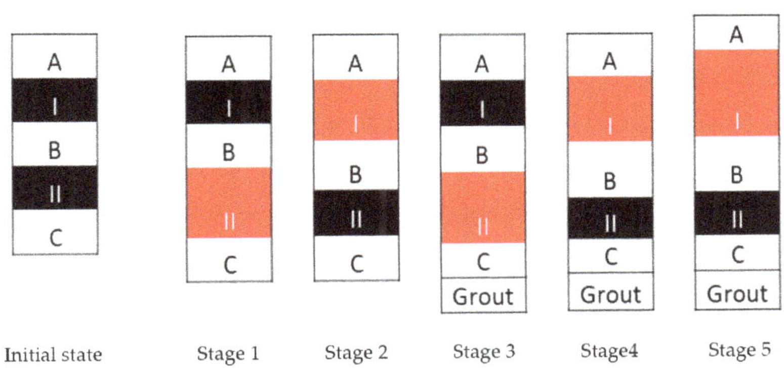

Figure 7. Locations of double supercells installed, separating the piles in Parts A, B and C (supercells are marked as red).

Table 4. Procedure and purposes of the BDSLTs.

	Loading State		Purposes
	Upper Supercell	Lower Supercell	
Stage 1	Maintain	Loading	Bearing capacity of Part C (before grouting; possibly with 'soft toe' issues)
Stage 2	Loading	Unloading	Bearing capacity of Part B
Stage 3	Maintain	Loading	Bearing capacity of Part C (after grouting)
Stage 4	Loading	Unloading	Removing the gap between Parts B and C
Stage 5	Loading	Maintain	Bearing capacity of Part A

Note: 'Maintain' signifies 'no relative displacement between two parts'; 'unloading' signifies 'free axial displacement between two parts'.

3.6. Pile Interpretation

The BDSLT, with a single-loading device, simultaneously provided two load–settlement curves for the pile's upper and lower segments (see Figure 8). As different loads transform the mechanism of pile behaviour into upward and downward displacements, interpreting BDSLT is different from interpreting conventional SLT. Previous research has found a correction factor for transferring the upward friction to the downward friction [11]. Further, the pile's elastic shortening (ΔS) and self-weight (G_p) also affected interpreted results [11]. Considering these factors, the equivalent method provided by Local Code JT/T 738–2009 [12] involved converting two load–settlement curves into a single curve (see Figure 8 and Equations (1) and (2)).

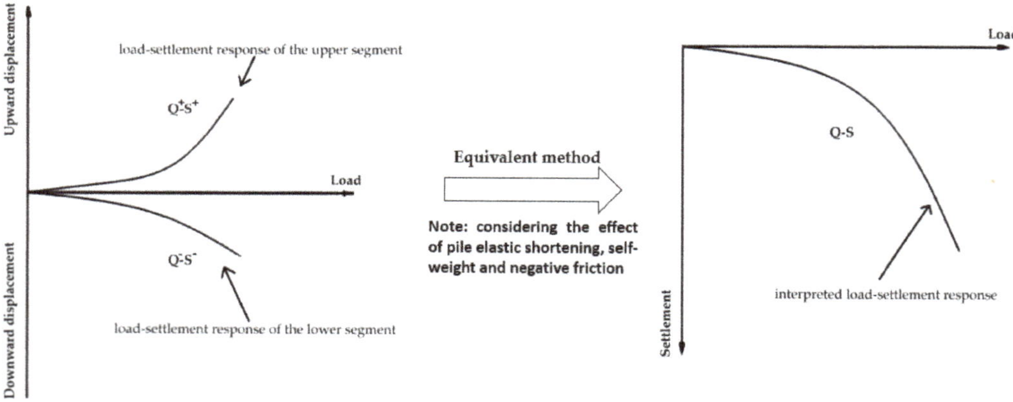

Figure 8. Q-s curve interpreted with equivalent method.

$$Q = Q_s + Q_e = k^+(Q^+ - G_p) + k^- Q^- \tag{1}$$

$$S = S^- + \Delta S = S^- + \Delta S_1 + \Delta S_2 = S^- + \frac{Q^- L}{E_p A_p} + \frac{(Q^+ - G_p)L}{2E_p A_p \gamma_m} \tag{2}$$

where

Q_s is the ultimate shaft bearing capacity;

Q_e is the ultimate end bearing capacity;

G_p *or* W is the pile's self-weight above the supercell;

Q^+ is the upward load on supercell (measured by the pressure gauge);

Q^- is the downward load on supercell;

k^+ is the pile's equivalent transform parameter above the supercell;

k^- is the pile's equivalent transform parameter below the supercell (which ignores the pile's load and displacement above supercell, $k^- = 1$);

S is the equivalent displacement on the pile head corresponding to Q;

S^- is the supercell's displacement;

ΔS is the pile's elastic compressive deformation;

ΔS_1 is the pile's elastic compressive deformation below the supercell due to the supercell's downward force;

ΔS_2 is the pile's elastic compressive deformation above the supercell due to the supercell's upward force;

L is the pile's length above the supercell;

E_p is the pile's elastic modulus;

A_p is the cross-section area of the pile; and

γ_m is the soil parameter ($\gamma_m = 0.8$ for clay and silt; $\gamma_m = 0.7$ for sand; $\gamma_m = 1.0$ for rock).

4. Analytical Solutions in Preliminary Design

Site investigation revealed that several analytical solutions (the semi-empirical method and the Alpha and Beta methods) were used to estimate total bearing capacity in the preliminary design, which increases construction safety. The Alpha and Beta methods were separately applied for pile shaft design in cohesive and cohesionless soils, respectively; however, unlike the Alpha and Beta methods, the semi-empirical method considered both site investigations and empirical data.

4.1. Semi-Empirical Method

Soil characteristics, pile dimensions and installation methods all influence soil–pile interface behaviours. Based on the site investigation and empirical database, the semi-empirical method was presented for the test piles' preliminary design in accordance with local Chinese standards (Technical Code for Building Pile foundation) [10]. As defined below, the estimated total bearing capacity (Q_{uk}), which comprised estimated shaft and end bearing capacity, can be computed by multiplying the ultimate resistance characteristic value (q_{sik} and q_{pk}) by the bearing area. Additionally, local Chinese standards (Technical Code for Building Pile Foundation) [10] specified the range of q_{sik} and q_{pk}, which were used in this research.

$$Q_{uk} = Q_{sk} + Q_{pk} = u\sum q_{sik} l_i + q_{pk} A_p \tag{3}$$

where

Q_{sk} is the estimated shaft bearing capacity;

Q_{pk} is the estimated end bearing capacity;

u is the perimeter of the cross-section area along the pile shaft;

l_i is the thickness of ith layer soil;

A_p is the cross-section area of the pile end;

q_{sik} is the ultimate shaft resistance characteristic value of ith layer soil; and

q_{pk} is the ultimate bottom resistance characteristic value of ith layer soil.

4.2. Alpha and Beta Methods

The Alpha (α) method appropriately estimated the shaft resistance of cohesive soil by defining the adhesion factor (α), as shown below [13]. Many existing studies have focused on the relationship between yield stress ratio and adhesion factor (α) for both displacement and non-displacement piles. Regarding non-displacement piles, Knappett and Craig (2012) [13] proposed a piecewise function for adhesion factors based on previous research [14–18]. Additionally, the Plastic Index determined the undrained shear strength over the pile's length [19].

$$Q_{sk} = \alpha \overline{s_u} A_s \tag{4}$$

$$s_u/\sigma'_{v0} = 0.11 + 0.0037PI \tag{5}$$

$$\alpha = 1 \qquad \text{for } s_u \le 30 \text{ kPa} \tag{6}$$

$$\alpha = 1.16 - \left(\frac{s_u}{185}\right) \qquad \text{for } 30 \text{ kPa} \le s_u \le 150 \text{ kPa} \tag{7}$$

$$\alpha = 0.35 \qquad \text{for } s_u \geq 150 \text{ kPa} \tag{8}$$

where

α represents the adhesion factors;

s_u is the undrained shear strength over the pile's length;

$\overline{s_u}$ is the average undrained shear strength over the pile's length;

A_s is the cross-section area of the pile shaft; and

σ'_{v0} *and* q' are the overburden pressure.

The Beta (β) method appropriately estimated the shaft resistance of cohesionless soil by the effective overburden pressure and factor β, which is determined by the coefficient of earth pressure and the interface friction angle. The testing results of the surface profile gauge revealed that pile surface roughness is much higher than particle size. Therefore, the internal friction angle obtained from the shear test can be considered the interface friction angle [20]. Equations (9) and (10) express the Beta method:

$$\tau_{int} = \beta\sigma'_{v0} = K \tan\delta'\sigma'_{v0} \tag{9}$$

$$K \approx K_0 = 1 - \sin\delta' \tag{10}$$

where

K is the coefficient of earth pressure;

K_0 is the coefficient of earth pressure at rest; and

δ' is the interface friction angle.

The end resistance of deep foundation is designed and regarded as a shallow foundation that considers the effect of soil property and foundation dimensions. Further, end bearing capacity (Q_{ult}) was obtained by multiplying the end resistance and cross-section area of the pile end. The formula for end resistance (q_{ult}) and end bearing capacity (Q_{ult}) is expressed with Equations (11) and (12):

$$q_{ult} = 1.3cN_c^* + q'N_q^* + 0.3\gamma BN_\gamma^* \text{ (For circular footing)} \tag{11}$$

$$Q_{ult} = q_{ult}A_{base} \tag{12}$$

where

N_c^*, N_q^* *and* N_γ^* are the Terzaghi's bearing capacity factors;

c is the cohesion strength of soil;

γ is the unit weight of soil;

B is the foundation's width; and

A_{base} is the cross-section area of the pile end.

5. Modified Closed-Form Analytical Solutions

The initial shear modulus (G_o or G_{max}), also known as small-strain shear modulus or low-amplitude shear modulus, is a fundamental characteristic of soil properties that can be measured using laboratory and field tests [21,22]; however, the plasticity index's high value yields a significant softening of the soil, which subsequently yields a decrease in shear modulus. Compared with the operational shear modulus (G or G_L), the initial shear modulus is limited in terms of predicting soil behaviour. Previous research has established a relationship between operational shear modulus (G) and initial shear modulus (G_{max}) to provide an analytical solution [23–26]. Randolph and Wroth (1978) [27] computed an elastic continuum solution in terms of operational shear modulus, in which they estimated the pile settlement in response to the anticipated loading. The basic closed-form solution developed by Randolph and Wroth [1,27] is expressed in Equations (13)–(15) below:

$$w_{(z)} = \frac{Q_{(z)}\left\{1 + \frac{4\eta \tanh[\mu(L-z)](L-z)}{\pi\lambda(1-v_s)\xi[\mu(L-z)]r_0}\right\}}{G_L r_0\left\{\frac{4\eta}{(1-v_s)\xi} + \frac{2\pi\rho_E \tanh[\mu(L-z)](L-z)}{\xi[\mu(L-z)]r_0}\right\}} \tag{13}$$

$$\frac{Q_{(z)}}{Q_b} = \frac{\left\{ \frac{4\eta}{(1-v_s)\xi} + \frac{2\pi\rho_E \tanh[\mu(L-z)](L-z)}{\xi[\mu(L-z)]r_0} \right\}}{\left\{ 1 + \frac{4\eta \tan h[\mu(L-z)](L-z)}{\pi\lambda(1-v_s)\xi[\mu(L-z)]r_0} \right\}} \quad (14)$$

$$w_{(z)} = w_b \cosh[\mu(L-z)] \quad (15)$$

Niazi and Mayne (2014) extended this theory in reference to BDSLT (see Table 5) and then applied the initial shear modulus to the soil stiffness reduction model in terms of the seismic cone penetration test (SCPTu) and CPTu-pile capacity correlations [7]. In the present study, the operational shear modulus (G or G_L) was computed with the modified closed-form analytical solutions, as informed by the load–settlement responses of Test Piles 1, 2 and 3. Due to the limitations of site investigation, the initial shear modulus (G_o) was determined using CPT cone resistance (q_c) (see Equation (16)) [21].

Table 5. Modified closed-form analytical solutions for two cases of O-cell pile loading.

	Related Formula
Total pile load	$Q_t = Q_{t1} + Q_{t2} = \left(Q_{s1} + W_{buoyant} + Q_{s2}\right) + (Q_{s3} + Q_b) = [\Sigma(f_{p1}\pi d_{s1} L_1) + \gamma_{buoyant\ (upper\ pile\ shaft)} \pi d_{s1}^2 L_1/4 + \Sigma(f_{p2}\pi d_{s2} L_2)] + [\Sigma(f_{p3}\pi d_{s3} L_3) + q_b \pi d_b^2/4]$
Upper shaft response	$w_1 = \frac{(Q_{s1} + W_{buoyant})\zeta_1 \mu L_1}{2\pi G_{L1} \rho_{E1} \tanh(\mu L_1) L_1}$
Middle shaft response	$w_2 = \frac{Q_{s2}\zeta_2 \mu L_2}{2\pi G_{L2} \rho_{E2} \tanh(\mu L_2) L_2}$
Lower shaft response	$w_3 = \frac{(Q_{s3} + Q_b)\left[1 + \frac{4\eta_3 \tanh(\mu L_3) L_3}{\pi\lambda_3 (1-v_{s3})\xi_3 \mu L_3 r_{o3}}\right]}{G_{L3} r_{o3} \left[\frac{4\eta_3}{(1-v_{s3})\xi_3} + \frac{2\pi\rho_{E3} \tanh(\mu L_3) L_3}{\zeta_3 \mu L_3 r_{o3}}\right]}$

where

Q_t is the applied compressive load at the pile head;
Q_s is the portion of Q_t, resisted by pile shaft;
$W_{buoyant}$ is the upper shaft segment's buoyant weight;
Q_b is the portion of Q_t, resisted by pile end;
f_p is the unit shaft resistance;
d_s is the pile shaft diameter;
L is the segment's length;
q_b is the unit base resistance;
d_b is the pile base diameter;
w is the axial segment displacement;
ζ is the average radius of influence in the surrounding soil mass affected by shearing stresses around the pile;
μL is the pile compressibility for the segment;
G_L is the operative shear modulus;
ρ_E is the modulus variation factor;
η is the factor for underreamed piles that take greater loads at pile base;
v_s is Poisson's ratio of soil; and
r_o is the radius of pile shaft.

$$G_0 = 62{q_c}^{1.15} \quad (16)$$

The ratio for the initial shear modulus (G_o or G_{max}) and operational shear modulus (G or G_L) is explained with the nonlinear stiffness reduction model, including Method 1 (Equation (17)) and Method 2 (Equation (18)). Method 1 uses several parameters to present the stiffness ratio (G_L/G_{Lmax}), such as by analysing the sensitivity of pseudo-strain (w/d), ratio of length and diameter (L/d), ratio between lengths of pile shaft segments (L-ratio) and Poisson ratio of soil (ν_s), which is supported by a robust database. Due to the limited data available for Method 1, it is presented here for completeness and is thus not used.

Method 2 establishes the nonlinear relationship between the pseudo-strain and stiffness ratio. Compared with Method 1, Method 2 solely considers the pseudo-strain sensitivity. It determines a series of parameters (A, B, C, n), as shown in Equations (19)–(21), that predicts Test Pile 4′s load–settlement response.

$$\text{Method 1}: \frac{G_L}{G_{L\,max}} = j\left[\frac{w(\%)}{d}\right]^k \left(\frac{L}{d}\right)^l (L - ratio)^m (\nu_s)^n (\xi)^o (\rho_E)^p \leq \left(\frac{G_L}{G_{L\,max}}\right)_{lim} \quad (17)$$

$$\text{Method 2}: \frac{G_L}{G_{L\,max}} = \frac{1}{A\left[\frac{w(\%)}{d}\right]^n + B\left[\frac{w(\%)}{d}\right] + C} \leq \left(\frac{G_L}{G_{L\,max}}\right)_{lim} \quad (18)$$

6. Results and Discussion

6.1. Pile Interpretation

Considering the pile's self-weight and negative friction, the load–settlement response of the upper part (Part A) is interpreted. It displays a linear relationship in load–settlement response, and the settlement is considerably lower than the 40 mm criterion. Therefore, according to conventional requirements, the bearing resistance does not reach the failure load [28]. Regarding the middle part (Part B), most of the interpreted load–settlement responses displayed a significant reduction (this stage's settlement change is over five times the former stage) at the failure load (see Figure 9b). Since the middle parts of Test Piles 2 and 4 had similar dimensions and overburden pressures, the increment of sand layer thickness significantly improved bearing capacity.

Overall, this case study indicates that pile settlement and diameter ratios of about 1% and 4% are expected to fully mobilise the pile shaft and base resistances, respectively. These observations align with those made by Al-Atroush et al. (2020) and Ong (2005) [29,30] and further justify the acceptance of the criteria adopted, in which the maximum pile settlements are limited to less than 40 mm; this effectively translates to 2.6% and 4% of the pile with diameters of 1.5 m (Piles 1 and 3) and 1 m (Piles 2 and 4), respectively.

The site investigation results reveal that the soft underlying stratum exists in most test piles, which might not satisfy the BDSLT reaction system's requirements (see Section 3.2). Previous research has found that bored pile post-grouting technology significantly improves the end bearing capacity and reduces the settlement [31,32]. Therefore, grouting technology is required to fully mobilise the end bearing capacity of the lower part (Part C). The testing result indicates great improvements after grouting at the pile end (see Table 6 and Figure 10). Compared with the clay-bearing stratum, the sand-bearing stratum with a higher level of hydraulic conductivity provides a better grouting improvement. Further, the higher overburden pressure yields a more considerable grouting improvement (see Test Piles 1 and 3 in Table 5).

Table 6. Interpreted end bearing capacities of test piles.

Pile No.	Pile Length (m)	Bearing Stratum of Part C	End Bearing Capacity (kN)		Improved Rate
			Before Grouting	After Grouting	
1	45.6	Sand	2959	8100	173.7%
2	50	Sand, clay	1600	2636	64.8%
3	60	Sand	1857	7500	303.9%
4	53	Clay	2741	4800	75.1%

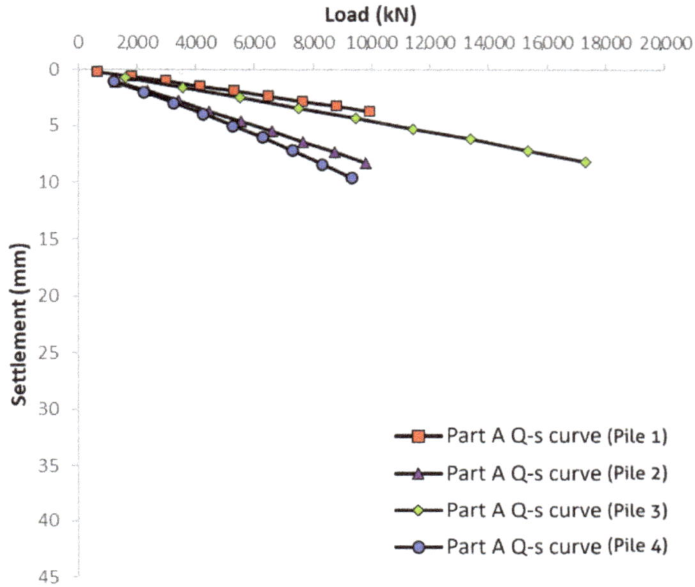

(**a**)

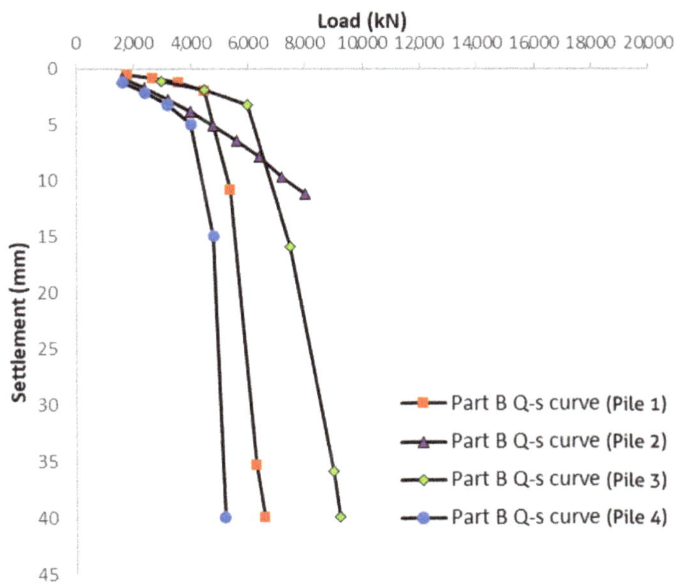

(**b**)

Figure 9. Interpreted load–settlement response of (**a**) upper part (Part A) (**b**) middle part (Part B).

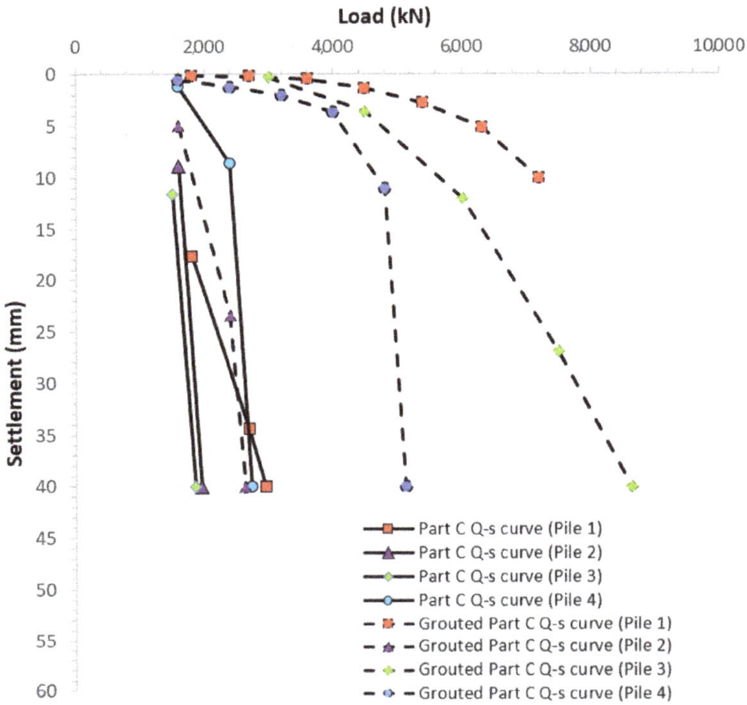

Figure 10. Interpreted load–settlement response of lower part (Part C).

6.2. Conventional Analytical Solutions

Two kinds of estimated bearing capacities were obtained in the preliminary design through the semi-empirical method and the Alpha and Beta methods (see Table 7). The predicted bearing capacity obtained from the latter two is not precise enough without considering the effect of polymer suspension. Polymer mixed with water stabilises the borehole during pile construction and affects the full mobilisation of pile resistance. Interface characteristics along the pile shaft are thus significantly influenced. Conversely, compared with the interpreted ultimate bearing capacity, the semi-empirical method provided an estimated shaft bearing capacity with higher fitness, as it considered pile surface roughness and soil stiffness; however, the relationship between shaft resistance and interface parameters was ambiguous. Therefore, establishing an algorithm is crucial, as is quantitatively analysing pile behaviour in the preliminary design.

Table 7. Estimated and interpreted results with test piles.

No.	Type of Theoretical Determination	Qs (Part A) kN	Qs (Part B) kN	Qs + Qb (Part C) kN
1	Semi-empirical method	11,283.34	5157.24	3865.20
	Alpha method and Beta method	2988.03	2959.35	9849.99
	Supercell test result	>9965	4500	2959
2	Semi-empirical method	8563.23	3090.32	2225.38
	Alpha method and Beta method	3975.08	3199.18	2522.05
	Supercell test result	>9798	>8000	1956

Table 7. *Cont.*

No.	Type of Theoretical Determination	Qs (Part A) kN	Qs (Part B) kN	Qs + Qb (Part C) kN
3	Semi-empirical method	15,735.61	5617.92	3268.80
	Alpha method and Beta method	8332.52	6600.02	15,715.64
	Supercell test result	>17316	6000	1857
4	Semi-empirical method	9090.51	3593.98	4231.10
	Alpha method and Beta method	2985.57	2135.71	3826.58
	Randolph method	**>8665**	**4196**	**2400**
	Supercell test result	>9329	4000	2400

6.3. Analytical Solution of the Randolph Method

According to the load–settlement responses of the test piles and site investigation outcomes, the operational shear modulus of test piles was determined using modified closed-form analytical solutions. The modulus reduction factor (G_L/G_{Lmax}) versus percent pseudo-strain (w/d [%]) demonstrated a nonlinear relationship (see Figure 11). The modulus reduction factors reduced in proportion to the increase in segment displacement, and a higher sand stratum thickness yielded a greater modulus reduction factor for upper pile segments (see Figures 3 and 11a). In contrast, the expected behaviours of the lower and middle segments provided outcomes with a high level of fitness, which might be due to the beneficial use of polymer slurry.

Further, the factors (A, B, C, n) of the nonlinear stiffness reduction model were deducted with curve fitting and then back-analysed within Test Piles 1, 2 and 3. By applying these factors to Equation (18), Equations (19)–(21) were obtained:

$$Upper\ part: \frac{G_L}{G_{L\ max}} = \frac{1}{17.25\left[\frac{w(\%)}{d}\right]^{0.969} + 1.0\left[\frac{w(\%)}{d}\right] + 0.8} \quad (19)$$

$$Middle\ part: \frac{G_L}{G_{L\ max}} = \frac{1}{4.32\left[\frac{w(\%)}{d}\right]^{0.805} + 9.86\left[\frac{w(\%)}{d}\right] + 3.26} \quad (20)$$

$$Lower\ part: \frac{G_L}{G_{L\ max}} = \frac{1}{135.1\left[\frac{w(\%)}{d}\right]^{0.859} + 1.16\left[\frac{w(\%)}{d}\right] + 0.81} \quad (21)$$

The operational shear modulus of Test Pile 4 was determined according to the curve-fitting results of the nonlinear stiffness reduction model (Equations (19)–(21)), and then the load–settlement response (see Figure 12) was predicted as informed by the modified closed-form analytical solutions. The result indicates a high fitness between the interpreted and predicted load–settlement response. For the middle and lower parts, ultimate bearing capacity was exactly predicted, and analytical solutions led to overestimated results with large settlement (>60 mm); however, due to the dispersed data of the test pile's upper part (see Figure 11a), the predicted load–settlement response might not be reliable; this indicates that further investigation is required.

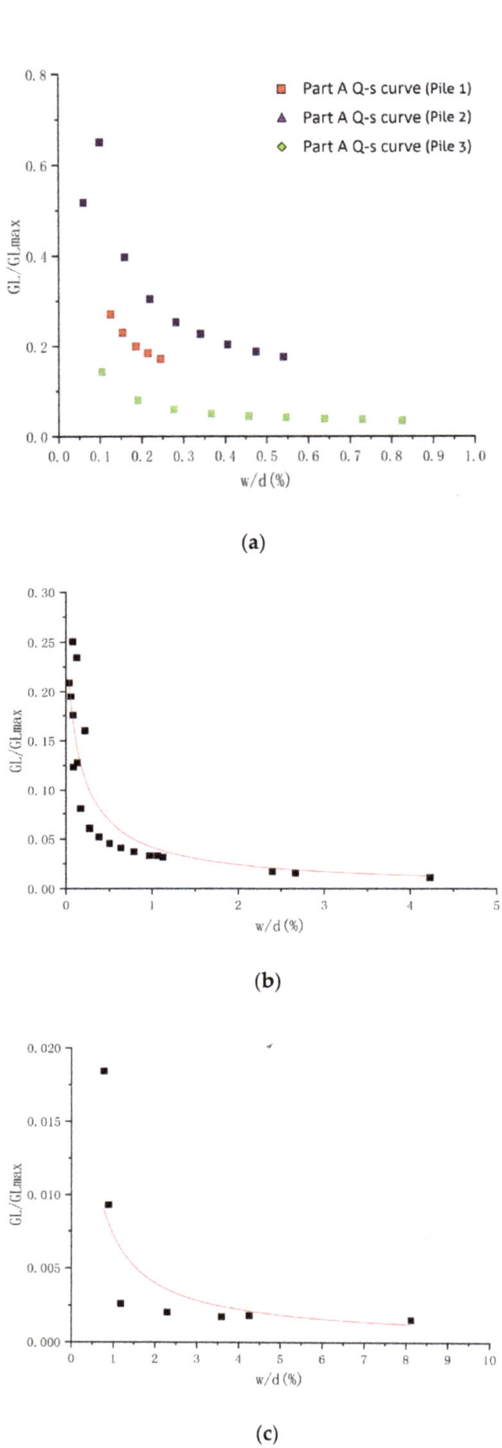

Figure 11. Modulus reduction factor (G_L/G_{Lmax}) versus per cent pseudo-strain (w/d [%]) for (**a**) upper part, (**b**) middle part and (**c**) lower part.

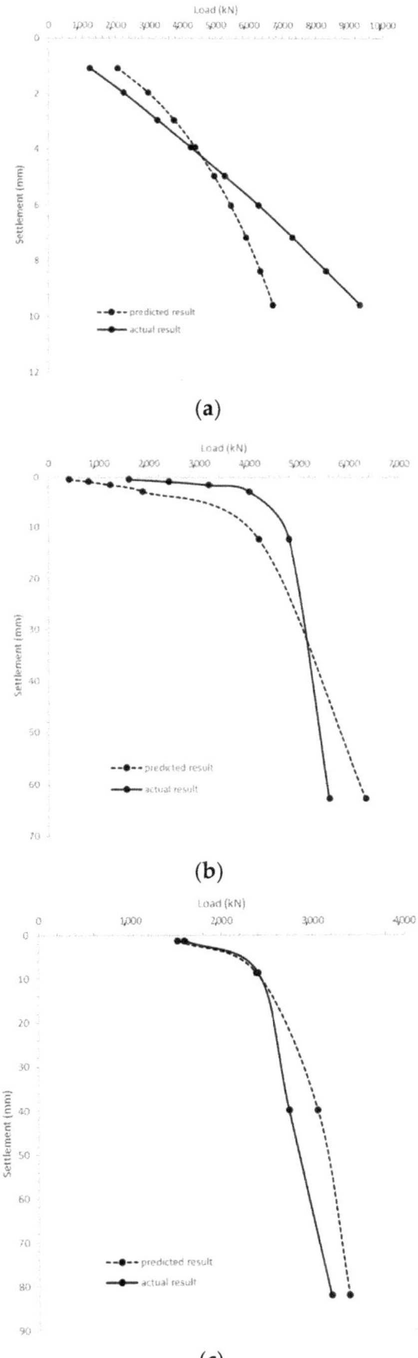

Figure 12. Predicted load–settlement responses based on modified closed-form analytical solutions for (**a**) upper part, (**b**) middle part and (**c**) lower part.

7. Conclusions and Recommendations

This article explored long-drilled pile behaviour in response to BDSLTs, and it validates theoretical research based on pile interpretation and analytical solutions. With two loading supercells welded to the single test pile, the BDSLT is thus more beneficial than just a conventional single-loading device in terms of providing more reliable test data. The load–settlement response obtained was interpreted with consideration of the negative friction effect and the pile's elastic deformation and self-weight, which contributes to greater accuracy. The interpreted results indicate that the sand stratum provides more resistance than the clay stratum, and that grouting improvement is significant in the sand stratum. Therefore, construction project site selection conducted in sand stratum will contribute to higher pile resistance and better grouting improvement. Ultimately, the base grouting effect is more beneficial for controlling pile settlement than increasing the working pile base capacity, especially when long friction-dominant piles are constructed in high ground water that might create 'soft toe' issues. Base grouting can solve these soft toe issues by enabling improved pile and rock contact.

However, as the preliminary design's analytical solutions are based on assumed factors and the availability of empirical data, their prediction accuracy is not reliable, especially if the Alpha and Beta methods are used. Additionally, the polymer applied for wet construction will affect the pile resistance, which might also contribute to decreased reliability; however, when the modified closed-form analytical solutions and interpreted load–settlement response are implemented carefully while considering all noted influencing factors, especially when the behaviour of soil stiffness is corrected, predicting the capacity of long bored pile foundations could be accomplished more accurately. Using the BDSLT also reduces the pile load testing time, as the supercells can be easily and reliably incorporated into the pile body during construction. BDSLT also ensures construction safety, as a large kentledge set-up can be avoided, especially if the pile load tests are conducted on soft grounds.

Author Contributions: Conceptualization, R.W. and D.E.L.O.; methodology, R.W., D.E.L.O. and J.Z.; software, R.W.; validation, R.W.; formal analysis, R.W. and S.L.; investigation, R.W. and S.L.; resources, S.L.; data curation, R.W.; writing—original draft preparation, R.W., D.E.L.O. and J.Z.; writing—review and editing, R.W., D.E.L.O. and J.Z.; visualization, R.W. and E.O.; supervision, D.E.L.O. and E.O.; project administration, D.E.L.O., S.L. and E.O.; funding acquisition, D.E.L.O. All authors have read and agreed to the published version of the manuscript.

Funding: This research received no external funding.

Institutional Review Board Statement: Not applicable.

Informed Consent Statement: Not applicable.

Data Availability Statement: The original experimental data are available on request from the corresponding author.

Conflicts of Interest: The authors declare no conflict of interest.

References

1. Randolph, M.F.; Wroth, C.A. Simple approach to pile design and the evaluation of pile tests. In *Behavior of Deep Foundations*; Lundgren, R., Ed.; ASTM International: Philadelphia, PA, USA, 1979; p. 484. [CrossRef]
2. Osterberg, J. New device for load testing driven piles and drilled shafts separates friction and end. In *Piling and Deep Foundations, Proceedings of the International Conference on Piling and Deep Foundations*; Burland, J., Mitchell, J., Eds.; CRC Press: London, UK, 1989; Volume 1, p. 421.
3. Choi, Y.; Lee, M.H.; Nam, M.S.; Kim, T.H.; Stuedlein, A.W. Development and implementation of a high-pressure, double-acting, bi-directional loading cell for drilled shafts. *Geotech. Test J.* **2016**, *39*, 196–205. [CrossRef]
4. Hayes, J. The landmark Osterberg cell test. *Deep Found.* **2012**, 45–49.
5. Zheng, Y.J.; Dai, X.; Li, L.X. Analysis of an O-cell pile test in Jinan with FEM. *Appl. Mech. Mater.* **2012**, *170–173*, 33–36. [CrossRef]
6. Baca, M.; Rybak, J. Pile base and shaft capacity under various types of loading. *Appl. Sci.* **2021**, *11*, 3396. [CrossRef]
7. Niazi, F.S.; Mayne, P.W. Axial pile response of bidirectional O-cell loading from modified analytical elastic solution and downhole shear wave velocity. *Can. Geotech. J.* **2014**, *51*, 1284–1302. [CrossRef]

8. Baca, M.; Brząkała, W.; Rybak, J. Bi-directional static load tests of pile models. *Appl. Sci.* **2020**, *10*, 5492. [CrossRef]
9. Baca, M.; Brzakala, W. Numerical modeling of pile installation influence on surrounding soil. In Proceedings of the International Multidisciplinary Scientific GeoConference: SGEM, Sofia, Bulgaria, 29 June—5 July 2017; Volume 17, pp. 619–626. [CrossRef]
10. *JGJ94–2008*; Technical Code for Building Pile Foundations. Ministry of Housing and Urban–Rural Development of the People's Republic of China: Beijing, China, 2008. (In Chinese)
11. Dai, G.; Gong, W. Application of bi-directional static loading test to deep foundations. *J. Rock Mech. Geotech. Eng.* **2012**, *4*, 269–275. [CrossRef]
12. *JT/T 738–2009*; Static Loading Test of Foundation Pile—Self-Balanced Method. Ministry of Transport of the People's Republic of China: Beijing, China, 2009. (In Chinese)
13. Knappett, J.; Craig, R.F. *Craig's Soil Mechanics*, 9th ed.; CRC Press: London, UK, 2019. [CrossRef]
14. Karlsrud, K.; Hansen, S.; Dyvik, R.; Kalsnes, B. NGI's pile tests at Tilbrook and Pentre—Review of testing procedures and results. In *Large-Scale Pile Tests in Clay*; Thomas Telford Publishing: London, UK, 1993; pp. 405–429.
15. Semple, R.M.; Rigden, W.J. Shaft capacity of driven pipe piles in clay. In *Analysis and Design of Pile Foundations*; ASCE: Reston, VA, USA, 1984; pp. 59–79.
16. Skempton, A.W. Cast in-situ bored piles in London clay. *Geotechnique* **1959**, *9*, 153–173. [CrossRef]
17. Stas, C.V.; Kulhawy, F.H. *Critical Evaluation of Design Methods for Foundations under Axial Uplift and Compression Loading*; Electric Power Research Institute: New York, NY, USA, 1983; Volume 3771.
18. Weltman, A.; Healy, P. Piling in boulder clay and other glacial tills. DOE and CIRIA Piling Development Group report PG5. *Int. J. Rock Mech. Min. Sci.* **1978**, *16*. [CrossRef]
19. Skempton, A.W. The post-glacial clays of Thames estuary at Tilbury and Shell-Haven. In Proceedings of the 3rd International Conference on Soil Mechanics and Foundation Engineering; ICOSOMEF: Zurich, Switzerland, 1953; Volume 1, pp. 302–308.
20. Wang, R.; Ong, D.E.L.; Peerun, M.I.; Jeng, D.S. Influence of surface roughness and particle characteristics on soil–structure interactions: A state-of-the-art review. *Geosciences* **2022**, *12*, 145. [CrossRef]
21. Anagnostopoulos, A.; Koukis, G.; Sabatakakis, N.; Tsiambaos, G. Empirical correlations of soil parameters based on cone penetration tests (CPT) for Greek soils. *Geotech. Geol. Eng.* **2003**, *21*, 377–387. [CrossRef]
22. Elbeggo, D.; Éthier, Y.; Karray, M.; Dubé, J.S. Valeurs de la Vitesse de Propagation des Ondes de Cisaillement dans les Argiles de la Mer de Champlain au Canada Predates par des Corrélations Empiriques. In Proceedings of the GeoMontreal Conference, Mont Tremblant, QC, Canada, 3–6 October 2013.
23. Amir-Faryar, B.; Aggour, M.S.; McCuen, R.H. Universal model forms for predicting the shear modulus and material damping of soils. *Geomech. Geoengin.* **2017**, *12*, 60–71. [CrossRef]
24. Ishibashi, I.; Zhang, X. Unified dynamic shear moduli and damping ratios of sand and clay. *Soils Found.* **1993**, *33*, 182–191. [CrossRef]
25. Kallioglou, P.; Tika, T.; Pitilakis, K. Shear modulus and damping ratio of cohesive soils. *J. Earthq. Eng.* **2008**, *12*, 879–913. [CrossRef]
26. Vardanega, P.J.; Bolton, M.D. Stiffness of clays and silts: Normalizing shear modulus and shear strain. *J. Geotech. Geoenviron. Eng.* **2013**, *139*, 1575–1589. [CrossRef]
27. Randolph, M.F.; Wroth, C.P. Analysis of deformation of vertically loaded piles. *J. Geotech. Geoenviron. Eng.* **1978**, *104*, 1465–1488. [CrossRef]
28. *JGJ/T 403–2017*; Technical Specification for Static Loading Test of Self-Balanced Method of Building Foundation Piles. Ministry of Housing and Urban–Rural Development of the People's Republic of China: Beijing, China, 2017. (In Chinese)
29. Al-Atroush, M.E.; Hefny, A.; Zaghloul, Y.; Sorour, T. Behavior of a large diameter bored pile in drained and undrained conditions: Comparative analysis. *Geosciences* **2020**, *10*, 261. [CrossRef]
30. Ong, D.E.L. Pile Behaviour Subject to Excavation-Induced Soil Movement in Clay. Ph.D. Thesis, National University of Singapore, Singapore, 21 April 2005.
31. Zhao, F. Mechanism research on the improvement of bearing capacity of single pile by pile-base post grouting technique and estimation of bearing capacity. *MATEC Web Conf.* **2015**, *25*, 03008. [CrossRef]
32. Bagui, S.K.; Puri, S.K.; Rao, V.; Dinesh, B.C.; Das, A. Estimation of base grout quantity for cast in situ piles based on field test results. *Adv. Bridge Eng.* **2020**, *1*, 4. [CrossRef]

geosciences

MDPI

Article

Behavior of a Large Diameter Bored Pile in Drained and Undrained Conditions: Comparative Analysis

Mohamed Ezzat Al-Atroush [1,*], Ashraf Hefny [2], Yasser Zaghloul [3] and Tamer Sorour [4]

1 Department of Engineering Management, College of Engineering, Prince Sultan University, Riyadh 11543, Saudi Arabia
2 Department of Civil & Env. Engineering, United Arab Emirates University, Al Ain 15258, UAE; a.hefny@uaeu.ac.ae
3 Department of Civil Engineering, Higher Institute of Engineering, Shorouk City, Cairo 11865, Egypt; y.zaghloul@sha.edu.eg
4 Department of Civil Engineering, Faculty of Engineering, Ain Shams University, Cairo 11865, Egypt; tamer.sorour@eng.asu.edu.eg
* Correspondence: mezzat@psu.edu.sa; Tel.: +966-506362379

Received: 24 June 2020; Accepted: 6 July 2020; Published: 8 July 2020

Abstract: Despite the difficulties in obtaining the ultimate capacity of the large diameter bored piles (LDBP) using the in situ loading test, this method is the most recommended by several codes and design standards. However, several settlement-based approaches, alongside the conventional capacity-based design approach for LDBP, are proposed in the event of the impossibility of performing a pile-loading test during the design phase. With that in mind, natural clays usually involve some degree of over consolidation; there is considerable debate among the various approaches on how to represent the behavior of the overconsolidated (OC)stiff clay and its design parameters, whether drained or undrained, in the pile-load test problems. In this paper, field measurements of axial loaded to failure LDBP load test installed in OC stiff clay (Alzey Bridge Case Study, Germany) have been used to assess the quality of two numerical models established to simulate the pile behavior in both drained and undrained conditions. After calibration, the load transfer mechanism of the LDBP in both drained and undrained conditions has been explored. Results of the numerical analyses showed the main differences between the soil pile interaction in both drained and undrained conditions. Also, field measurements have been used to assess the ultimate pile capacity estimated using different methods.

Keywords: large diameter bored pile; finite element method; load transfer; failure mechanism; overconsolidated stiff clay; two dimensional axisymmetric; drained and undrained conditions

1. Introduction

The in situ full-scale loading test is the most recommended methodology by several international design standards to determine the ultimate capacity of the large diameter bored piles (LDBP). However, loading of this class of piles till reaching apparent failure is practically seldom. The pile load-settlement performance at failure state is a challenge, especially for the large diameter bored piles, as failure is difficult to be achieved through the full-scale loading tests [1–3]. Nevertheless, many would agree that the ultimate pile bearing capacity can be defined as the load at which a considerable increase in the pile settlement occurs under sustained or slight increase of the applied load [1], i.e., the pile plunges.

Numerical analysis has recently become a powerful method and reliable tool widely used for solving many geotechnical problems. Many authors have discussed numerical studies related to soil-structure interaction and the axially loaded single pile [2–7]. Despite that rarity of the available loading tests on LDBP that achieved apparent failure was the main reason that the measurements of the well-documented Alzey bridge case history have been utilized in many numerical studies performed by several researchers, e.g., [3]; and [7].

Unfortunately, given the conclusions drawn in those studies, the findings of these numerical analyses did not indicate an agreement with field measurements at the state of failure. For instance, a back-analysis study [7] has been performed for the mentioned LDBP load test (Alzey bridge case study), and it was concluded that it is necessary to perform more back analyses of pile load tests to achieve better agreement with field measurements and give general recommendations. In another numerical study [3] conducted to simulate the same field case study, it was concluded that the finite element models performed were able to predict the deformation under the working load accurately but were not capable of simulating the load-deformation behavior adequately beyond the working load.

Such a disagreement between the calculated and the measured values may exist because of many numerical factors. A highly sensitive numerical model is often required to simulate the behavior of the LDBP at the failure state. The most expected responsible factors for the disagreement between the field measurements and the numerical results at higher loads may be due to one or all of the following factors; mesh size and dependency, model sensitivity to geometry dimensions, nonlinear analysis methodologies (convergence criterion), constitutive soil model and the adopted properties, and finally the construction effect represented in both interface element and drainage condition.

Natural clays usually involve some degree of over consolidation (OC) because of the process of mechanical unloading such as erosion, excavations, and changes in groundwater pressures. Uniform undrained cycles of loading on the overconsolidated clays lead to an accumulation of shear, induced pore pressures and can cause failure even when the magnitude of the shear stress is a fraction of the monotonic undrained shear strength of the clay [8]. Thus, the coupling of shear and volumetric behavior is required for accurate modeling of overconsolidated clays under cyclic loading [9] even at small or medium stress (or strain) levels. In general, the soil is a multiphase material; its stress, strain, and strength are represented by pressure dependency with coupling between shear and volumetric behavior [10]. For example, during drained shearing, dense sands and highly over-consolidated clays tend to dilate, whereas, loose sands and normally consolidated clays tend to contract.

The tendency of soil to either dilate or contract in shear often defines if the key design parameters will be either the drained shear strength (c' and φ') or the undrained shear strength (s_u) [11]. Porous materials like soils have different design properties under drained and undrained conditions. The existence of either a drained or an undrained condition in a soil depends on: soil type, geological formation (i.e., fissures, etc.), and rate of loading.

The literature review shows that there is a considerable debate on how to model or consider the behavior of the OC stiff clay and its design parameters in the pile axial loading problems. Several numerical studies have utilized the drained condition to define the behavior of the overconsolidated stiff clay [3,7]. This was explained by the micro-fissures associated with the OC stiff clay that may provide avenues for local drainage, soil along fissures has softened (increased water content) and is softer than intact material. However, on the other side, some field and experimental studies (i.e., [12–14]) mainly investigated the behavior of piles installed in overconsolidated stiff clay using the undrained design parameters.

This paper discusses the response of the large-diameter bored piles installed in predominately OC stiff clay. Two numerical models are established to simulate both drained and undrained behavior of an axial loaded to failure large diameter bored pile load test (Alzey Bridge Case Study). Whereas the in situ test achieved the failure, the sensitivity of those numerical models is an issue. The results of the numerical models are assessed by comparing the settlement and ultimate pile capacity obtained with the field measurements. This evaluation also highlights the main differences of LDBP behavior in both drainage conditions. The results of these numerical results are used to evaluate the estimated pile ultimate capacity using different international codes and design methods.

2. In Situ Pile-Loading Test

A full-scale, and well-instrumented pile-loading test has been performed by Sommer [12] for a single large diameter bored pile (LDBP) with length and diameter of 9.50 m and 1.30 m, respectively. This pile was constructed in a homogeneous over-consolidated stiff clay soil layer. The consistency limits were determined as 0.80 and 0.20 for liquid limit and plastic limit, respectively. Also, soil water content was obtained as 22%. The groundwater table was 3.5 m below the ground surface. Average unconfined compressive strength (Q_u) along borehole length was obtained as 300 kN/m^2, as given in Figure 1.

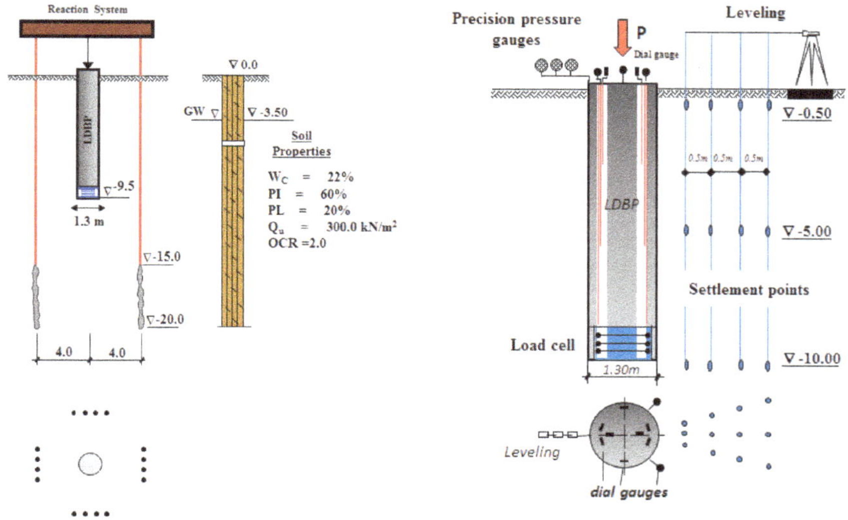

(**a**) (**b**)

Figure 1. Large diameter bored pile load test [12]. (**a**) Test arrangement and typical soil profile with mechanical properties (Modified from [12]). (**b**) Measuring devices and instrumentations (Modified from [12]).

Components of the reaction system used in the large-diameter pile-loading test are given in Figure 1a. The system consisted of steel girders held by 20.0 m length of burden anchors. In order to ascertain that anchors' positions would not affect the pile-bearing and skin friction results, the compression anchors were allocated at a horizontal distance equals three times of pile diameter (4.0 m), from the pile central loading axis (Figure 1a). Also, the anchors were vertically extended to 15 to 20 m depth below the ground surface. Figure 1b presents the instrumentations and measuring devices that have been used in the loading test. Pile was equipped with an end bearing prefabricated concrete base containing a load cell to measure the transferred stresses at the pile base level. The load cell was placed immediately under the pile base. The difference between the load cell measurement and the applied load gave the transferred load by friction. In addition, ground settlement points with steel rods of 25 mm diameter have been used to measure the settlement of the soil near the pile. These settlement points were protected by plastic pipes and installed at a horizontal distance of 0.5 m, 1.0 m, 1.5 m, and 2.0 m from the pile's shaft. Three different levels were assigned for the settlement points. These levels were −0.50 m below ground surface, at the middle of the pile length at the level of −5.00 m, and under the pile base at the level of −10.00 m (Figure 1b). Furthermore, pile settlement was measured using dial gauges at the head level and also monitored using a concreted precision leveling device.

Figure 2 shows the in situ measurements of the loading test. Pile settlement values measured using the dial gauges under each applied loading increment; also, the respective values of bearing stress, unit skin friction, and the total applied pressure are given in the same figure.

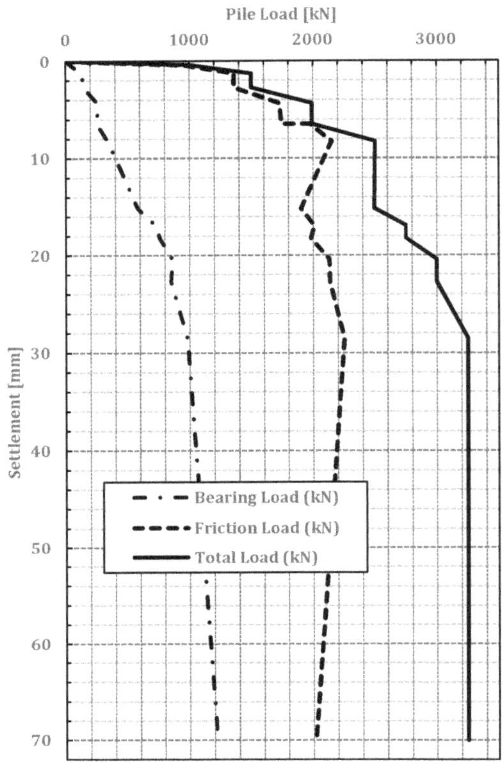

Bearing Stress	Skin Friction	Applied Presure
kN/m²	kN/m²	kN/m²
110	35	1130
10%	90%	100%
	S = 2.75 mm	
190	45	1500
13%	87%	100%
	S = 6.46 mm	
440	49	1880
23%	77%	100%
	S = 15.18 mm	
640	55	2260
28%	72%	100%
	S = 22.75 mm	
740	58	2450
30%	70%	100%
	S = 28.54 mm	
920	520	2450
38%	62%	100%
	S = 70.0 mm	

Figure 2. In situ measurements of the well-instrumented large diameter bored pile (Modified from [12]).

3. Numerical Modelling

The established models should simulate the behavior of the loaded to failure, large diameter bored pile, of the Alzey bridge case study, so that the sensitivity of the established model is very important to achieve good agreement between the numerical and field results at the failure state. In this regard, several sensitivity analyses are conducted to adopt the optimum mesh size and geometry dimensions that can accurately represent the simulated case study.

3.1. Methodology

Two-dimensional axisymmetric models are used in this study. For the established models, the external boundaries are supported to avoid the numerical instability (singularity) of the finite element model. The left and right sides are taken as fixed in the lateral-direction, and free to settle in the vertical direction. Fixed supports are employed at the bottom boundary in both the horizontal and vertical directions. Conversely, the top boundary is considered free.

The analysis is divided into three stages; the first stage represents the initial stresses of the soil before pile implementation. The second stage starts with changing the pile volume to concrete material as a replacement of soil material. At this stage, rigid interface elements are used to connect pile and soil mesh elements to avoid any numerical instability (singularity) [15], and pile self-weight is considered at this stage. The calculated deformations of the first and second stages of analysis are discarded, in order to start accounting for pile settlement due to the applied loads only. Interface elements are activated in the third stage of analysis, and the rigid interface elements are deactivated. The load is applied to the pile head in the third stage using incremental loading steps to simulate the pile-loading sequence with the same field loading test steps (see Figure 2).

The load transfer mechanism of the LDBP is investigated, using the numerical models, by determining the pile bearing load at each loading increment utilizing the obtained bearing stress at the pile base level. Pile friction resistance is calculated by deducting the bearing load from the total applied amount. Thus, the relation between pile settlement and ultimate bearing, friction, and total capacities can be constructed.

Fundamental to note that in this evaluation, the finite element analysis criteria mainly aspires to achieve the failure load by applying load with a value greater than the ultimate load measured in the field study [12], to allow the finite element solver to achieve the failure according to the employed convergence criteria. The maximum load where convergence can be obtained in the numerical model is considered as the failure load. Furthermore, failure is also indicated by the apparent large settlement that is expected to be induced at the ultimate load, similar to the field-loading test results.

3.2. Mesh Dependency

Three axisymmetric 2D finite element models with different levels of mesh refinement are established with quadratic 8-noded high order elements, using Midas GTS NX package, to investigate the dependency of the mesh size and its effect on the analysis results. As shown in Figure 3a the first model is established with a coarse mesh of size ranging from 0.325 m (around the pile) to 1.0 m (at the boundaries). Medium to fine mesh with size varying from 0.21 m to 0.40 m is adopted in the second model (Figure 3b). The third model is generated with a very fine mesh of size of 0.10 m near the pile and gradually increasing to 0.20 m at boundaries location (Figure 3c).

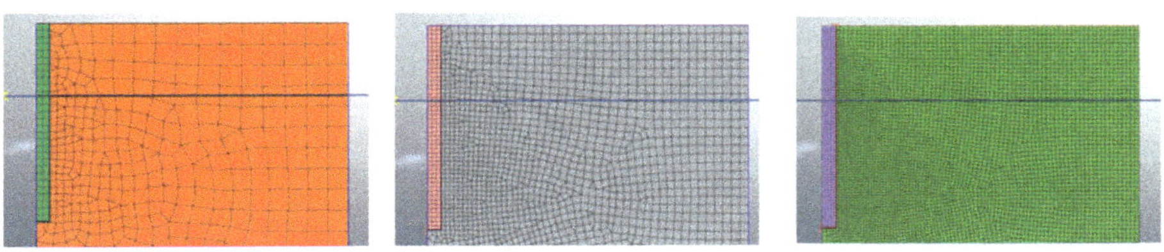

(a) (b) (c)

Figure 3. Finite element meshes with different sizes. (**a**) Coarse mesh, (**b**) medium mesh, (**c**) fine mesh.

Noteworthy to address that the three finite element models are established in equal geometry dimensions (15 m width and 16 m depth). Also, the same constitutive model and soil parameters are adopted according to a previous study [3], which means that the mesh size was the only variable in this step of evaluation. During the analysis, it was observed that analysis time is significantly increased with decreasing mesh size. Results of pile settlement and ultimate total, bearing and friction resistances of the three models are obtained as given in Figure 4, based on the methodology explained in the previous Section 3.1.

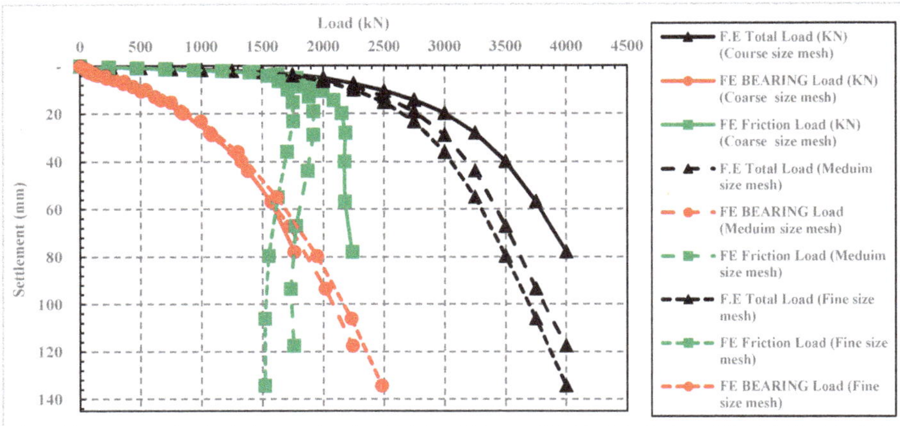

Figure 4. Comparison between pile settlement, total, bearing, and friction resistances results of the three finite element models with different levels of mesh refinement.

Figure 4 highlights the significant effect of mesh size on pile load settlement results. As shown, the pile settlement increases with decreasing mesh size, and the model with fine mesh achieved the highest settlement values. This comparison also shows that up to working load (1500 kN) minor effect of the mesh size was observed on the results of the pile load settlement. The same observation is also exhibited in the results of the total, bearing, and friction resistance. As shown in Figure 4, at the applied load of 1500 kN, almost equal values are obtained using the three numerical models with different mesh sizes, for both bearing and friction resistances.

At higher loads, the results of the three models showed that full friction mobilization is occurred at almost equal settlement value near about 1.5% of the pile diameter (see Figure 4). In addition, the number of the mesh elements created at the pile base level affected bearing resistance results of the three models. In the coarse mesh model, there are only two elements at the pile base, while, five elements are created at the pile base of the fine mesh model. The highest bearing resistance is obtained from the model with a fine mesh (see Figure 4). This was explained by Wehnert [7], as at each mesh element, stresses are calculated at the Gauss points; accordingly, reducing the element size increases the number of these points, whereas stresses are calculated, allowing for much precise stress distribution. Besides, Wehnert [7] recommended that at least two or three elements at the pile base are required to get rid of the mesh dependency effect on pile bearing resistance results.

The mesh size also has a significant effect on the pile failure mechanism, as demonstrated in Figure 5. The formed plastic points under the last load increment are more dense and smooth in the fine mesh model results (Figure 5c) compared to the results of the coarse mesh model (Figure 5a).

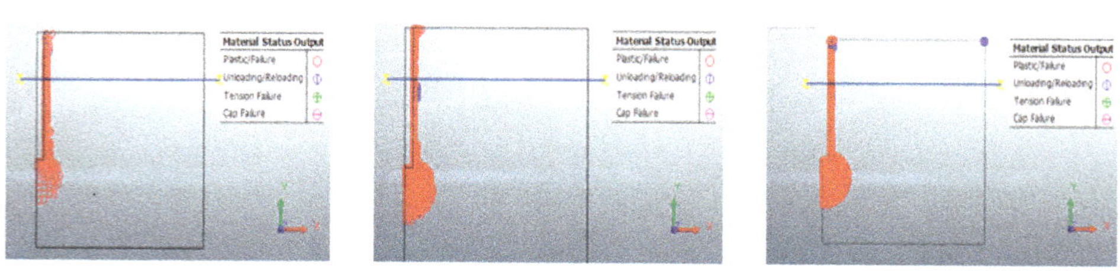

(a) (b) (c)

Figure 5. The shape of the formed plastic points at the last load increment of the three finite element models with varying sizes of mesh. (**a**) Coarse mesh, (**b**) medium mesh, (**c**) fine mesh.

In conclusion, to justify the economic computational feasibility of the numerical model, mesh with medium to fine size (0.21 m) is to be adopted in the upcoming investigations, as it allows to have three elements at the pile base level. Also, minor differences in pile settlement and ultimate capacities results are observed between numerical results of the model with a medium to fine size mesh compared to the other with very fine size mesh.

3.3. Geometry of the Numerical Model

Ten analysis attempts are carried out with different geometry dimensions (widths and depths) to investigate the effect of boundaries position on the analysis result. The first five finite element models (Figure 6) are established with equal depths of 16 m, and different widths of 10 m (~8D), 15 m (~12D), 20 m (~16D), 25 m (~20D), and 30 m (~24D), to assess the geometry width effect on the resulted pile settlement. Simultaneously, optimum depth of geometry is explored using five numerical models (Figure 7) with equal widths of 15 m, and different depths of 16 m (~1.5L), 25 m (~2.5L), 30 m (~3L), 40 m (~4L), and 45 m (~4.5L).

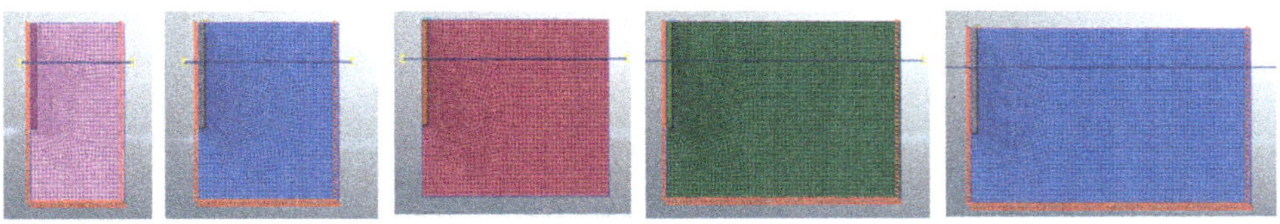

(a) (b) (c) (d) (e)

Figure 6. Five finite element models with different widths. (**a**) 10 m × 16 m, (**b**) 15 m × 16 m, (**c**) 20 m × 16 m, (**d**) 25 m × 16 m, (**e**) 30 m × 16 m.

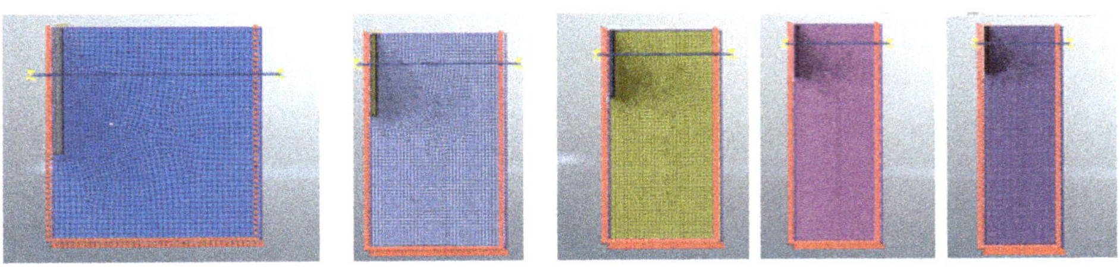

(a) (b) (c) (d) (e)

Figure 7. Five finite element models of with different depths. (**a**) 15 m × 16 m, (**b**) 15 m × 25 m, (**c**) 15 m × 30 m, (**d**) 15 m × 40 m, (**e**) 15 m × 45 m.

Figure 8a,b presents the relation between the obtained pile settlement and the geometry width and depth, respectively. It can be seen that the obtained pile settlement almost becomes with constant value when model geometry width and depth are considered as 20 m (~16D) and 40 m (~4L), respectively. Therefore, these dimensions will be adopted in the upcoming analyses.

Based on the sensitivity analyses performed, it can be concluded that to achieve the maximum expected settlement, the numerical model should be established with high order medium to fine mesh and geometry dimensions of 20 m × 40 m to ensure that boundaries positions will not affect the results at the analysis zone. In addition, the pile should be represented by a fine mesh with a size of 0.216 m, and at least three mesh elements have to be adopted at its base.

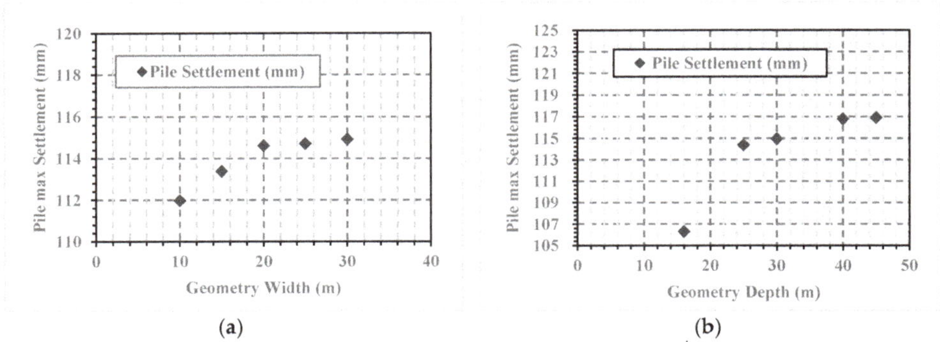

Figure 8. Relations between maximum obtained pile settlement and model geometry. (**a**) Model geometry width. (**b**) Model geometry depth.

In this evaluation, the finite element analysis criteria mainly aspire to achieve the failure load. As explained before, the maximum load where convergence can be obtained in the numerical model is considered as the failure load. For that purpose, and based on several analyses attempts performed, it is highly recommended to utilize the secant stiffness nonlinear convergence method. Moreover, it provides much stable analysis because of its ability to find convergence when the stability of analysis is an issue, as in case of approaching failure loads, or when large strains are expected [16]. Also, the maximum number of iterations per increment, and the maximum number for bisection level, are essential parameters when large strains are expected, and its recommended to be taken with high value. This may increase the computational usage, although the analysis solver would be able to complete the analysis and find a convergence, even at the last loading increment (ultimate load).

4. Drained Condition

Based on the sensitivity analyses performed, the numerical model shown in Figure 9 has been established. To satisfy both the accuracy of the results and the analysis time requirements, a compromise solution has been adopted as follows. A fine mesh with a size of 0.216 m was considered around and below the pile element (10 m width and 19.5 m depth), and gradually, soil mesh size is increased to be 0.65 m at the boundary locations.

The established model has been utilized in a drained numerical analysis, using the soil and pile parameters given in Tables 1 and 2, respectively. Results of this study showed that an excellent agreement was obtained between the in situ measurements and finite element results in both the pile load settlement and load transfer relationships when Modified Mohr-Coulomb (MMC) constitutive model is used to simulate the overconsolidated stiff clay soil behavior. Also, the large induced pile settlement at the failure state has been accurately determined. Figure 10 shows a comparison between field measurements and the obtained results of pile bearing, friction, the total load using the calibrated FE model (drained condition). More details of the drained analysis are given in [10].

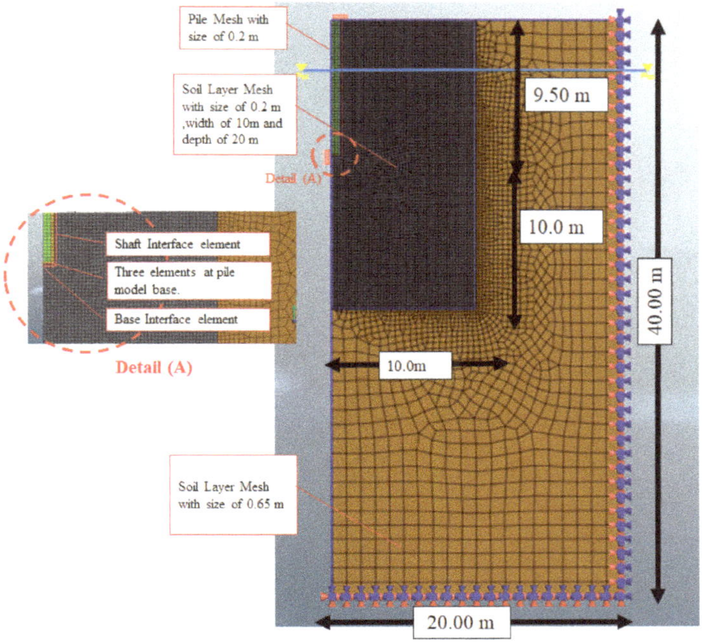

Figure 9. (2D) axisymmetric finite element model established based on the results of the sensitivity analysis performed (After [10]).

Table 1. Over-consolidated stiff clay soil (drained parameters) (After [10]).

Soil Parameter	MMC	Unit
Type of material behavior	Drained	
Soil weight above/Below phr. Level ($\gamma_{unsat} \backslash \gamma_{sat}$)	20	kN/m^3
Poisson's ratio (ν)	0.3	-
Secant stiffness (E_{50}^{ref})	45,000	kN/m^2
Oedometer stiffness ($E_{oed}{}^{ref}$)	45,000	kN/m^2
Unloading-reloading stiffness (E_{ur}^{ref})	90,000	kN/m^2
Power of stress level (m)	0.5	-
Unloading-reloading Poisson's ratio (N_{ur})	0.2	-
Reference pressure (P_{ref})	100	kN/m^2
Cohesion (c)	20	kN/m^2
Friction angle (φ')	22.5	Degree (°)
Dilatancy angle (ψ)	0.1	Degree (°)
Lateral earth pressure coefficient (K_0)	0.80	-
Overconsolidation Ration (OCR)	2.0	-
Soil Tensile Strength	0.0	kN/m^2

Table 2. Large diameter pile structural parameters (After [10]).

Pe Material	Concrete	Unit
Young Modulus ($E_{elastic}$)	24248711	kN/m^2
Poisson's Ratio (μ)	0.20	-
Unit weight (γ_c)	24.0	kN/m^3

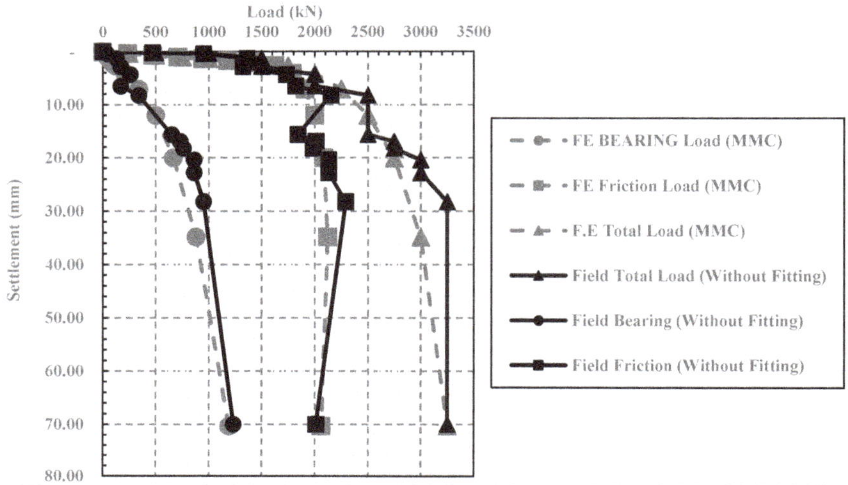

Figure 10. Comparison between field measurements and the obtained results of pile bearing, friction, the total load using the calibrated FE model [10].

5. Undrained Condition

Based on the results presented in the drained analysis study [10], the Modified Mohr Columb constitutive model is used to simulate the overconsolidated soil in the undrained condition, as well. The influences of the other parameters, such as mesh size and model geometry, etc., are filtered out and taken with same values adopted in the calibrated drained analysis. The soil drainage condition and the corresponding changes in the soil and interface parameters are the only variables in this investigation.

Referring to the field study [12], soil undrained shear strength of 300 kN/m^2 was determined using the unconfined compression test for cylindrical samples taken from the tested pile location. Therefore, the undrained cohesion is taken as 150 kN/m^2 in this analysis.

The undrained soil modulus of elasticity of the overconsolidated stiff clay soil is calculated using Duncan and Buchignani method [17] utilizing the plasticity index, over consolidation ratio, and undrained soil cohesion (60%, 2.0, and 150 kN/m^2). Besides, similar to the drained analysis attempt, tangential Young's modulus E_{50} and unloading modulus E_{ur} are taken as equal, and two times of E_{eod} respectively.

Soil undrained Passion's ratio is exercised with 0.495 instead of 0.50 to avoid any numerical errors. Modified Mohr-Coulomb constitutive model is also used to simulate the undrained soil condition, and the drainage option is adopted with the third class [Undrained c] (Undrained stiffness\Undrained strength). Soil tensile strength is taken equal zero, to prevent any expected tensile stresses along the pile shaft, and to ensure that failure will occur because of shear. The lateral earth pressure coefficient is calculated as 1.00, according to Equation (1). Table 3 summarizes the soil parameters adopted in the undrained analysis. Pile material is defined with the same properties mentioned in the drained analysis (Table 2).

$$K_0 = \upsilon/(1 - \upsilon) \quad (1)$$

where, v: Soil Poisson's ratio (undrained)

Table 3. Over consolidated stiff clay undrained soil parameters.

Soil Parameter	Modified Mohr-Coulomb (MMC)	Unit
Type of material behavior	Un-Drained	
Soil weight ($\gamma_{unsat} \backslash \gamma_{sat}$)	20	kN/m^3
Poisson's ratio (ν_u)	0.495	
Secant stiffness (E_{50}^{ref})	52,000	kN/m^2
Oedometer stiffness ($E_{oed}{}^{ref}$)	52,000	kN/m^2
Unloading-reloading stiffness (E_{ur}^{ref})	10,400	kN/m^2
Reference pressure (P_{ref})	100	kN/m^2
Power of stress level (m)	0.50	
Undrained Cohesion (c_u)	150	kN/m^2
Friction angle (φ_u)	zero	Degree (°)
Dilatancy angle (ψ)	zero	Degree (°)
Lateral earth pressure coefficient (K_0)	1.0	-
Soil Tensile Strength	0.0	kN/m^2

The soil unit skin friction was measured in the field [12] through two different pile-loading tests, for two piles with different lengths of 9.50 m, and 13 m. It was stated that the average measured skin friction (α*cu) is 53 kN/m^2, which represented almost about 36% of the undrained shear strength (150 kN/m^2) of the clay soil around the pile shaft [12].

Two interface elements are used to simulate the reduction in soil shear strength along the pile shaft. First is the shaft interface that used to simulate the interaction between the soil and pile shaft. The second is the base interface, which represents the pile base interaction with the beneath soil.

A reduction factor (R) of 36% is considered to define the undrained adhesion of the shaft interface. Also, a reduction factor equals 1.0 is taken for the base interface, as given in Table 4. Normal (K_n) and shear (K_t) stiffness moduli are automatically calculated based on the adopted soil module of elasticity [15]. Fundamental to state that values of normal and shear stiffness modulus differ from those taken in the drained analysis [10], which is attributed to the differences in Poisson's ratio and soil modulus of elasticity values in the undrained condition. Furthermore, interface tensile strength is taken as zero, to ensure that failure will occur because of shear not because of tension. Also, the tension cut option is considered in analysis options to avoid any tension results.

Analysis is performed with the same sequence that was presented before in Section 3.1. During the analysis, it was noted that the solver failed to find convergence at a load of 3300 kN and warned with failure.

Figure 11a compares between field measurements and the obtained pile settlement results using both drained and un-drained analyses. Perfect agreement can be seen from this figure between the results of the undrained numerical model and the field measurements. Also, failure is represented by the large apparent settlement result obtained at the last load increment. However, failure occurred at a load of 3300 kN instead of 3250 kN (failure load according to field measurements and drained analysis). It was also observed that a slightly greater settlement result is obtained at the last loading increment of the undrained analysis (77 mm). On the other side, the relation between total load, pile friction, and bearing resistances with settlement obtained using finite element undrained analysis are compared with field measurements in Figure 11b.

Table 4. Interface elements parameters.

Parameter	Shaft Interface	Base Interface	Unit
Interface nonlinearity	Coulomb Friction	Coulomb Friction	
Interface Adhesion (c_a)	53	150	kN/m^2
Interface Friction angle ($\varnothing_i$)	zero	zero	Degree (°)
Interface Dilatancy angle (ψ_i)	zero	zero	Degree (°)
Shear stiffness modulus (k_t)	125,217	34,7826	kN/m^3
Normal stiffness modulus (k_n)	1,377,390	3,826,086	kN/m^3
Tensile Strength	0	0	kN/m^2
Reduction Factor (R)	0.36	1	-

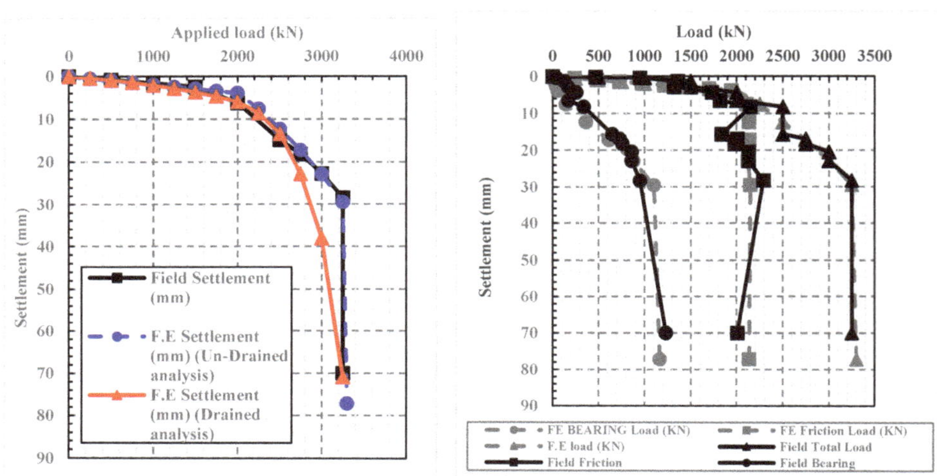

(**a**) (**b**)

Figure 11. Comparison between field measurements and results of the undrained numerical analysis (**a**) pile settlement results. (**b**) Pile load transfer results.

It can be seen from Figure 11b that good agreement is obtained between the undrained numerical results and the field measurements. Similar to the drained analysis results, the load is predominantly transferred by friction at the initial loads, and up to the working load of 1500 kN. Full mobilization of the friction resistance is achieved at an applied load of 2100 kN. After this load, the friction resistance becomes a constant value. In contrast, at higher loads, pile bearing resistance is obviously increased to achieve its maximum value at the load of 3250 kN. At a load of 3300 kN, a sudden substantial increase in pile settlement results, also a slight increase in the bearing resistance results is noted.

Figure 12 shows a comparison between field measurements, drained, and undrained analysis results of the total, friction, and bearing resistances. Good agreement is obtained between field measurements and both drained and undrained analyses result at the ultimate state, when a reduction factor of 0.36 and 1.0 are adopted in undrained and drained analyses, respectively. However, the obtained ultimate capacity from undrained analysis (3300 kN) is slightly higher than that obtained from the drained analysis (3250 kN).

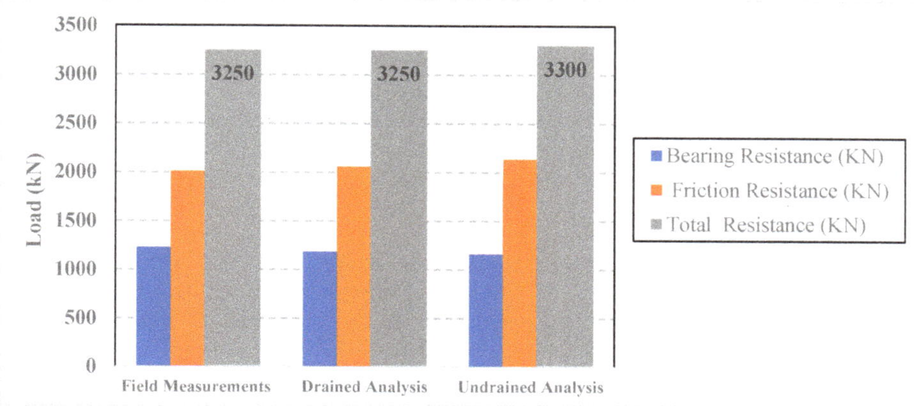

Figure 12. Comparison between field measurements and drained\un-drained analyses results.

6. Load Transfer Mechanism of Large Diameter Bored Pile (Drained\Un-Drained)

Figure 13a,b present the axial load distribution along pile shaft length for both drained and undrained analyses, respectively. Pile axial load distribution is calculated at each loading increment by multiplying the normal vertical stress in the shaft times the pile cross-sectional area. Worthwhile noting that pile load values at the base level are calculated by the integration of bearing stresses at the pile base.

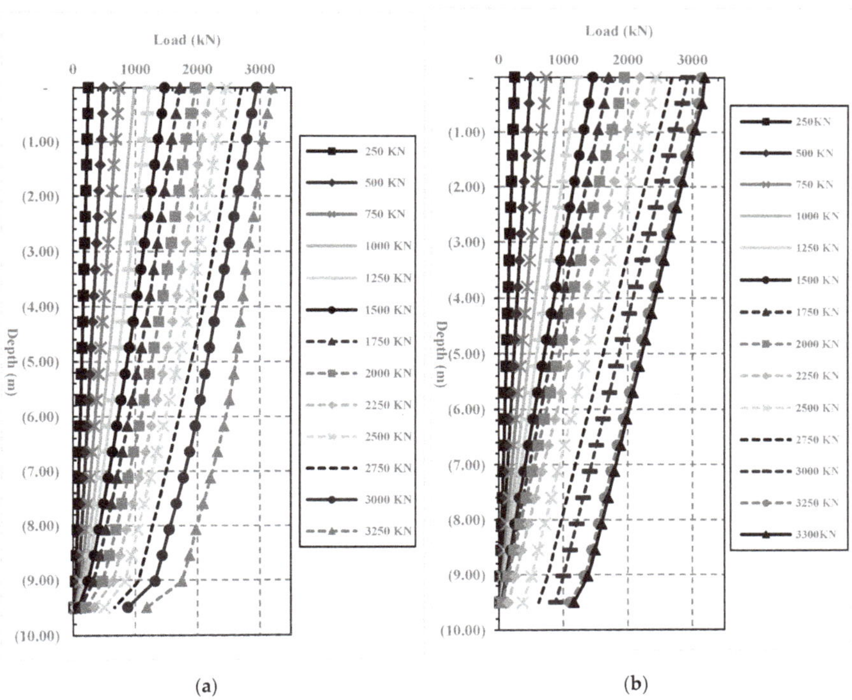

Figure 13. Results of the pile axial load distribution. (**a**) Drained analysis. (**b**) Un-drained analysis.

It can be seen from Figure 13a,b that the obtained pile axial load equals the applied load at the head level and decreases with depth to achieve its smallest value at the base level. This behavior is obvious in both drained and undrained analyses results. However, the rate of axial load decrease with depth is different, in both types of analysis. This is ascribed to the variance in the friction resistance of the soil in the two different conditions (drained/undrained). According to the Mohr-coulomb criterion, in the drained state the amount of load transferred by friction is increasing with depth because of the increase of the overburden pressure ($|\tau| = \sigma_h \tan\varphi_i + c_i$). In contrast, for the undrained condition, the amount of load transferred by friction is constant with depth ($|\tau| = c_u$).

Tangential stresses of the interface elements in the vertical direction (Y-Axis) are obtained from the calibrated numerical models at each load increment along the interface length. The relations between interface tangential stresses and pile length are presented in Figure 14a,b for both drained and undrained analyses, respectively. These tangential stresses represent the distribution of the soil unit skin friction along the pile shaft length at each load increment. In general, skin friction results are consistent with the result of pile load distribution (Figure 13), and indeed the obtained skin friction values are increased with the depth in the drained condition, and also they are constant with the depth in the undrained condition.

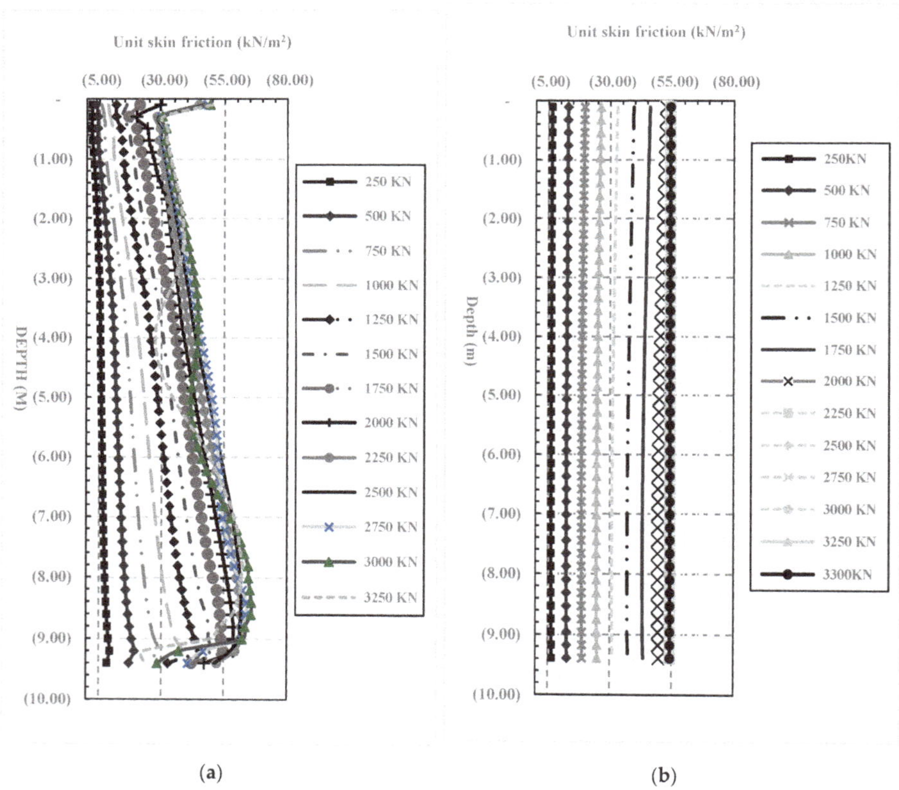

Figure 14. Results of the average values of unit skin friction obtained using numerical analysis. (**a**) Drained analysis [10]. (**b**) Un-drained analysis.

The pile load transfer behavior presented before in Figures 10 and 11b showed that the full friction mobilization occurred after load increment of 2000 kN in both drained and undrained cases. These results are also consistent with the skin friction results shown in Figure 14a,b. As the obtained shear stress values after this load almost become equal to the soil shear strength (according to MC criterion), which confirmed that full friction mobilization occurred after this load in both drained and undrained conditions.

For the drained condition (Figure 14a), the value of the unit skin friction at the ground surface equals the value of the adopted effective cohesion (c′), and gradually increased with the depth with slope equals (tan Ø′). It was also noticed that at the last three loading increments, a significant increase in skin friction occurs at the ground surface level. The same observation was also reported in [18] and [19], as a substantial increase was observed in the lateral earth pressure coefficient at the ground surface in several performed axial pile-loading static tests. This was attributed to the dilation effects near the ground surface, where the confining pressure is low compared to deeper depths. Figure 14a also shows that unit skin friction values tend to decrease after the load increment of 2000 kN at the last three applied load increments (2500, 3000, and 3250 kN). This decrease is attributed to the arching action [10,20,21]. On the other side, unit skin friction is obtained in undrained conditions, with value, equals the undrained adhesion. This value is increasing with applied load increases to reach its max value of the adopted undrained adhesion and remains with the same value until the end of the loading test.

The disadvantage of the undrained total stress analysis (Undrained Model C) is that no distinction is made between effective stresses and pore pressures. Hence, all output referring to effective stresses should now be interpreted as total stresses, and all pore pressures are equal to zero [15]. Therefore, no change in vertical and horizontal stress results are noted in this model, and consequently, no arching action is observed in the results of the undrained analysis.

Regarding the bearing resistance of the LDBP, Sommer [12], calculated the bearing capacity factor (N_c) in the undrained condition by dividing the value of the ultimate bearing stress measured at the failure load by the undrained cohesion value for the bearing soil (150 kN/m^2). Hence, the N_c bearing capacity factor was determined as 6.0.

Sommer also compared this result with Whitaker and Cooke [13] and Breth [14], loading test results that were performed in similar frankfurter and Londoner stiff clay soil. N_c factor was obtained with a value of (6.90) by Whitaker and Cooke [13], using a pile-loading test of a large diameter bored pile (0.94 m diameter). Also, a higher value of (9.50) was obtained by Breth [14] using the pile-loading test of a small diameter bored pile (0.42 m diameter). On the other side, with the same calculation procedure, the N_c bearing capacity factor is obtained as 6.0, using finite element undrained analysis results of the bearing stress at the ultimate pile load, which agrees with calculations presented in the field study [12].

Fundamental observation should be highlighted in this regard; the value of bearing stress at the pile base level is affected by the pile's diameter. Consequently, the bearing capacity factor (N_c) will also be influenced by the pile diameter. These findings are also in contradiction with several codes and design standards, such as DIN 4014 [22] and ECP202/4 [23]. As, they estimate only an absolute value of bearing stress under large diameter bored piles bases only based on the predicted settlement value at failure (Specific settlement-based criteria [5%D]), without considering any effect for the pile diameter on the bearing stress values.

7. Comparative Analysis

In case of impossibility to perform pile-loading test at the design phase, codes and design standards proposed several settlement-based design approaches for the large diameter bored piles (LDBP) alongside the conventional capacity-based design approaches. German standards (DIN) [22] and Egyptian code (ECP) [23] approaches are examples of those settlement-based methods. On the other side, Meyerhof [24] method is one of the commonly used capacity-based design approaches.

In this section, field measurements of the Alzey bridge case study will be used to assess the calculated ultimate capacity of the LDBP using two different methods of both capacity-based and settlement- based design approaches. According to the available data in the field study [12], Meyerhof [24] capacity-based method and ECP [23] settlement-based design approaches have been chosen. Other methods, such as AASHTO (LRFD) [25], require soil testing results such as SPT or CPT test results, which were not provided in the field study.

In both DIN [22] and ECP [23] methods, full friction mobilization is estimated to be achieved at a settlement of 1% of pile diameter. Also, at a settlement value of 5% of pile diameter bearing mobilization is expected. For cohesive soil, ECP provides estimated values for both soil unit skin friction and bearing resistance corresponding to the predicted settlement values. Also, it recommends ignoring soil friction resistance at the first two meters below the ground surface and at a distance equals pile diameter (D) above the pile base level as well. Using this effective length and the pile perimeter (O), pile friction capacity can be determined. Furthermore, the pile bearing load is calculated by multiplying the estimated value of bearing stress by pile cross-sectional area.

Fundamental to note that the ECP method did not consider any effect for the cohesive soil type, class, or degree of consolidation on the suggested undrained bearing stresses values. Besides, these suggested values are recommended for any large diameter bored pile, whatever its diameter. Subsequently, the relations between pile settlement, friction load, bearing load, and total ultimate load are plotted and compared with the field measurements in Figure 15.

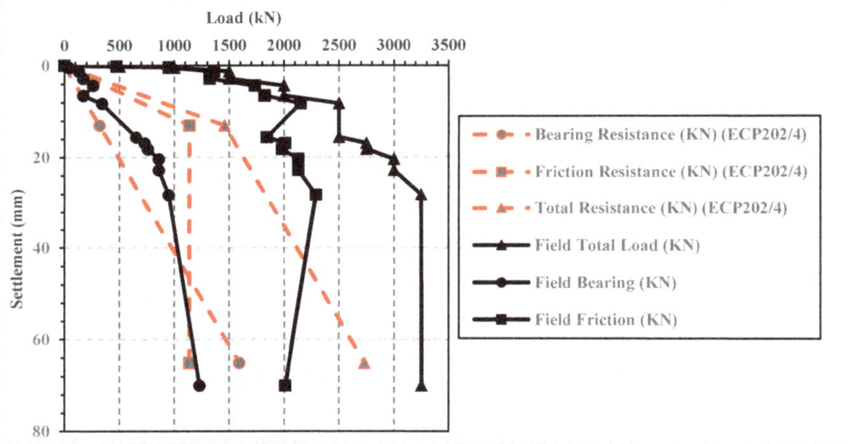

Figure 15. Comparison between field measurements of Alzey case history, and calculated values of pile settlement, friction, bearing, and total ultimate load using a settlement-based method [23].

Significant differences are evident between the calculated capacities using the settlement based method [23] and field measured values, as shown in Figure 15. At a settlement of 1% of pile diameter (13 mm), the calculated pile friction capacity (1139 kN) represents about 60% of that measured in the field loading test (2000 kN). Conversely, the calculated pile bearing ultimate capacity at pile settlement percentage of 5% the pile diameter (65 mm), is greater than the measured pile ultimate bearing resistance with a difference of about 27%. This means that this settlement-based method estimates a higher value for ultimate bearing stress (1200 kN/m^2) than the field measured one (920 kN/m^2). Despite that, the acquired pile ultimate capacity using the settlement-based method (2731 kN) is lower than the field measured value (3250 kN) with a difference of about 15%.

On the other hand, Meyerhof [24] conventional capacity-based design approach (Equation (2)) is used to calculate the ultimate pile capacity for the same large diameter bored pile in both drained and undrained conditions. Meyerhof stated that immediately after pile casting, the shaft adhesion is closely given by the undrained shear strength of clay. However, at later stages and particularly at the end of the foundation construction, the shaft resistance of piles will be governed by the effective drained shear strength parameters (c and φ) of remolded clay failing close to the shaft. The corresponding effective unit skin friction in the homogeneous clay may then be taken, as shown in Equation (2). Besides, in saturated homogeneous clay under undrained conditions, Meyerhof highlighted that the value of N_c below the critical depth varies with the sensitivity and deformation characteristics of the clay from about (5) for very sensitive brittle normally consolidated clay to about (10) for insensitive stiff overconsolidated clay, and value of (9) is frequently used for bearing capacity estimates of bored piles.

Using the soil and pile parameters presented before in Tables 1 and 2 (Drained parameters), bearing, friction, and total ultimate capabilities are calculated as 1980 kN, 1820 kN, and 3800 kN, respectively. Also, using the undrained soil and interface parameters (Tables 3 and 4), the undrained bearing, friction, and total ultimate capabilities are determined as 1791 kN, 2095 kN, and 3886 kN, respectively, using Equation (3) [24].

$$P_{ult} = A_s\,(c_a + K_s\gamma\,\frac{L}{2}\,\tan\,\delta) + A_b\,(cN_c + \gamma LN_q + \gamma\,\frac{D}{2}\,N_\gamma) \quad (2)$$

$$P_{ult} = A_s\,(c_a) + A_b\,(c\,N_c) \quad (3)$$

where,

c_a: soil adhesion per unit area;

δ: friction angle of the soil on the shaft.

K_s: the at-rest earth pressure coefficient on the shaft.

γ: soil unit weight

N_c, N_q, and N_γ: factors of bearing capacity, depend on ϕ and the embedment depth ratio L/B.

Figure 16 compares between field measurements, numerical results, and the calculated ultimate resistances using both the settlement-based [23] and the capacity-based [24] methods.

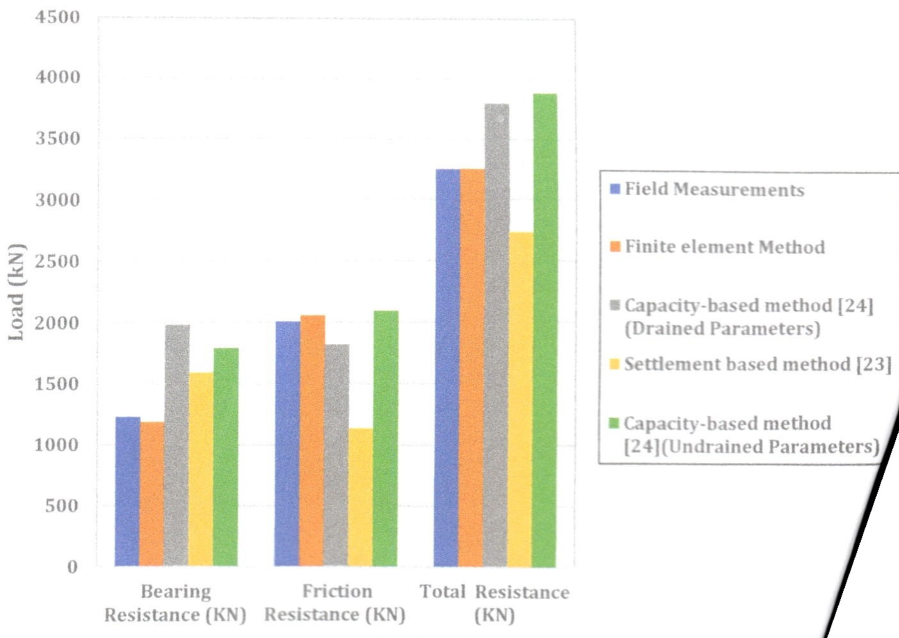

Figure 16. Comparison between field measurements and the calculated friction, bearing, a ultimate resistance using both the settlement-based [23] and the capacity-based [24] method

The ultimate drained capacity determined using the capacity-based method [24] field measurement, with a percentage of about 17%. This difference is attributed to th value of pile-bearing resistance (1980 kN) compared to field measured bearing resistance (1230 kN). The acquired bearing resistance from the Meyerhof equation represents 60% greater than field measurements. Nevertheless, the calculated pile friction resistance using the Meyerhof formula is lower than field measurement (12%). In addition, the ultimate undrained capacity obtained using Meyerhof undrained formula (Equation (3)) is near equal the drained capacity determined using Equation (2) (2.2% difference), which is consistent with the results obtained using the numerical analyses (Figure 12). Moreover, the implemented settlement-based [23] method underestimates the large diameter pile capacity compared to field measurement (15% difference), and utilized capacity-based method [24] overestimates the ultimate capacity of the pile (difference of about 17%). However, the finite element model using Modified Mohr-Coulomb constitutive model success to simulate the behavior of large-diameter bored pile and good agreement is obtained between the obtained ultimate capacity and field measurements.

8. Conclusions

Based on the numerical study conducted, the following conclusions have been drawn:

1. Numerical analysis is capable of predicting not only the working capacity but also the ultimate capacity of the large diameter bored piles (LDBP). Also, the large induced pile settlement at the failure state can be determined using the finite element method, when the appropriate soil constitutive model is carefully selected. Also, sufficient sensitivity analyses should be carried out to assess the quality of the generated mesh and to justify the economic computational feasibility of the established model.
2. Very good agreements were obtained between field measurements of Alzey bridge case study and both undrained and drained analyses results, even, at the failure state, when reduction factor of 0.36, and 1.0 of interface strength were adopted in analyses, respectively.
3. Both drained and undrained numerical results revealed that about 90% of the total applied load was predominantly transferred by friction at the initial loading increments (working loads), and only about 10% of the total applied load was carried by pile bearing resistance. However, at the ultimate load, friction resistance is 62%, and bearing resistance increased to 38% of the total pplied load.

both drained and undrained analyses, full friction mobilization occurred at a value of nent near equals 1.0% of the pile diameter; also, full bearing mobilization occurred at a settlement near equals 5.5% of the pile diameter.

sted Nc value of (9.0) may be only valid for the small diameter bored piles (less than ever, for large diameter bored piles, they are dependent on pile diameter.

aring stress below the large diameter pile base is affected by the pile diameter. codes and design standards proposed constant bearing stress to be used at a value (i.e., 5% D) to estimate the ultimate capacity of the LDBP, irrespective vithout any discrimination for any class of the cohesive soils.

method overestimated the ultimate pile capacity of the Alzey case LDBP, ased method underestimates the ultimate pile capacity of the Alzey

d methodology, M.E.A.-A. and A.H.; software, M.E.A.-A.; validation, writing—original draft preparation, M.E.A.-A., Y.Z. and T.S.; A.H.; visualization, M.E.A.-A.; supervision, A.H. All authors the manuscript.

d by the Structures and Materials (S&M) Research Lab uthors would like to express their sincere gratitude to ns, and continuous encouragement to perform this research.

Conflicts of Interest: The authors declare no conflict of interest.

References

1. Fellenius, B.H. *What Capacity to Choose from the Results of a Static Loading Test*; Deep Foundation Institute, Fulcrum: Hawthorne, NJ, USA, 2001; pp. 19–22.
2. Lee, J.H.; Salgado, R. Determination of Pile Base Resistance in Sands. *J. Geotech. Geoenviron. Eng. ASCE* **1999**, *125*, 673–683. [CrossRef]
3. El-Mossallamy, Y. Load-settlement behaviour of large diameter bored piles in over-consolidated clay. Numerical models in geomechanics. In Proceedings of the 7th International Symposium on Numerical Models in Geomechanics, NUMOG VII, Graz, Austria, 1–3 September 1999; pp. 443–450.
4. Davis, R.O.; Change, K.C.; Mullenger, G. Modeling of axially loaded piles: Comparisons with pile test load. In *Computer and Physical Modeling in Geotechnical Engineering*; PASCAL database: Rotterdam, The Netherlands, 1989.
5. Baars, S.V.; Niekerk, W.V. Numerical modelling of tension piles. In *Beyond 2000 in Computational Geotechnics*; Taylor & Francis: Boca Raton, FL, USA, 1999; pp. 237–246.
6. Meissner, H.; Shen, Y.L.; Van Impe, W.; Vogt, C. *Punching Effects for Bored Piles*; Deep Foundation Institute, and Auger Piles: Hawthorne, NJ, USA, 1993.
7. Wehnert, M.; Vermeer, P.A. Numerical Analyses of Load Tests on Bored Piles. Numerical models in geomechanics. In Proceedings of the 9th Numerical Models in Geomechanics, NUMOG IX, Ottawa, ON, Canada, 25–27 August 2004.
8. Malek, A.M.; Azzouz, A.S.; Baligh, M.M.; Germaine, J.T. Behavior of Foundation Clays Supporting Compliant Offshore Structures. *J. Geotech. Geoenviron. Eng. ASCE* **1989**, *115*, 615–636. [CrossRef]
9. Gu, C.; Wang, J.; Cai, Y.; Sun, L.; Wang, P.; Dong, Q. Deformation characteristics of overconsolidated clay sheared under constant and variable confining pressure. *Soils Found.* **2016**, *56*, 427–439. [CrossRef]
10. Ezzat, M.; Zaghloul, Y.; Sorour, T.; Hefny, A.; Eid, M. Numerical Simulation of Axially Loaded to Failure Large Diameter Bored Pile. *Int. J. Geotech. Geol. Eng. World Acad. Sci. Eng. Technol.* **2019**, *13*, 289–303.
11. Coutinho, R.Q.; Mayne, P.W. *Geotechnical and Geophysical Site Characterization 4*, 1st ed.; CRC Press, Taylor & Francis: Boca Raton, FL, USA, 2012; p. 5.
12. Sommer, H.; Hammbach, P. Großpfahlversuche im Ton für die Gründung der Talbrücke Alzey. *Der Bauingenieur* **1974**, *49*, 310–317.
13. Whitaker, T.; Cooke, R.W. An investigation of the shaft and base resistance of large bored piles in London Clay. In *Large Bored Piles*; Institution of Civil Engineers, ICE Virtual Library: London, UK, 1966; pp. 7–49.
14. Breth, H. Das Tragverhalten des Frankfurter Tons bei im Tiefbau auftretenden Beanspruchungen. In *Mitteilungen der Versuchsanstalt für Bodenmechanik und Grundbau der TH Darmstadt*; für Bodenmechanik und Grundbau: Versuchsanst, Germany, 1970.
15. MIDAS. *GTS NX User Manual, Analysis Reference Chapter 4 Materials*; Section [2]; Plastic Material Properties; MIDAS Teotech: Gyeonggi-do, Korea, 2009.
16. *Midas GTS User Supplied Subroutine*; MIDAS Information Technology Co. Ltd.: Seongnam, Korea, 2009; Chapter 3; p. 62.
17. Duncan, J.M.; Buchignani, A.L. *An Engineering Manual for Settlement Studies*; Berkeley, Department of Civil Engineering, University of California: Oakland, CA, USA, 1987.
18. O'Neill, M.W.; Reese, L.C. *Drilled Shaft: Construction Procedures and Design Methods*; Federal Highway Administration: Washington, DC, USA, 1999.
19. Rollins, K.M.; Clayton, R.J.; Mikesell, R.C.; Blaise, B.C. Drilled Shaft Side Friction in Gravelly Soils. *J. Geotech. Geoenviron. Eng. ASCE* **2005**, *131*, 987–1003. [CrossRef]
20. Franke, E. Großbohrpfähle. In *Vorträge der Baugrundtagung 1970 in Düsseldorf*; 5.167 Essen; Deutsche Gesellschaft für Erd-und Grundbau e.V.: Essen, Germany, 1970.
21. Touma, F.T.; Reese, L.C. Behavior of bored piles in sand. *J. Geotech. Eng. Div. ASCE* **1974**, *100*, 749–761.
22. *DIN 4014, German Association for Earthworks and Foundation Engineering*; Deutsches Institute fur Normung: Berlin, Germany, 1990.
23. ECP 202/4. *Egyptian Code for Soil Mechanics—Design and Construction of Foundations*; Part 4, Deep foundations; The Housing and Building Research Center (HBRC): Cairo, Egypt, 2005.

24. Meyerhof, G.G. Bearing Capacity and Settlement of Pile Foundations. *J. Geotech. Eng. Div. Am. Soc. Civ. Eng.* **1976**, *102*, 197–228.
25. *AASHTO, LRFD Bridge Design Specification*, 2nd ed.; American Association of State Highway and Transportation Officials: Washington, DC, USA, 1998.

geosciences

MDPI

Article

Data-Driven Field Observational Method of a Contiguous Bored Pile Wall System Affected by Accidental Groundwater Drawdown

Elizabeth Eu-Mee Chong [1,*] and Dominic Ek-Leong Ong [1,2]

1 Lecturer, Centre for Sustainable Technologies, Swinburne University of Technology Sarawak Campus, Kuching 93350, Malaysia; d.ong@griffith.edu.au

2 Senior Lecturer, Griffith University, Nathan, Queensland 4111, Australia

* Correspondence: echong@swinburne.edu.my

Received: 4 June 2020; Accepted: 10 July 2020; Published: 13 July 2020

Abstract: This paper presents the use of a 700 mm-diameter contiguous bored pile (CBP) wall for a main basement deep excavation project with cut-and-cover tunnel. Due to the presence of cement grout columns between piles behind the CBP wall, the main basement was considered to be 'impermeable'. However, site observations have shown that installation of ground anchors have unintentionally punctured the water tightness of the wall, creating leakages through the CBP wall and the possibility of localized groundwater lowering, as evidenced by the relatively large settlements. In the absence of cement grout columns at the cut-and-cover tunnel section, immediate groundwater drawdown was observed with the excavation rate. Settlement induced by the excavation and groundwater drawdown only slowed down upon the casting of skinwall to prevent groundwater from flowing through the wall. The accidental groundwater leakage led to small wall deflection. The ratio of maximum settlement to maximum deflection is atypical to those reported in the literature. The analysis also revealed that corner effect is significant with smaller settlement registered at the corners of the wall.

Keywords: deep excavation; contiguous bored pile wall; capping beam; groundwater; wall permeability

1. Introduction

The rapid development of Kuching City, in the state of Sarawak, Malaysian Borneo, in recent years has seen the needs for more underground structures to be constructed. Deep excavation projects often cause stresses to the surrounding buildings and structures. Among the factors contributing to these stresses are groundwater imbalance, soil movement, and construction sequence and activities.

Many studies of deep excavation in clay have been well documented over the years [1–9], but studies of deep excavation in sand are scarce. Moreover, most of the deep excavation projects are often constructed by using diaphragm wall as retaining system as opposed to the contiguous bored pile (CBP) wall used in this study. A database by Long [10] showed that diaphragm wall, sheetpile wall, and contiguous wall are widely used throughout the world although the first two constitute most of the database. One of the characteristics of CBP wall is the gap between successive piles that allow for more economical design. These gaps would allow groundwater to flow through the wall causing groundwater drawdown. In overconsolidated clays in London, Powrie et al. [11] acknowledged the possibility of groundwater flowing through the wall that caused a reduction in long-term pore pressures behind the retaining walls. Through simulations using flowtank, Richards et al. [12] demonstrated the groundwater table immediately behind the wall goes from near vertical to near horizontal, when groundwater were allowed to be discharged. These gaps are often sealed to prevent groundwater flow and despite the measures taken to ensure that all gaps are sealed, leaking may still occur, leading to possibly further drawdown. Leakages are discontinuities in the wall and is experienced in geotechnical applications such as vertical cut-off walls [13], thus reflecting the case presented in this paper. On the other hand, in examining leakages in jet-grouted cut-off walls, Pan et al. [14] stated that the flow rate of the leakages increases with depth and exposed length of the wall, and decreases when more jet-grouted rows with larger diameter and closer spacing were used. These observations in general support the field observations made herein where the rate of leakage reduced when the skinwall (water barrier) was constructed.

As not many deep excavation projects in Kuching have been documented, this gave rise to an opportunity to study the performance of the soil-structure responses of a three-level basement of a shopping mall and hotel complex with a cut-and-cover tunnel situated in the heart of Kuching city. This paper discusses the unique performance of a deep excavation project with CBP wall tied back with two layers of ground anchors. Detailed site information, the subsoil conditions, construction sequence, field measured data, and site observations will be discussed. The effect of ground anchor installation and dewatering on the soil movement around the deep excavation project will be explained in further detail. The three-dimensional corner effect is also presented.

2. Project Background and Subsoil Conditions

This deep excavation project in Kuching City is made up of two main sections, that is, the main basement area which covers approximately 13,000 m^2; and a sloping cut-and-cover tunnel of 120 m long. The longest and widest sections of the main excavation area are 96 m and 155 m, respectively. The reduced level (RL) for the existing ground ranges from RL + 3 m to RL + 3.5 m. The ground floor level was built to RL + 3.6 m. There are three levels of basement, i.e., Basement B1 (RL − 0.6 m), Basement B2 (RL − 3.6 m), and Basement B3 (RL − 6.6 m).

The site layout, together with various geotechnical instrumentations, is shown in Figure 1. The project site is surrounded by century-old colonial era buildings, which consist of two-storey masonry shophouses to the West and Northwest, and one- to two-storey concrete buildings to the Northeast, East, and Southwest of the site. The distance of the closest building to the center of the CBP wall is less than 2 m. Most of these shophouses and buildings were built in the early to mid-20th century. The shophouses are suspected to be constructed on timber bakau piles, as evidenced by the remnants of the demolition of the existing buildings to make way for the current project shown in Figure 2. The timber bakau piles are estimated to be about 6–9 m in length, as reported by Goh and Mair [15] for a project constructed in Singapore at about the same time under the British empire. Owing to the shallow depth, the shophouses are considered as 'floating' on the timber bakau piles. At the southern side of the site is an open field.

The main basement was bounded by eight walls consisting of six CBP walls, namely Walls A to F, as indicated in Figure 1 and two existing reinforced concrete (RC) walls. It was constructed with a bottom-up method and the total depth of excavation averaged about 10 m. A 700-mm diameter CBP with center-to-center spacing of 750 mm was used as the retaining wall and tied back with two levels of temporary ground anchors installed at a horizontal spacing of 2.25 m. At the concave corners, double strutting was used instead of ground anchors. The use of temporary ground anchors and corner struts provided unobstructed working space for excavation and basement construction. The length of the CBP wall is 14 m, which leaves the CBP wall embedment length to be about 4 m. To seal the gaps between the successive bored piles, 200-mm diameter cement group columns (CGC) were installed until the depth where hard strata could be found. The details of the CBP wall and CGC can be seen in Figure 3.

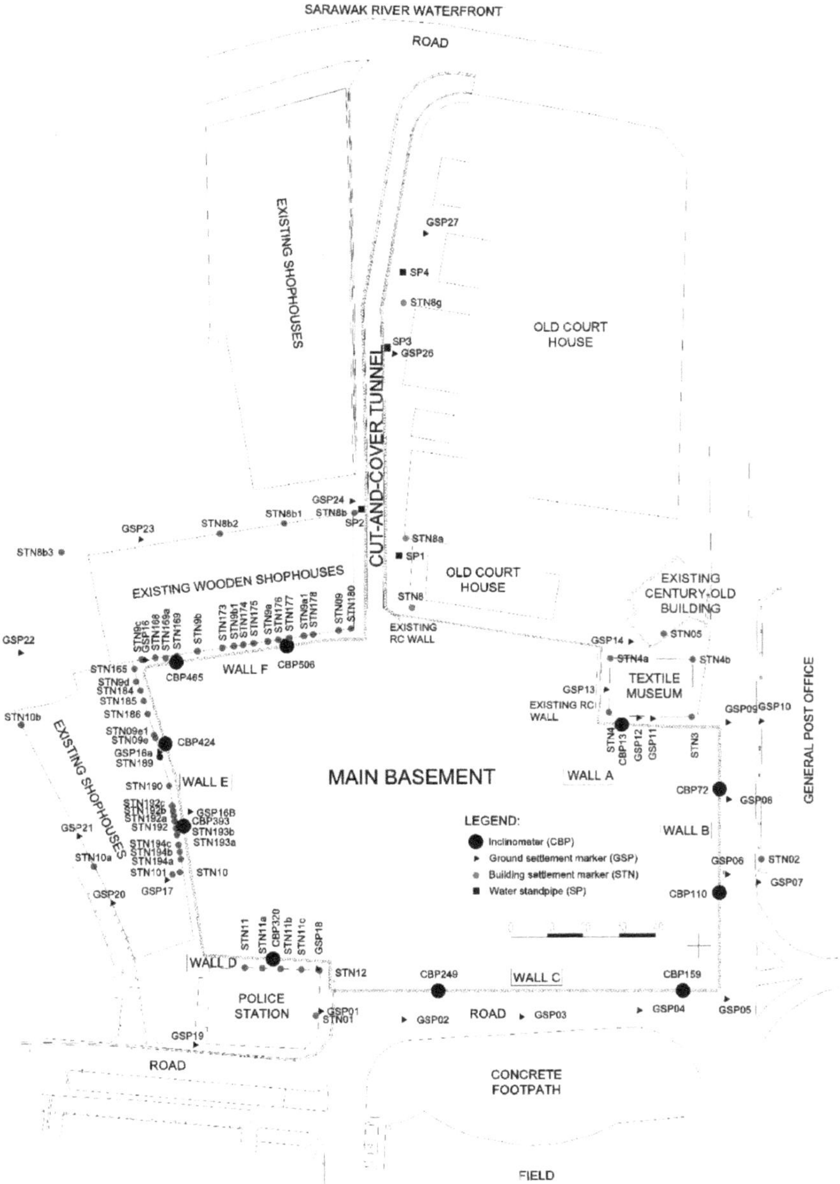

Figure 1. Site layout plan of the main basement and cut-and-cover tunnel with geotechnical instrumentation.

Figure 2. Timber bakau piles used as support system for the demolished buildings.

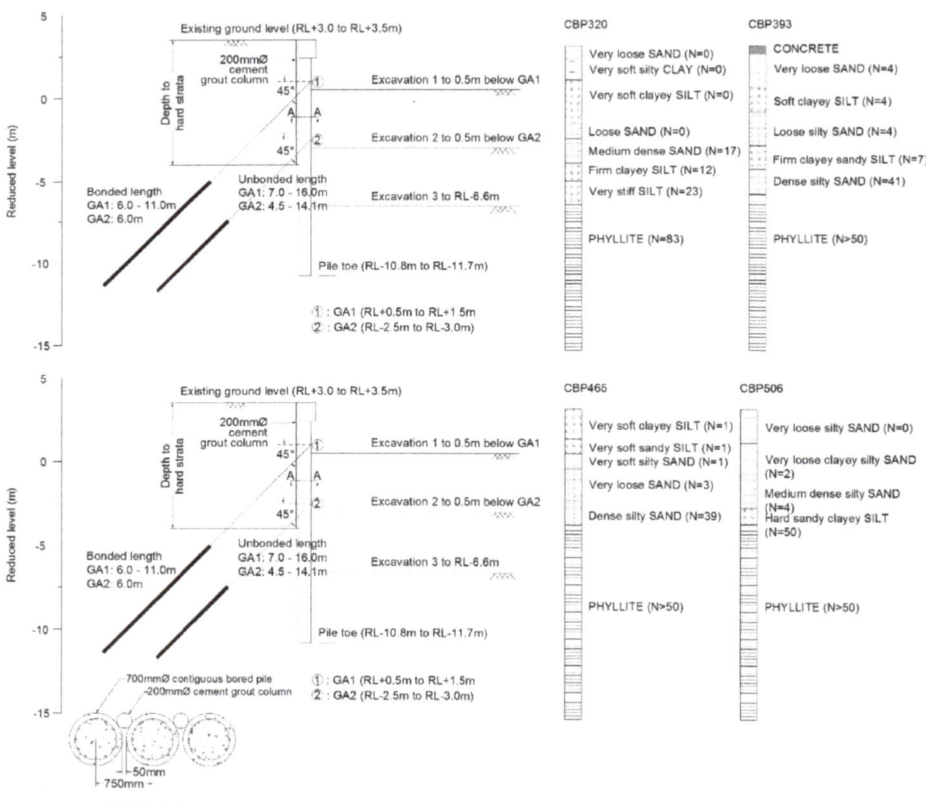

Figure 3. Cross section of contiguous bored pile (CBP) wall details and soil profiles for different wall sections.

The cut-and-cover tunnel is one of the two vehicle exits from the basement parking area. Due to this, the depth of the excavation varies along the chainages of the tunnel from 7 m deep to ground level.

A similar 700-mm diameter CBP wall lined up both sides of the tunnel and temporary strutting were available at the top of the CBP wall. The absence of cement grout columns that was supposed to plug the gaps between CBPs has created a 'permeable' wall condition along this section. Wooden shophouses occupy one side of the tunnel, while a raised Old Court House sits on the other side of the tunnel. The closest distance from the buildings to the tunnel is 0.6 m. Due to such a close proximity, a few bored piles could not be installed as the drilling machine could not be maneuvered to the location and it was replaced with soldier bored piles instead. A river is located perpendicularly approximately 50 m from the end of the cut-and-cover tunnel. Groundwater table was generally encountered at about 0.7 m to 1.0 m below the ground level.

2.1. Soil Conditions

Figure 3 shows the cross sections of the CBP wall and selected soil profiles along Walls D, E, and F. The site is characterized by loose sand interbedded with firm clayey silt with localized pockets of clay and peat. Figure 4 shows the soil properties of standard penetration test (SPT) 'N', particle content, water content, plastic limit and liquid limit of the construction site obtained from 28 boreholes drilled during the soil investigation works. The first 5 m have relatively low 'N' values, approximately 1–15. This is followed by 15 < N < 50 in the next 5–7 m, i.e., about 8 m from existing ground level. The sand contents are quite high, ranging from 60% to 90% for the top 6 m, as shown in Figure 4b and reduced to 50% at the depth of approximately 8 m. This is consistent with the soil profiles where hard strata in the form of metamorphic phyllite can be found at the depth of approximately 6–8 m below the existing ground level with occasional metasandstone. Despite the fact that the rock quality designation (RQD) of phyllite is less than 25% upon coring, the strength of phyllite as a mass block is relatively undisturbed and exhibited high strength [16]. Due to the presence of rock at shallow depths, the CBP walls are consistently socketed into the predominantly phyllite bedrock.

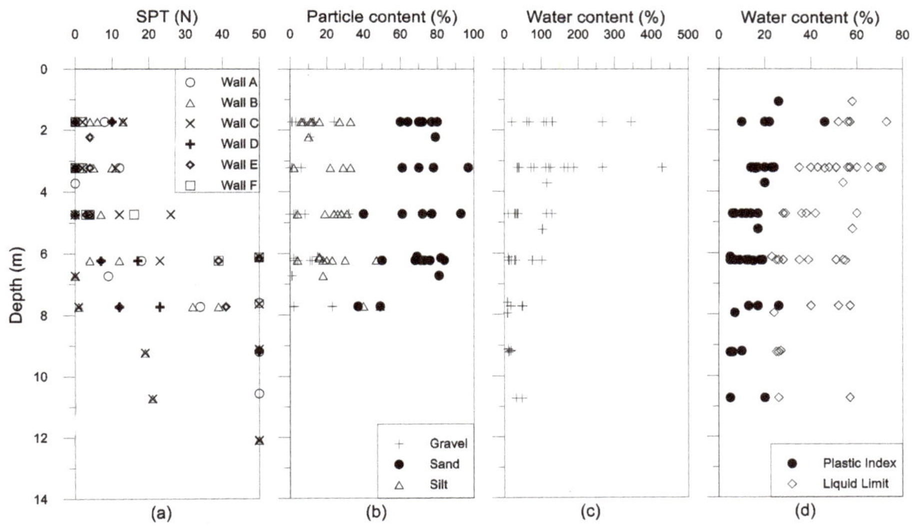

Figure 4. Soil properties obtained from borehole logs: (**a**) standard penetration test (SPT) 'N'; (**b**) particle content; (**c**) water content; (**d**) plastic limit and liquid limit.

2.2. Geotechnical Instrumentation and Monitoring

In an effort to monitor the performance of the CBP wall during construction, geotechnical monitoring instruments such as inclinometers, building and ground settlement markers, and water standpipe were set up surrounding the excavation site and monitored throughout the whole construction. The layout of the instrumentation plan is shown in Figure 1.

Before the commencement of bored pile installation, there were 45 and 41 numbers of building and ground settlement markers installed surrounding the site, respectively. These settlement points were measured weekly throughout the construction. When relatively large settlements were observed during the construction, additional 40 building markers were installed along Walls E and F, and in the vicinity of the cut-and-cover tunnel. For these sections, the settlement monitoring was carried out biweekly. Unfortunately, 27 of the total settlement markers had to be either replaced or discontinued due to construction activity and disturbance from public.

A total of 10 inclinometers were installed in the CBP wall to measure the lateral movement of the CBP wall. No inclinometer was available at the existing RC wall. Each inclinometer measured at about 20 m length, enabling an extension of 6 m after the pile toe to capture any toe movement, as was reported by Hsieh et al. [17]. A weekly reading was taken for monitoring of wall movement.

Four 50-mm diameter open perforated water standpipes were available along the cut-and-cover tunnel. Monitoring the groundwater level through the open water standpipes is considered sufficient due to the high permeability of the sandy soils. The water levels were taken on a weekly basis at these water standpipes.

2.3. Construction Sequence

Throughout the construction of the basement, three major construction stages could be identified: installation of CBP wall, main excavation works, and substructure constructions. When the contractor took over the site, there was a huge basin at the center of the site left by the demolition of existing buildings. Hacking, relocating, and removing the existing infrastructures within the site perimeter were the initial site preparatory works.

Due to the large excavation size, the excavation progressed clockwise from Wall A to Wall F, followed by excavation at the cut-and-cover tunnel. The construction began with the installation of 700-mm diameter bored piles using concrete grade 40 MPa. An alternate sequence of concreting the bored piles prevented the surrounding soil from collapsing and allowed the concrete to set in the steel casings used. Subsequently, the 50-mm gap between the piles was sealed using 200-mm diameter cement grout columns of varying depths to the bedrock. The pile heads were hacked about 1.5 m from ground level to enable the construction of capping beam. The first stage of excavation was typically about 2–3 m from the existing ground level until 0.5 m below the first level of temporary ground anchor (GA-1), between RL0.0 m to RL + 1.0 m to allow room for ground anchor drill auger to work. Using the rotary wash drilling method, temporary 150-mm diameter ground anchors were installed at 45° angle to provide horizontal restraint for the wall. The ground anchors were made of unravel PC strand system and tested to 1.25 times of the working load. Upon insertion of the ground anchor into the borehole, cement grout of 30 MPa was pumped in through tremie pipe. The ground anchors had 6 m fixed length and approximately 10 m free length, varying from section to section. Once the concrete has set, the ground anchors were pre-stressed to 60 T and the locked off load was 66 T or 110% of the working load to consider for slippage and creep within the tendon [18]. Upon completion, the second stage of excavation took place after the pre-stressing until 0.5 m below the second level of ground anchor (GA-2), which is between RL − 3.0 m to RL − 3.5 m. The same ground anchor installation process is repeated for GA-2. The final stage is excavation to formation level at RL − 6.6 m. The access earth ramp for the construction machineries was located close to CBP249, hence GA-2 installation was done much later. The duration from the start of excavation until the completion of excavation works for each wall section took about three months, except CBP249.

In the middle of the main basement below the formation level, a rectangular pit measuring 15 m × 20 m × 5 m was excavated for septic treatment plant (STP). It was assumed that the construction of the STP has no implication on the soil movement due to (a) being sufficiently far from the CBP walls which is out of the influence zone of an excavation, (b) RC walls were used for the STP, and (c) the whole STP is constructed in mainly phyllite. The construction of substructure began following a dormant period after excavation. After the completion of each level floor slab, skinwalls was concreted over the CBP wall as permanent seal to prevent seepage of groundwater into the basement. Prior to the concreting of skinwalls, the temporary ground anchors were cut off and the anchor heads removed.

For the cut-and-cover tunnel, the construction began with the installation of bored piles which was continued from the main basement. Due to the nature of the cut-and-cover tunnel being an exit ramp, the length of the bored piles varied from 11.5 m at the start of the tunnel to 8.5 m at the end of the tunnel. However, no cement grout column was installed behind the CBP wall. The pile heads were hacked to enable the tunnel cover slab construction. As the excavation progressed, temporary strutting at the top of the piles were installed for the first 90 m of the tunnel. The remaining length of the tunnel did not require temporary strutting due to shallower excavation depths. Construction of skinwalls followed after the concreting of the tunnel slab. The excavation sequence is summarized in Table 1 for main basement and Table 2 for the cut-and-cover tunnel.

Table 1. Construction sequence for main basement.

Stage	Construction Activity
1	Installation of CBP Wall
2	Casting of Pile Capping Beam
3	Excavate to 0.5 m below GA-1 (B1) (RL0.0m to RL + 1.0 m)
4	Installation of the first level of ground anchor (GA-1) (RL + 0.5 m to RL + 1.5 m)
5	Stressing of the first level of ground anchor (GA-1)
6	Excavate to 0.5 m below GA-2 (B2) (RL − 3.0 m to RL − 3.5 m)
7	Installation of the second level of ground anchor (GA-2) (RL − 2.5 m to RL − 3.0 m)
8	Stressing of the second level of ground anchor (GA-2)
9	Excavate to the formation level (B3) (RL − 6.6 m)

Table 2. Construction sequence for cut-and-cover tunnel.

Stage	Construction Activity
1	Installation of CBP Wall
2	Excavation 1 to RL + 0.5 m
3	Excavation 2 to RL − 3.0 m
4	Installation of temporary struts
5	Excavation 3 to RL − 3.7 m
6	Excavation 4 to RL − 4.3 m
7	Construction of skinwall

3. Site Observations

The dense array of field instrumentations allowed the project to be monitored from many aspects, such as the vertical and horizontal soil movements and the groundwater loss. The allowable building settlement was restricted to 10 mm while the ground settlement was limited to 50 mm. Skempton and MacDonald [19] suggested the allowable settlement for isolated foundations and raft foundations in sands to be 50 mm and 50–75 mm, respectively. Since timber bakau piles were used to support the shophouses along Walls E and F, these buildings were considered to be 'floating' raft foundation that will partially settle with the ground.

3.1. Building and Ground Settlements

To facilitate the discussion on the overall response of the building and ground to the construction activities, only the settlement profiles near six out of 10 inclinometer sections will be presented, while the overall settlement for various construction activities will be summarized. Most of the settlement markers were closely located parallel to the CBP walls, due to the location of the existing buildings.

Generally, CBP installation did not contribute much settlement, except for CBP320 and CBP465. Through an empirical method by Clough and O'Rourke [3], the upper bound settlement for diaphragm wall installation in granular soils is 0.12% of the wall depth, equivalent to 16.8 mm in the present case. Gaba et al. [20] suggested the surface movement at wall for CBP wall installation in stiff clay 0.04% of wall depth, which translates to approximately 5.6 mm for the present case. From Table 3, the ground settlement during the CBP installation for the entire site ranges from 0.3 mm to 17.6 mm. If Clough and O'Rourke's limit were to be used, ground settlement at CBP320 and CBP465, which measured at 17.6 mm and 13.2 mm (both 24% of the total settlement), respectively, would have exceeded the limit. The ground settlement near CBP72, CBP110 and CBP249 shows a rather substantial percentage of settlement, 27%, 22% and 20%, respectively. However, the magnitude of those settlements is relatively smaller than CBP320 and CBP465.

Table 3. Summary of building and ground settlement during CBP wall installation and main excavation works.

	Activity	CBP Wall Section									
		13	72	110	159	249	320	393	424	465	506
CBP wall installation	Building settlement during the activity (mm)	−0.1	NIL	NIL	NIL	NIL	1.4	−11.0	−5.9	−14.0	−4.4
	% of total settlement	1	NIL	NIL	NIL	NIL	−44	6	11	25	8
	Ground settlement during the activity (mm)	−0.3	−6.4	−7.4	−3.7	−4.3	−17.6	−4.3	−9.6	−13.2	NIL
	% of total settlement	1	27	22	15	20	24	2	12	24	NIL
Main excavation works	Building settlement during the activity (mm)	−6.7	NIL	0	NIL	NIL	4.2	−98.3	−32.8	−20.2	−18.7
	% of total settlement	16	NIL	0	NIL	NIL	−131	55	62	36	34
	Ground settlement during the activity (mm)	−7.1	−3.1	−8.4	−10.5	−16.9	−17.2	−93.3	−49.3	−17.9	NIL
	% of total settlement	12	13	26	43	77	24	53	61	32	NIL

As expected, the main excavation works had caused much of the settlement to occur, especially along Walls E and F. The building settlement near CBP393, CBP424, CBP465, and CBP506 recorded settlement from 18.7 mm to 98.3 mm, much greater than the allowable building settlement of 10 mm. For these sections, the ground settlement markers also recorded similar magnitude to the building markers, ranging from 17.9 mm to 93.3 mm. It can be deduced that the buildings (shophouses) which were founded on timber bakau piles settled together with the ground upon removal of the overburden stresses. Essentially, this means the buildings are 'floating' on the timber bakau piles. A separate discussion on these wall sections will be made in the next section. For CBP249, the main excavation works contributed to 16.9 mm ground settlement (77% of the total settlement). The high percentage of settlement was due to the longer construction period as CBP249 was located near the temporary access ramp. Figure 5 shows the overall building and ground settlement profiles for CBP13 and CBP320. There is a remarkable difference between the building and ground settlement. The latter is settling much more than the building settlement. This finding suggested that the buildings near to these two sections were built on solid foundation that were able to tolerate the soil movement. The ground, however, settles when the horizontal stress increases.

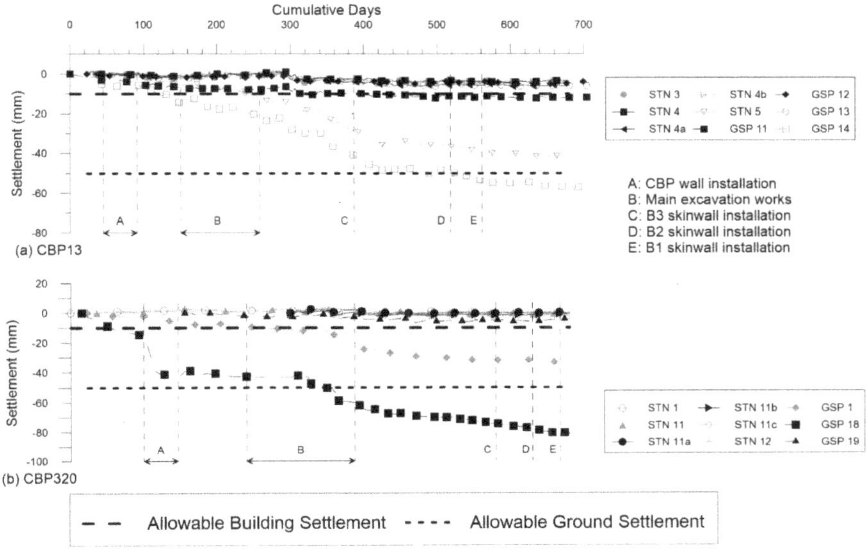

Figure 5. Overall building and ground settlement profiles for (**a**) CBP13 and (**b**) CBP320.

3.2. *Specific Observations Along Walls E and F*

As mentioned in the preceding section, the building and ground settlement along Walls E and F were relatively large during the main excavation works. Figure 6 presents the measured settlements near four inclinometer sections along Walls E and F that had experienced rather substantial movement during the main excavation works and warrant for a closer examination. The main excavation works consist of Stages 3 to 9, that is, the three stages of excavation and ground anchor installation and pre-stressing in between the excavation works. These activities contributed up to 98.3 mm and 93.3 mm settlement, accounting for 55% and 53% of the total building and ground settlement, respectively. Based on the settlement markers that were ideally lined perpendicularly to the inclinometer section (marked by black color line in the figure), STN193b near CBP393 recorded 117.4 mm building settlement at Day 400. For other inclinometer sections, the largest settlement for CBP424, CBP465 and CBP506 measured at 60.9 mm, 43.6 mm and 13.8 mm, respectively. There were other settlement markers that recorded larger settlement than reported here but these settlement markers were not in a perpendicular direction behind the inclinometer sections. Upon the alarming settlement observation, additional settlement markers, mostly building settlement markers, were installed and increased monitoring of twice weekly were practiced.

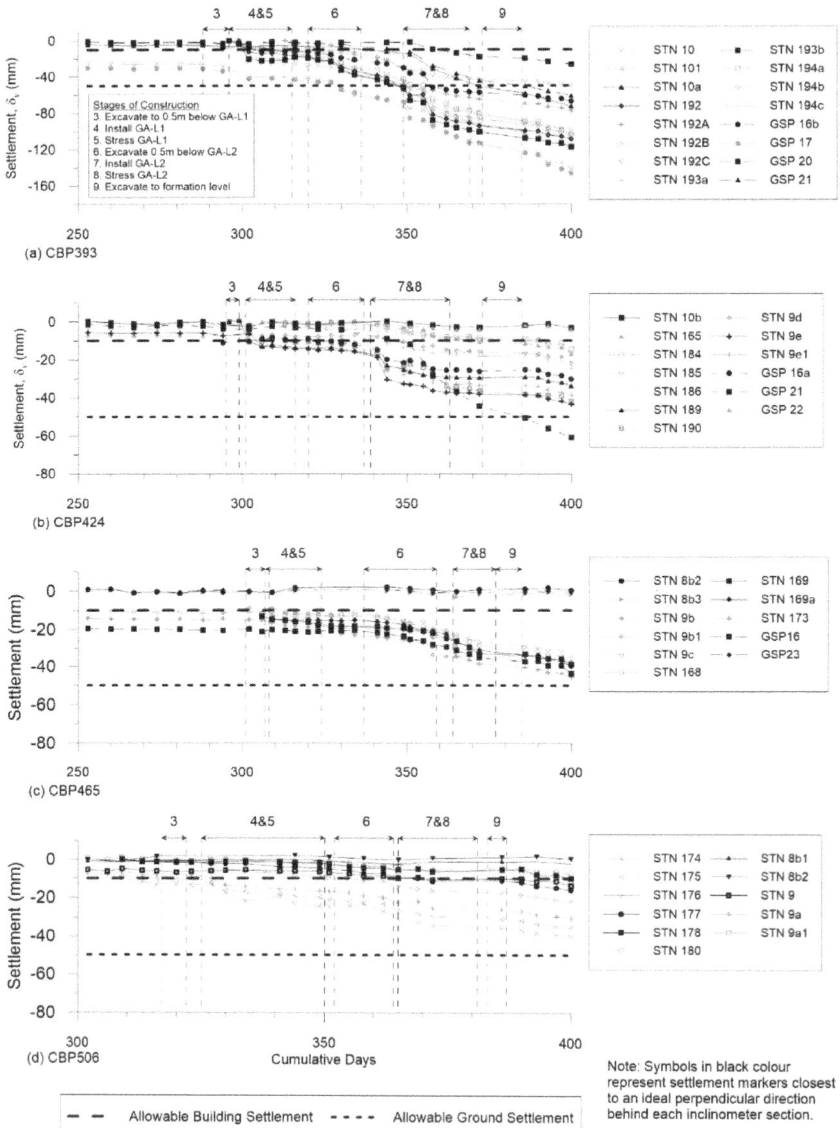

Figure 6. Specific building and ground settlement profiles throughout the main excavation works at (**a**) CBP393; (**b**) CBP424; (**c**) CBP465; and (**d**) CBP506.

It was observed that Stages 4, 5, 7, and 8, which are the ground anchor works, contributed to large and immediate settlement. Table 4 summarizes the percentage of settlement during the ground anchor works and excavation works. The ground anchor works had caused building settlement of up to 47% and ground settlement of up to 37%. Meanwhile, the excavation works caused smaller percentage of settlement, except for CBP506 with 80% and CBP465 with 53%. These two outliers could be ignored as the final settlement was only 1.5 mm, and thus the large percentage of settlement was caused by fluctuation in the data. The installation of GA-1 caused smaller settlement compared to GA-2, while the second stage of excavation had the largest settlements. Rotary wash drilling method had been used in the ground anchor installation works. Due to the use of water during drilling process, the soil particles may possibly be washed out from the holes and thus causing large settlements. Kempfert and Gebreselassie [21] reported that for an excavation project in soft soil with soldier piles and tied back with a single layer of ground anchor spaced at 0.90 m, it recorded 60–72% settlement that occurred during anchor installation. They attributed the appreciable settlement to fresh cement–bentonite slurry in the drill hole and the vibration arising from the anchor installation method.

Table 4. Summary of the total building and ground settlement percentage during ground anchor works and excavation works.

Activity	Stages	Settlement Category	Percentage of Total Settlement during the Duration (%)			
			CBP393	**CBP424**	**CBP465**	**CBP506**
Ground anchor works	4&5	Building	7–15	0–37	1–7	0–8
		Ground	0–5	2	0	No data
	7&8	Building	9–33	12–47	1–18	2–19
		Ground	0–34	10–37	11	No data
Excavation works	3	Building	1–3	0	1–6	0–4
		Ground	0–4	0	1–15	No data
	6	Building	2–18	2–7	2–80	4–53
		Ground	0–11	7	14	No data
	9	Building	1–8	1–3	1–18	0–4
		Ground	0–8	4	8	No data

Another possible reason for the relatively large settlements is groundwater loss. An acute observation revealed continuous water stain marks below the ground anchor positions for most of the wall sections, especially Walls D, E, and F, as can be seen in Figure 7a. The CBP wall was considered to be watertight as cement grout columns were used to seal off the gaps in between successive bored piles for the entire wall. This was confirmed through the dry CBP walls at the corner of the main basement (right side of Figure 7a), where corner struts were used instead of ground anchors. However, as the ground anchor center-to-center spacing is 2.25 m, it coincides with the cement grout columns that provides water proofing. The ground anchor drilling process (Figure 7b) cut through the grout columns and 'punctures' the water-tightness of the CBP wall. Clough and O'Rourke [3] described several cases where potential water flow may lead to ground settlement, among them are flow through wall flaw as shown in Figure 8. Although the ground anchors were grouted immediately after the insertion of ground anchor, fresh concrete requires curing time to effectively seal the gap. Site observations indicated that at the first opportunity when the puncture occurred, the groundwater showed immediate response by seeping through the drill hole as shown in Figure 7c. As the groundwater table was generally about 1.0m below the existing ground level, which is higher than GA-1 level, the water stain marks occurred for both GA-1 and GA-2 levels. This response further confirms the previously discussed Figure 6, where the large and immediate settlement could be observed during ground anchor works as discussed earlier. The groundwater loss issue ceased when skinwall was constructed. Unfortunately, water standpipes were not available at the main basement, as it was thought to be unnecessary due to the presence of cement grout column.

Figure 7. (**a**) Groundwater leakage from where ground anchors were installed; (**b**) drilling of ground anchor in between bored piles; and (**c**) close-up of the water leakage from the ground anchor.

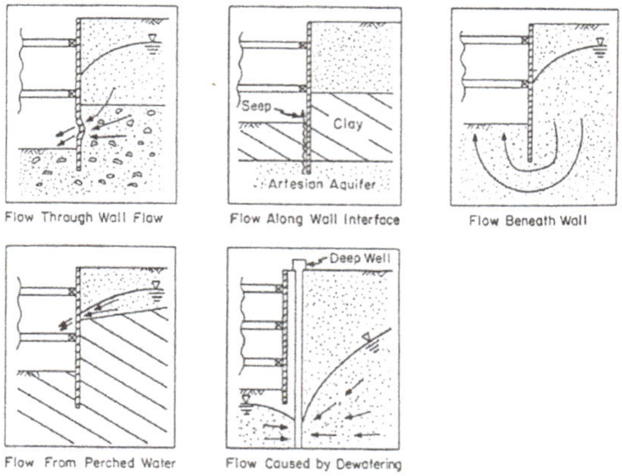

Figure 8. Potential water flow which may lead to ground settlement [3].

Figure 9 shows the relationship between the measured maximum building and ground surface settlement (δ_{vm}) and the excavation depth (H) for all excavation stages. The δ_{vm} generally increase as H increases, especially for Walls D, E, and F. However, there are also a good spread of data that has registered very small δ_{vm}, allowing the data to fall within the range of 0% to 1.23%H, for both building and ground settlement. Included in Figure 9b are the δ_{vm} /H from databases such as Clough and O'Rourke [3] for excavation in stiff clays, residual soils, and sands; and Moormann [22] for non-cohesive soils; and some data from excavation in sand [23–25] for comparison. These case histories had relatively small settlement compared to the present study, generally below 0.33%*H*. As previously discussed, the flaw in the CBP wall upon ground anchor installation had caused the groundwater loss and significant settlement had been observed. Due to some settlement markers being located further away from the wall or on buildings that has more solid foundation, a big range of δ_{vm}/H is observed. Zhang et al. [26] reported a quite similar experience to the present study, where δ_{vm}/H was 0.9% and attributed the large ground surface settlements to significant groundwater drawdown.

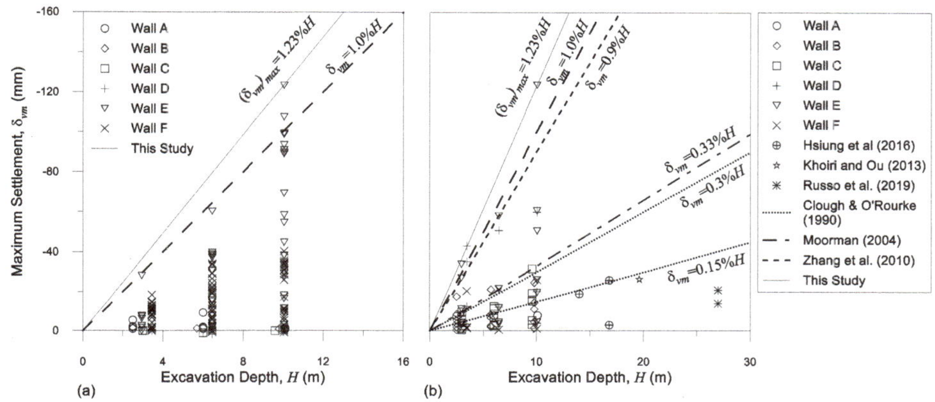

Figure 9. Relationship between maximum settlement and excavation depth for all the excavation stages for (**a**) building; and (**b**) ground.

3.3. *Groundwater Levels*

The groundwater monitoring was only available at the cut-and-cover tunnel where the CBP walls were considered to be 'permeable' due to the absence of cement grout column. During the early stages of construction, it was observed that the soils in the gaps between the pile were constantly damp, indicating slow groundwater leakage through the wall. Figure 10 displays the measured groundwater level and the corresponding settlement profiles for SP1, SP2, SP3, and SP4. It is observed that when the excavation was carried out, the groundwater level behind the wall were immediately drawndown. During the quiescent period for skinwall preparatory works, the groundwater level was noted to maintain at the final excavation levels consistently for SP1, SP2, and SP3. This demonstrated the groundwater loss through the gaps in between the piles experimented by Richards et al. [12]. Prior to concreting the foundation slab, submersible pumps were placed in depressed pump sumps to extract the seepage water on the excavation floor so that concreting can be carried out in the dry. Similarly, the casting of skinwall was carried out in dry condition. Casting of skinwall cut off the groundwater flow through the wall and re-established the groundwater to its original level. Although the settlement was less than 5 mm, the settlement profiles for the three water standpipes showed a similar trend where during the excavation induced groundwater lowering, the settlement occurred at higher rate and slowed down considerably upon the construction of skinwall. This observation also strengthens the earlier observation where groundwater lowering induced by ground anchor installation works had caused relatively large settlement to occur along Walls E and F. At the end of the tunnel, the groundwater level was observed to drawdown gradually with the excavation as shown in Figure 10d. Similarly, with the settlements near the other three standpipes, minimal settlement was observed. Unfortunately, due to SP4 being destroyed during construction activity, no data could be obtained after skinwall construction.

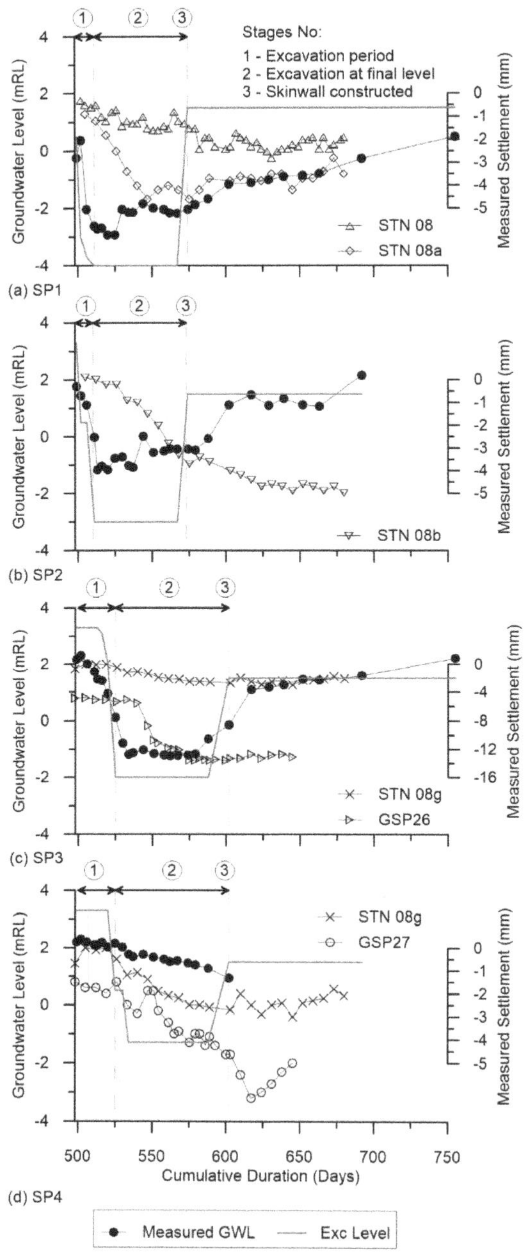

Figure 10. Measured groundwater and settlement profiles for (**a**) SP1; (**b**) SP2; (**c**) SP3; and (**d**) SP4.

3.4. CBP Wall Deflections

Before further examination of the horizontal wall deflection (δ_h), there is a need to establish if the wall toe movement is significant. Hwang et al. [27] suggested that the magnitude of the wall deflection is affected by the wall toe movements and further argued that wall deflection simply cannot be moving outward, due to the limit of strut elongation. Figure 11 shows the progression of the wall toe movements at all excavation depths. The first stage of excavation hardly recorded any movement except for CBP159. The second stage of excavation sees a little more movement, but it is within 0.5 mm from the wall. The final stage of excavation caused the wall toe to move inward with the largest movement of 2 mm. With such small measurements, the wall toe movements were considered to be insignificant. This observation further strengthens the fact when the bored piles are socketed into phyllite by about 4 m, the mass block of phyllite layer provided sufficient strength to restrict the wall movement. Furthermore, it is possible to have an outward toe movement in this case, as over-stressing the ground anchor will cause the outward toe movement.

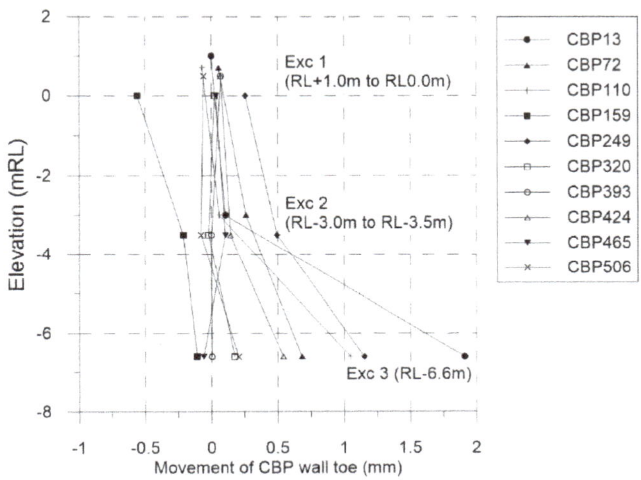

Figure 11. Progressive wall toe movements at all excavation depths.

Figure 12 shows the horizontal wall deflections for excavation Stages 3, 6, and 9. It is observed that the wall behaved in cantilever mode at Stage 3, where the largest movement occurred at the top of the wall. This movement is expected as the soil near the top has been excavated for pile head hacking and subsequently to 0.5 m below GA-1. The maximum horizontal deflection (δ_{hm}) for Stage 3 occurred at CBP13 with 4.04 mm. After the installation of GA-1, the deflection profile changed to prop mode, where the top of the wall moved towards the retained soil. For CBP13, CBP393, and CBP424, the effectiveness of ground anchors were demonstrated when the wall is pulled into the soil by the ground anchor as indicated by the negative deflection value [28]. Although the wall deflection profiles for CBP159 and CBP249 continued the cantilever mode, it was due to CBP159 being situated in a localized pocket of peat which exerts higher pressure onto the wall, while CBP249 was located near the temporary access ramp with heavy machineries passing by. The largest deflection for Stage 6 occurred at CBP249, measuring 7.9 mm. The deep inward movement for all the sections continued to Stage 9, although CBP393 exhibited otherwise. The largest deflection occurred at CBP249, measuring 14.3 mm.

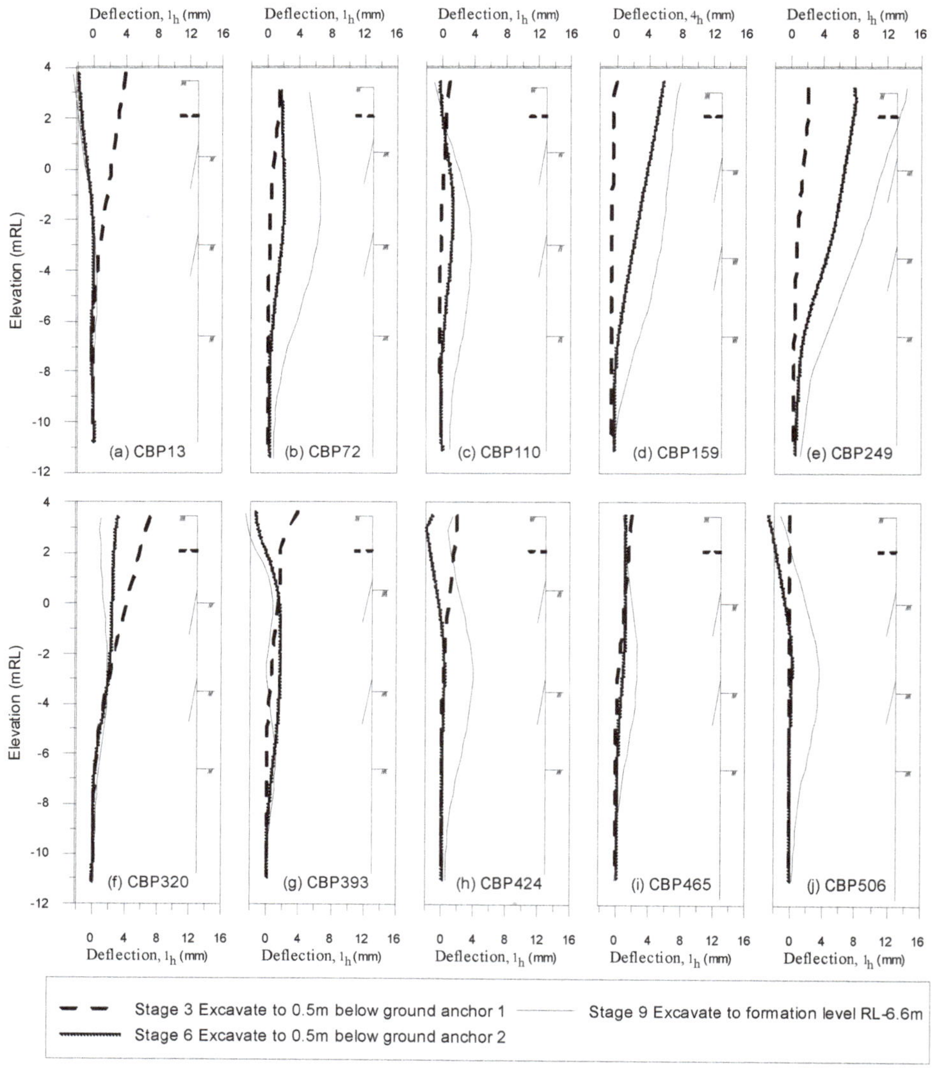

Figure 12. Horizontal wall deflections for excavation Stages 3, 6, and 9.

Most of the δ_{hm} occurred near or higher than the excavation surface, similar to observations in clay [29,30]. From Figure 13, the depth to the maximum lateral wall deflection, H_m, is generally close to the excavation depth. Apart from a few exceptions, the data falls within $H_m = H$ to $H_m = H–4.5$, indicating H_m to occur higher than the excavated surface. For CBP506 at Stage 3, the maximum deflection was recorded at 18-m depth, but it was caused by fluctuation of very small wall deflection. At Stage 9, most of the maximum deflections occurred near the depth of the second excavation. Owing to the shallow depth to hard strata, some of the Stage 9 excavation are done at the phyllite layer. Therefore, smaller deflections were observed near to the surface of final excavation.

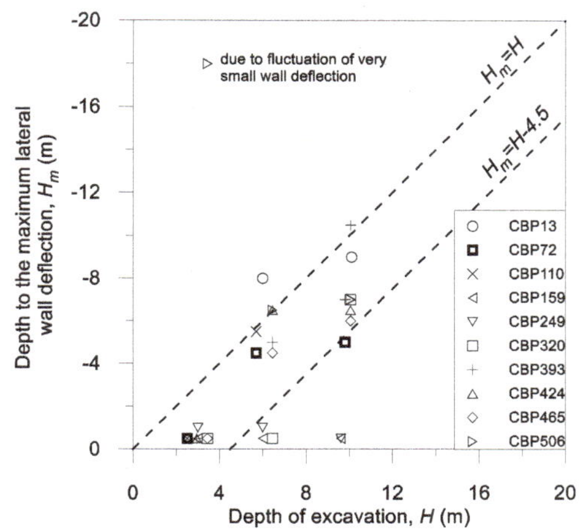

Figure 13. Relationship between location of maximum horizontal wall deflection and excavation depth.

Figure 14 presents the relationship between the δ_{hm} and H. The entire δ_{hm} lie below 0.2%H. The small deflection experienced in the present study is likely due to the groundwater drawdown caused by the puncturing of walls during ground anchor installation. Relief of pore water pressure in the ground due to the accidental dewatering has resulted in less lateral earth pressure to be exerted on the CBP wall. Other reported δ_{hm}/H for sandy soils tend to be higher, within the range of 0.28 [23,31] to 1.33 [32], which did not observe any groundwater drawdown. Moormann [22] reported δ_{hm}/H = 0.5% for excavation in sand. Besides, the shallow depth of phyllite layer have led to smaller deflections as well.

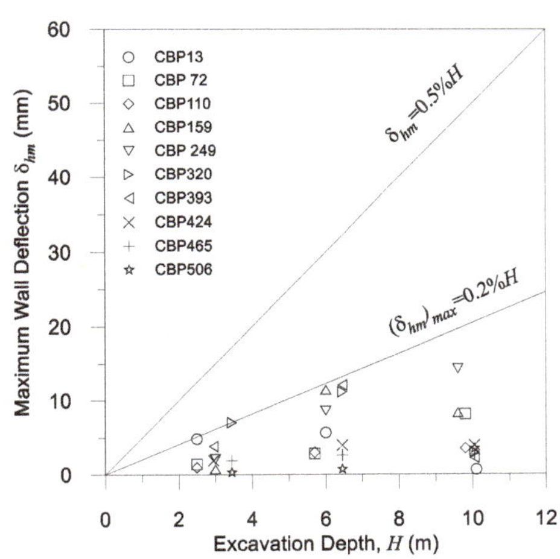

Figure 14. Relationship between maximum wall deflection and excavation depth for all the excavation stages.

3.5. Relationship between Maximum Wall Deflections and Maximum Settlements

As has been established in the present study, the building settlements were observed to be relatively large with small wall deflection. Figure 15a examines the normalized relationship of maximum building settlement and maximum wall deflection. A typical relationship is δ_{vm}/δ_{hm} = 0.5–1.0 [29]. In the present study, the δ_{vm}/δ_{hm} ratio is 14.8. The upper bound of the ratio is contributed by the large building settlements around CBP393. Ignoring these values, the maximum δ_{vm}/δ_{hm} ratio will be 6, still significantly larger than the typical value of 0.5–1.0. Zhang et al. [26] reported δ_{vm}/δ_{hm} of 10 and attributed the excessive ground surface settlements to significant groundwater drawdown due to under-drainage mechanism. Therefore, drawing upon the data from the water standpipes at the cut-and-cover tunnel, it is deduced that groundwater drawdown, caused by ground anchor installation that punctured the wall, has led to significant settlement. The same observation is also extended to the ground settlement where Figure 15b illustrates the maximum δ_{vm}/δ_{hm} ratio is 14.5. The comparable δ_{vm}/δ_{hm} ratios between normalized building and ground settlement was due to shophouses being built on 'floating' timber bakau piles that follows the ground settlement.

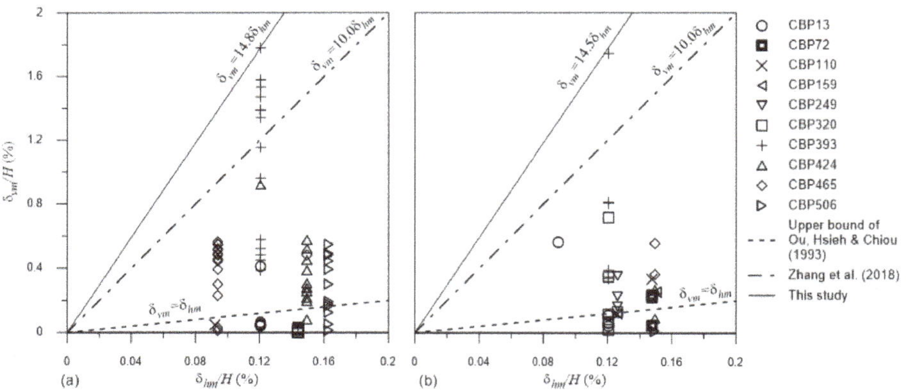

Figure 15. Relationship of maximum (**a**) building settlement and maximum wall deflection; and (**b**) ground settlement and maximum wall deflection.

3.6. Corner Effect on the Maximum Settlements

Higher concentration of settlement markers along Walls E and F will be presented for discussion of corner effect. Figure 16a shows the relationship of building settlement (δ_v) normalized with δ_{vm} and distance to corner (d') normalized with wall length (L) for Wall E. At d'/L = 0.0–0.3, no settlement markers were available. For all the three stages of excavation, the settlement data concentrated within d'/L = 0.3–0.5 (centre of the wall) are larger than those at d'/L = 1.0 (corner). This implies the corner effect is present through the corner restraint of the CBP wall.

Along Wall F, the corner effect for all three stages of excavation displayed the corner effect prominently, as observed from Figure 16b. It is to be noted that the larger settlements occurred from d'/L = 0.0 to d'/L = 0.5 and the left wall corner is located at d'/L = −0.3. The slight shift in the d'/L towards the left corner could be explained by the irregular geometry of the excavation where d'/L = 0 is 80° concave corner while d'/L = 1.0 is a 90° convex corner, which would have lesser earth pressure on the wall, and thus less settlement.

It seems the groundwater lowering due to puncturing of the wall does not seem to diminish the corner effect. Furthermore, the groundwater leakage was observed uniformly for Walls E and F, thus any effect from groundwater lowering on the corner restraint is not significant.

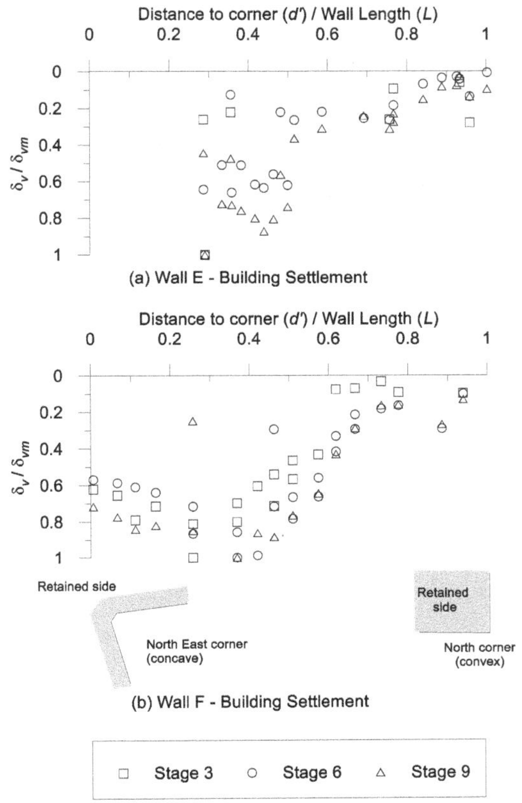

Figure 16. Normalized relationship of maximum settlement and distance to corner for (**a**) building settlement for Wall E; and (**b**) building settlement for Wall F.

4. Conclusions

This paper presents the construction and monitoring of a deep excavation project in the heart of Kuching City. The site background and observations have been thoroughly discussed. Based on the field measurement results, the following conclusions can be drawn:

1. Generally, CBP installation works did not cause much settlement to occur. The main excavation works had caused substantial building and ground settlement up to 98.3 mm and 93.3 mm, respectively. This accounts for 55% of the total building settlement and 53% of the total ground settlement. Due to the similarity in magnitude of the settlement, the buildings that are founded on timber bakau piles along Walls E and F are said to be settling together with the ground.
2. Relatively large settlements were observed along Walls E and F, owing to ground anchor works that had (a) caused the soil particle to be washed out during the drilling process; and (b) punctured the CBP walls with cement grout column causing continuous groundwater loss. The ground anchor installation and pre-stressing works contributed up to 47% and 37% of the total building and ground settlement, respectively. The δ_{vm}/H is 1.23% for both building and ground settlement, much larger than most case histories.
3. For the cut-and-cover tunnel where no CGC were available, the groundwater level was immediately drawn according to the excavation rate. Settlement rate slowed down upon the casting of skinwall. The construction of skinwall had effectively sealed off the site and

prevented groundwater from leaking into the site, as evidenced at both the main basement and cut-and-cover tunnel.

4. The toe movement is minimal, indicating that the wall is indeed properly socketed into phyllite. The initial deflection profile showed cantilever behavior and subsequently changed to prop mode after the installation of ground anchors, with the location of maximum deflection within $H_m = H$ to $H_m = H$–4.5. The δ_{hm} for the final stage of excavation is 14.3 mm, while the δ_{hm}/H is less than 0.2%.
5. Due to the relatively large δ_{vm} and small δ_{hm}, the δ_{vm}/δ_{hm} ratio is 14.8 and 14.5 for building and ground, respectively, far greater than those reported in the literature. Again, this further supports the observation where the settlement had been caused by groundwater loss. The similar δ_{vm}/δ_{hm} ratios between building and ground pointed to the fact that the building and the ground were settling together, as the buildings had been built on 'floating' timber bakau piles.
6. Corner effect is apparent for Walls E and F, although the larger *d'/L* occurred within 0 to 0.5. The imbalance in the plane-strain condition is most likely due to the irregular geometry of the excavation that consists of a concave and a convex corner.

Author Contributions: Conceptualization, E.E.-M.C. and D.E.-L.O.; formal analysis, E.E.-M.C.; investigation, E.E.-M.C.; data curation, E.E.-M.C.; writing—original draft preparation, E.E.-M.C.; writing—review and editing, D.E.-L.O.; supervision, D.E.-L.O.; funding acquisition, D.E.-L.O. All authors have read and agreed to the published version of the manuscript.

Funding: This research was funded by Rakyat Elite Sdn. Bhd.

Acknowledgments: The authors wish to express their gratitude to Bina Puri Construction Sdn. Bhd. and KTA (Sarawak) Sdn. Bhd. for their assistance in the collection of field data and other information.

Conflicts of Interest: The authors declare no conflict of interest. The funders had no role in the design of the study; in the collection, analyses, or interpretation of data; in the writing of the manuscript, or in the decision to publish the results.

References

1. Blackburn, J.T.; Finno, R. Three-dimensional responses observed in an internally braced excavation in soft clay. *J. Geotech. Geoenviron. Eng.* **2007**, *133*, 1364–1373. [CrossRef]
2. Burland, J.B.; Simpson, B.; St. John, H.D. Movements around excavations in London clay. In Proceedings of the 7th European Conference on Soil Mechanics and Foundation Engineering, Brighton, UK, 10–13 September 1979; Balkema: Rotterdam, The Netherlands, 1879; pp. 13–19.
3. Clough, G.W.; O'Rourke, T.D. Construction Induced Movements of In-situ Walls. In Proceedings of the Design and Performance of Earth Retaining Structures, New York, NY, USA, 18–21 June 1990; pp. 439–470.
4. Hashash, Y.M.A.; Whittle, A.J. Ground movement prediction for deep excavations in soft clay. *J. Geotech. Eng.* **1996**, *122*, 474–486. [CrossRef]
5. Hsiung, B.-C.; Yang, K.-H.; Aila, W.; Ge, L. Three-dimensional effects of a deep excavation on wall defections in central Jakarta. *Tunn. Undergr. Space Technol.* **2018**, *72*, 84–96. [CrossRef]
6. Ng, C.W.W.; Lings, M.L. Effects of Modeling Soil Nonlinearity and Wall Installation on Back-Analysis of Deep Excavation in Stiff Clay. *J. Geotech. Eng.* **1995**, *121*, 687–695. [CrossRef]
7. Ng, C.W.W.; Yan, R.W.M. Three-dimensional modelling of a diaphragm wall construction sequence. *Géotechnique* **1999**, *49*, 825–834. [CrossRef]
8. Ou, C.-Y.; Hsieh, P.-G.; Lin, Y.-L. A parametric study of wall deflections in deep excavations with the installation of cross walls. *Comput. Geotech.* **2013**, *50*, 55–65. [CrossRef]
9. Tan, Y.; Wei, B. Observed Behaviours of a Long and Deep Excavation Constructed by Cut-and-Cover Technique in Shanghai Soft Clay. *J. Geotec. Geoenviron. Eng.* **2012**, *138*, 69–88. [CrossRef]
10. Long, M. Database for Retaining Wall and Ground Movements Due to Deep Excavations. *J. Geotech. Geoenviron. Eng.* **2001**, *127*, 203–224. [CrossRef]
11. Powrie, W.; Chandler, R.J.; Carder, D.R.; Watson, G.V.R. Back-analysis of an embedded retaining wall with a stabilizing base slab. *Proc. Inst. Civ. Eng. Geotech. Eng.* **1999**, *137*, 75–86. [CrossRef]

12. Richards, D.J.; Wiggan, C.A.; Powrie, W. Seepage and pore pressures around contiguous pile retaining walls. *Géotechnique* **2016**, *66*, 523–532. [CrossRef]
13. Croce, P.; Modoni, G. Design of jet-grouting cut-offs. *Proc. ICE Ground Improv.* **2007**, *11*, 11–19. [CrossRef]
14. Pan, Y.; Liu, Y.; Chen, E.J. Probabilistic investigation on defective jet-grouted cut-off wall with random geometric imperfections. *Géotechnique* **2019**, *69*, 420–433. [CrossRef]
15. Goh, K.H.; Mair, R.J. The horizontal response of framed buildings on individual footings to excavation-induced movements. *Geotech. Asp. Undergr. Constr. Soft Ground Viggiani* **2012**, 895–902. [CrossRef]
16. Ong, D.E.L.; Choo, C.S. Sustainable construction of a bored pile foundation system in erratic phyllite. In Proceedings of the ASEAN Australian Engineering Congress, Kuching, Sarawak, Malaysia, 25–27 July 2011.
17. Hsieh, P.-G.; Ou, C.-Y.; Lin, Y.-K.; Lu, F.-C. Lessons Learned in Design of an Excavation with the Installation of Buttress Walls. *J. Geoengin.* **2015**, *10*, 63–73. [CrossRef]
18. Puller, M. *Deep Excavations: A Practical Manual*; Thomas Telford Publishing: London, UK, 1996.
19. Skempton, A.W.; MacDonald, D.H. The allowable settlements of buildings. *Proc. Inst. Civ. Eng.* **1956**, *5*, 727–768. [CrossRef]
20. Gaba, A.R.; Simpson, B.; Powrie, W.; Beadman, D.R. *Embedded Retaining Walls—Guidance for Economic Design*; Ciria: London, UK, 2003; Volume 156.
21. Kempfert, H.G.; Gebreselassie, B. Effect of anchor installation on settlement of nearby structures in soft soils. In Proceedings of the International Symposium on Geotechnical Aspects of Underground Construction in Soft Ground, Tokyo, Japan, 19–20 July 1999; pp. 665–670.
22. Moormann, C. Analysis of wall and ground movements due to deep excavations in soft soil based on a new worldwide database. *Soils Found.* **2004**, *44*, 87–98. [CrossRef]
23. Hsiung, B.-C.B.; Yang, K.-H.; Aila, W.; Hung, C. Three-dimensional effects of a deep excavation on wall deflections in loose to medium dense sands. *Comput. Geotech.* **2016**, *80*, 138–151. [CrossRef]
24. Khoiri, M.; Ou, C.-Y. Evaluation of deformation parameter for deep excavation in sand through case histories. *Comput. Geotech.* **2013**, *47*, 57–67. [CrossRef]
25. Russo, G.; Nicotera, M.V.; Autuori, S. Three-Dimensional Performance of a Deep Excavation in Sand. *J. Geotech. Geoenviron. Eng.* **2019**, *145*, 05019001. [CrossRef]
26. Zhang, W.G.; Goh, A.T.C.; Goh, K.H.; Chew, O.Y.S.; Zhou, D.; Zhang, R. Performance of braced excavation in residual soil with groundwater drawdown. *Undergr. Space* **2018**, *3*, 150–165. [CrossRef]
27. Hwang, R.; Moh, Z.-C.; Wang, C.H. Toe movements of diaphragm walls and correction of inclinometer readings. *J. Geoengin.* **2007**, *2*, 61–71. [CrossRef]
28. Wong, I.H.; Poh, T.Y.; Chuah, H.L. Performance of Excavations for Depressed Expressway in Singapore. *J. Geotech. Geoenviron. Eng.* **1997**, *123*, 617–625. [CrossRef]
29. Ou, C.-Y.; Hsieh, P.-G.; Chiou, D.-C. Characteristics of ground surface settlement during excavation. *Can. Geotech. J.* **1993**, *30*, 758–767. [CrossRef]
30. Wang, Z.W.; Ng, C.W.W.; Liu, G.B. Characteristics of wall deflections and ground surface settlements in Shanghai. *Can. Geotech. J.* **2005**, *42*, 1243–1254. [CrossRef]
31. Hsiung, B.-C.B. A case study on the behaviour of a deep excavation in sand. *Comput. Geotech.* **2009**, *36*, 665–675. [CrossRef]
32. Ou, C.-Y.; Lai, C.-H. Finite-element analysis of deep excavation in layered sandy and clayey soil deposits. *Can. Geotech. J.* **1994**, *31*, 204–214. [CrossRef]

geosciences

Review

Review of Soil-Structure Interaction Based on Continuum Mechanics Theory and Use of High Performance Computing

Muhammad Rizwan Riaz [1,*], Hiroki Motoyama [2] and Muneo Hori [3]

1 Department of Civil Engineering, University of Engineering and Technology, Lahore 54890, Pakistan
2 Resilience Science Research Division, Kagawa University, Takamatsu City, Kagawa 760-8521, Japan; motoyama.hiroki@kagawa-u.ac.jp
3 Japan Agency for Marine-Earth Science and Technology, Yokohama City, Kanagaw 236-0001, Japan; horimune@jamstec.go.jp
* Correspondence: rizwan.riaz@uet.edu.pk

Abstract: Recent achievement of research on soil-structure interaction (SSI) is reviewed, with a main focus on the numerical analysis. The review is based on the continuum mechanics theory and the use of high-performance computing (HPC) and clarifies the characteristics of a wide range of treatment of SSI from a simplified model to a high fidelity model. Emphasized is that all the treatment can be regarded as the result of the mathematical approximations in solving a physical continuum mechanics problem of a soil-structure system. The use of HPC is inevitable if we need to obtain a solution of higher accuracy and finer resolution. An example of using HPC for the analysis of SSI is presented.

Keywords: seismic response analysis; soil structure interaction; continuum mechanics; structural mechanics; soil spring; high performance computing

Citation: Riaz, M.R.; Motoyama, H.; Hori, M. Review of Soil-Structure Interaction Based on Continuum Mechanics Theory and Use of High Performance Computing. *Geosciences* **2021**, *11*, 72. https://doi.org/10.3390/geosciences11020072

Academic Editors: Jesus Martinez-Frias, Dominic E.L. Ong and Salvatore Grasso

Received: 2 December 2020
Accepted: 2 February 2021
Published: 8 February 2021

Publisher's Note: MDPI stays neutral with regard to jurisdictional claims in published maps and institutional affiliations.

1. Introduction

Research related to the effects of soil-structure interaction (SSI) on structural seismic response has been a key issue in earthquake engineering for several decades. Numerous articles on SSI, which included extensive review papers, were published; see [1–4] in the early days of earthquake engineering. Regarding numerical analysis that uses finite element method (FEM) for the evaluation of SSI, the most influential work is Wolf [5].

In this review paper, we aim at summarizing recent advancements in numerical analysis related to SSI. To this end, we consider the following two fundamentals: (1) the continuum mechanics theory on which most rigorous treatment of SSI is made and (2) the use of high-performance computing (HPC) for large-scale seismic response analysis considering SSI. These two fundamentals are related to each other because the continuum mechanics theory leads to the governing equations for SSI which are given as a set of four-dimensional partial differential equations for the three components of a displacement vector function and the set cannot be numerically solved without using modern computers of fast CPU's (Central Processing Unit) and large computing memories. A soil-structure system of 100 m dimension would need an analysis model of $10^{6\sim9}$ degree-of-freedom (DOF) if the spatial resolution is 1~0.1 m and the duration of 10~100 s would need $10^{4\sim5}$ time steps if the temporal resolution is 0.01 s.

When computers were slow and could not solve problems of 1,000,000 DOF, we had to approximate the governing equation that was derived from the continuum mechanics theory to obtain an approximate solution of the governing equation. We dare say that structural mechanics theory for beam, plate or shell can be regarded as a sophisticated approximation of the continuum mechanics theory and an approximate theory can provide an approximate but accurate solution to the original continuum mechanics problem using a computer of limited performance. As for SSI, however, such an approximate but accurate solution could not be easily obtained because developing approximate treatment of soil was much more difficult than developing approximate treatment of structures. For instance, a set of soil-springs are the simplest approximate treatment of soil. This treatment works well for certain conditions but the applicability of this treatment is much more limited than the applicability of the beam, plate or shell theory for structures.

Based on the two fundamentals of the continuum mechanics theory and the use of HPC, we will review a methodology for numerically evaluating SSI. The progress of computer hardware and software is fast and the use of computers of higher performance will be an ordinary practice for solving larger scale problems DOF of which exceeds 1,000,000 or 10,000,000. Moreover, the cost for numerical computation is decreasing; computers and programs are available for the evaluation of SSI. We thus emphasize the significance of understating the mechanism of SSI from the viewpoint of the continuum mechanics theory and correctly computing the effects of SSI using an analysis model of high fidelity for a soil-structure system.

The contents of this review paper are as follows. Firstly, we review past review articles on SSI, focusing on the numerical study. Secondly, we present the most rigorous treatment of SSI based on the continuum mechanics theory. We present an interpretation of the ordinary treatment of SSI as a mathematical approximation of the rigorous treatment. Thirdly, we provide an example of using FEM enhanced with HPC for evaluating SSI by using a large-scale analysis model. We point out future works that are needed for further understanding and numerical analysis of SSI.

2. Literature Survey on Evaluation of SSI

The change in the dynamic response of the structure because of the presence of underlying soil and vice versa is called SSI. There are two aspects of SSI. The first is the modification of the free-field ground motion due to the presence of the structure and the second is the modification of the structural response due to the flexibility of the supporting soil. In the case of flexible structures founded on rigid ground, the effect of SSI on the structural response is not significant; and it has been ignored in the conventional structural analysis. For stiff and heavy structures, the soil flexibility can change the structural response significantly and ignoring SSI can lead to unsafe design [6]. This is the major target of evaluating the effects of SSI.

SSI was studied in the early twentieth century and it was found that the effects of SSI could influence structural seismic response and ought to be taken into consideration for the seismic resistance design. Observations made during large earthquakes in the past emphasized the need to correctly evaluate the effects of SSI on the seismic response analysis. The difficulty of evaluating the effects of SSI is the dependence of SSI on the characteristics of both the structure and the soil. As mentioned, the effects of SSI tend to be more significant for the case of heavy and stiff structures founded on weaker soils. The characteristics of input ground motion influence the effects of SSI on the structural seismic response as well. However, this influence is often overlooked.

The first attempt to study the phenomenon of SSI was made in Japan by Sezawa and Kanai in 1935 [7,8]. The theory of SSI was properly developed for the first time by Reissner, who studied the behavior of a circular foundation lying on an elastic half-space and subjected to vertical time-harmonic loading, in 1936 [9]. Since then, numerous analytical and numerical research studies have been conducted on the evaluation of SSI for a more accurate seismic response analysis; some recent studies are [10–14]. Several experimental works involving full or reduced scale tests of SSI problem have also been conducted during this period; see [15–18].

In general, two approaches are used for evaluating the effects of SSI. The first approach is the direct method in which the structure and soil are modeled and their response is solved simultaneously by carrying out the time integration of governing equations of the structure and soil models. The second approach is the substructure method (or the indirect method) in which the structure and soil are separated by introducing a rigid interface between them and their responses are solved individually in the frequency domain. A key issue of the substructure method is the dynamic impedances for the soil domain which is modeled as a spring and a dashpot.

2.1. Direct Method for Evaluating SSI

In the direct method, the structure and soil are considered as a system and an analysis model for them is constructed. A suitable governing equation is posed for the structure and soil and the finite element method (FEM) is used to solve the equation in the time domain. This approach is conceptually simple. The direct method involves the non-linear numerical analysis of a high-fidelity model of the structure which is constructed from three-dimensional geometry of the structure. However, it is often the case that some physical assumptions are made in posing the governing equation, such as the application of two-dimensional model and the application of the structural element models. Making physical assumptions is always acceptable with a condition that the assumptions made ought to be validated by the experiments or the observation.

A key issue of the direct method is the determination of the dimension (or size) of the soil domain and the artificial boundary conditions on the side surfaces of the soil domain. They are essential for accurately simulating the effect of the surrounding soil as well as to avoid the trapping of waves reflected from the structure inside the soil domain. For this reason, a standard option is to use a sufficiently large dimension for the soil domain so that the wave reflected from the structure is dissipated before reaching the boundary. Several studies were made to develop suitable artificial boundary conditions to reduce the dimension of the soil domain. The representative artificial boundary conditions are the introduction of the viscous boundary, consistent boundary, unified boundary, transmitting boundary, periodic boundary and the viscous spring boundary; see [19–24]. These boundary conditions act either as non-reflecting boundaries or as adsorbing layers, in order to avoid the reflections of the outward propagating waves from the soil domain.

Despite the presence of artificial boundaries, the direct method requires an analysis model of a significant volume of the soil domain for the accurate evaluation of SSI. Such an analysis model becomes significantly large if the shear wave velocity is 100 m/s and the temporal resolution is 0.1 s, the wavelength of the shear wave is 10 m and we need the spatial resolution of 1 m to accurately compute this wave. As finer temporal resolution (or responses in higher frequency) is required, the time increment, as well as the element size, decreases accordingly, which results in an analysis model of larger DOF and a larger number of time steps. For this reason, the direct approach with three-dimensional high-fidelity models has not been often used. With the advancement of computers, the works applying larger three-dimensional models have been increasing; see [25–28]. The characteristic of such works is that they applied the solid element models with non-linear property only to the soil domain, not to the structure domain. Recently, computational techniques enable us to apply the direct method for the evaluation of SSI, which includes the non-linear numerical analysis of a high fidelity model of a soil-structure system; see [29–31]. Still, the governing equation and the analysis model are simplified to obtain an approximate solution, by making certain physical assumptions for material properties of soil and concrete.

2.2. Substructure Method for Evaluation of SSI

In the substructure method, the structure and soil are considered separately; it is assumed that a rigid body foundation is put between the structure and soil. In this method, the seismic response analysis takes the following three steps: (1) the determination of the input ground motion, which is different from the free field ground motion due to the presence of the structure (termed as kinematic interaction); (2) the determination of the frequency dependent impedance functions for the foundation, by giving a unit harmonic excitation of displacement/rotation at a particular frequency to the foundation, computing the reactional force/moment of the soil and computing the ratio of the input and the reaction as an impedance function; and (3) the calculation of the dynamic response of the structure supported by the springs with the impedance functions determined in the second step, subjected to the input motion determined in the first step (termed as inertial interaction).

As for the determination of the input ground motion, analytical solutions were found in the substructure method for the cases of a structure and a half-space of soil with simple configuration and material properties. Research was carried out for a surface foundation and a piled or embedded foundation. A one-dimensional analysis that assumes uniform or stratified ground structure(s) and the vertical propagation of shear waves is used for the evaluation. Kinematic interaction is the lack of match of the foundation with the free field ground motion. It occurs because of the difference in foundation stiffness from that of the surrounding ground which causes reflection and refraction of the seismic waves approaching the soil foundation interface. For the case of the surface foundation, the kinematic interaction is ignored since it is assumed that only shear waves propagate in the vertical direction; as explained later, the presence of the structure locally influences the shear waves because the traction free boundary conditions on the ground surface are not satisfied on the bottom of the structure even though its weight is regarded as the force on the vertical direction. However, for the case of a stiff embedded foundation, the free field input ground motion is modified resulting in rotational motion for the foundation.

As for the determination of the frequency-dependent impedance functions, the mathematical problem of the soil domain is formulated as a mixed boundary value problem. Several analytical or empirical solutions were found for this problem; for instance, see [32–36] for the analytical solutions and [37] for the empirical solutions. These solutions were applied to model the soil domain as sway-rocking spring for a range of structures; see [38–42], as well as to develop numerical analysis programs such as CLASSI [43], FLUSH [44] and SASSI [45]. Further extending this approach, several more complex models, termed as lumped parameter models for SSI analysis, are also developed; see [46–48].

It should be noted that the applicability of these solutions is limited to determine the frequency-dependent impedance functions. This is because the solutions are for a simple circular or rectangular rigid foundation which is perfectly bonded to a homogeneous half space of linearly isotropic elasticity subjected to the excitation at the centroid of the foundation. The solutions could be used for a soil-structure system that is similar to the ideally simplified model. There were studies that showed significant differences between the observed data and the solutions that were synthesized by using the analytical solutions [49].

Other than the two basic approaches as mentioned above, a third hybrid approach based on macro-elements has also been developed recently. This approach combines the features of the substructure approach, that is, domain decomposition and finite element modeling. The soil and foundation system is replaced by a single element termed as macro-element located at the foundation level. The major assumption for this approach is also the same as for substructure approach, that is, the presence of the rigid body at foundation level. The first macro element was proposed by Nova & Montrasio [50] for a simplified case of footing and soil subjected to monotonic loading. Since then several macro element models have been proposed with different loading scenarios, foundation shapes and soil types; see for instance [51–55].

We have to point out that the frequency domain analysis of a structure is smart in the sense that it considerably reduces the numerical computation for the structural responses which are mainly governed by a few first modes. However, dynamic responses of three-dimensional grounds are different from the structures and the decomposition of the responses in the time domain is much more complicated; the use of a stratified ground model provides a clear perspective to the soil behavior but it might lead to over-simplifications compared with simplifications of structural modal analysis.

It should also be noted that while the applicability is limited, the use of simplified models of the soil domain is needed at the initial stages of numerical analysis to obtain overall soil responses or for the seismic probabilistic risk analysis which needs only rough estimates of ground motion amplification. In order to link the substructure method that uses a simplified model and the direct method that uses a high fidelity model, we have to maintain the consistency of the simplified model, that is, solving the same physical problem with different degrees of mathematical approximations. Losing the consistency would be fatal because it is impossible to evaluate the accuracy of the solution. Comparison with observed data is a straightforward way to examine the accuracy of the numerical analysis and it is easy for a simplified model to reproduce observed data by tuning a few parameters of the model. For the prediction, however, we have to understand the limitation of using the simplified model.

The co-existence of the conventional simplified model and the high fidelity model causes some confusion about the numerical analysis of evaluating the effects of SSI on the structural seismic responses; these two groups of the analysis models are constructed for solving different mathematical problems and hence it appears that the different mathematical problems correspond to different physical problems; see [56]. As shown in the next section, however, the continuum mechanics theory provides a single physical problem for a soil-structure system which is expressed in the form of a functional (or a Lagrangian) and various mathematical problems are derived from the functional by making suitable mathematical approximations. Unlike the physical assumptions which need to be validated by the experiments or the observation, we do not have to validate mathematical approximations. Instead, we examine the accuracy of the solutions which are obtained by solving the mathematically approximated problem with the exact solution. It should be noted that each mathematically approximated problem has its own exact solution but that exact solution is regarded as an approximate solution for the original problem which is not mathematically approximated.

3. Analysis of SSI Based on Continuum Mechanics Theory

The continuum mechanics theory, which is based on Newtonian mechanics, can predict mechanical behaviors of solid or fluid if its size is much larger than molecules or atoms that constitute the body. As mentioned, this theory leads to a set of coupled four-dimensional partial differential equations as the governing equation of the mechanical behavior, which cannot be solved without using modern fast and large computers. The governing equation is simple since it does not include terms of interactions. Indeed, the differential equations are for the temporal and spatial derivatives of a displacement function at one point and one instance and terms related to other points or times are not included. However, the interaction between one part and the other part of the body or between a certain time and other times are accurately evaluated by solving the differential equations; we should recall that Newton's second law is expressed in the form of a temporal differential equation, which is prescribed for each instance and does not include terms of the past or future. According to the continuum mechanics theory, therefore, the accurate evaluation of SSI is made just by solving the governing equation accurately, provided that initial and boundary conditions, as well as the configuration and material properties of the body, are prescribed accurately.

Based on the continuum mechanics theory, it is readily proved that some structural mechanics theories are a mathematical approximation of the continuum mechanics theory; see [57]. For instance, a beam theory which uses only Young's modulus and neglects Poisson's ratio is rigorously derived from the linear isotropic elasticity of the continuum mechanic theory (that uses both Young's modulus and Poisson's ratio) by making suitable mathematical approximations; an assumption of a one-dimensional stress-strain relation via Young's modulus is not needed. A plate theory is derived from the continuum mechanics theory in the same manner, that is, by making mathematical approximations.

The derivation of the continuum mechanics theory to structural mechanics theory uses a Lagrangian of the continuum. If a suitable Lagrangian is given to a system of a structure and soil, it can yield a wide range of mathematical problems; the most rigorous problem is to the variational problem of the Lagrangian without making any mathematical approximations and the simplest problem results in a mass spring system. The mathematical problems derived from the Lagrangian are different due to the difference in the mathematical approximations. Still, these problems are consistent since they provide approximate solutions to the common problem of the Lagrangian. For the simplified modeling, the consistency to the original continuum mechanics problem is of essential importance.

Using the Lagrangian, we show that the conventional soil-spring model can be objectively constructed from it, by applying mathematical approximations to displacement functions. No physical assumptions should be made in constructing the soil-spring model to maintain consistency with the original continuum mechanics problem.

3.1. Lagrangian of SSI

As the simplest example, we consider a linearly elastic body. A Lagrangian is formulated for a system of a structure and soil; it is a functional for a displacement function. The wave equation is derived from the Lagrangian as the governing equation for the displacement function when no mathematical approximations are made. Another governing equation can be derived from the Lagrangian when a displacement function of a particular form, which is regarded as a mathematical approximation, is used.

To formulate the Lagrangian, we denote a structure and soil by S and G, respectively; G is regarded as a finite region where ground motion is input at the bottom surface and suitable absorbing boundary conditions are posed at the side surfaces; see Figure 1. Assuming infinitesimally small deformation, we define the following Lagrangian for a displacement function, $\boldsymbol{u}(\boldsymbol{x},t)$:

$$\mathcal{L}[\boldsymbol{u}] = \int_{S+G} \frac{1}{2}\rho(\boldsymbol{x})\dot{\boldsymbol{u}}(\boldsymbol{x},t)\cdot\dot{\boldsymbol{u}}(\boldsymbol{x},t) - \frac{1}{2}\nabla\boldsymbol{u}(\boldsymbol{x},t) : \boldsymbol{c}(\boldsymbol{x}) : \nabla\boldsymbol{u}(\boldsymbol{x},t)\, dv, \quad (1)$$

where ρ and $\boldsymbol{c}$ are density and elasticity, $\dot{\boldsymbol{u}}$ and $\nabla\boldsymbol{u}$ stand for the temporal and spatial derivative of $\boldsymbol{u}$ and $\cdot$ and $:$ are inner product and second-order contraction, respectively; a bold character is used for a vector or tensor quantity. The Lagrangian problem of $\delta\int\mathcal{L}\,\mathrm{d}t = 0$ yields

$$\rho(\boldsymbol{x})\ddot{\boldsymbol{u}}(\boldsymbol{x},t) - \nabla\cdot(\boldsymbol{c}(\boldsymbol{x}) : \nabla\boldsymbol{u}(\boldsymbol{x},t)) = 0. \quad (2)$$

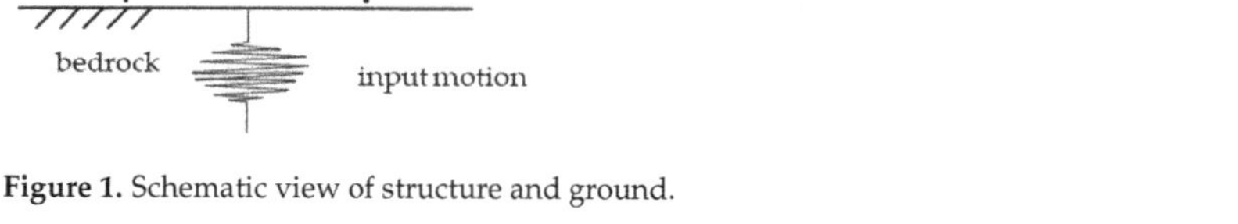

Figure 1. Schematic view of structure and ground.

This is the wave equation or the governing equation for $\boldsymbol{u}$. As is seen, Equation (2) does not have any terms for SSI in it. Still, the solution of Equation (2) accounts for the effects of SSI on the structural responses.

Now, we consider suitable mathematical approximations for $\boldsymbol{u}$, paying attention to the condition on the interface between S and G, denoted by I. The conventional analysis of SSI usually assumes that I is an infinitesimally thin and rigid body plate. The restriction ought to enforce the continuity of displacement and traction across I; see Appendix A.

3.2. Consistent Mass Spring Model Accounting for SSI

As the first step of reducing an HPC model, we study the case when a displacement function is described only in S and a displacement function in G is ignored. We consider the following displacement function:

$$\boldsymbol{u}(\boldsymbol{x},t) = g(t)\boldsymbol{\Psi}(\boldsymbol{x}) + u(t)\boldsymbol{\varphi}(\boldsymbol{x}) \text{ in } S \tag{3}$$

where g is a one-directional horizontal input ground motion on I, $\boldsymbol{\Psi}$ is a constant vector field the direction of which is parallel to the one-directional input ground motion and $\boldsymbol{\varphi}$ is the first mode of the structure, satisfying $\rho\omega^2\boldsymbol{\varphi} + \nabla\cdot(\boldsymbol{c}:\nabla\boldsymbol{\varphi}) = 0$ in S, with ω being the natural frequency of the first mode. Substitution of Equation (3) into $\mathcal{L}$ of Equation (1) yields

$$\mathcal{L} = \frac{1}{2}m^{\psi\psi}\dot{g}^2 + m^{\psi\varphi}\dot{g}\dot{u} + \frac{1}{2}m^{\varphi\varphi}\dot{u}^2 - \frac{1}{2}k^{\varphi\varphi}u^2 \tag{4}$$

where m^{AB} and k^{AB} are defined as $m^{AB} = \int_S \rho\boldsymbol{A}\cdot\boldsymbol{B}\,\mathrm{d}v$ and $k^{AB} = \int_S \nabla\boldsymbol{A}:\boldsymbol{c}:\nabla\boldsymbol{B}\,\mathrm{d}v$. For this $\mathcal{L}$ of Equation (4), $\delta\int\mathcal{L}\,\mathrm{d}t = 0$ leads to

$$m^{\varphi\varphi}\ddot{u} + k^{\varphi\varphi}u = -m^{\psi\varphi}\ddot{g}. \tag{5}$$

By posing suitable initial conditions for unknown u, we can uniquely obtain u.

Equation (5) coincides with the governing equation of a mass spring model; see Figure 2. By definition, $m^{\varphi\varphi} = \int_S \rho\boldsymbol{\varphi}\cdot\boldsymbol{\varphi}\,\mathrm{d}v$ and $k^{\varphi\varphi} = \int_S \nabla\boldsymbol{\varphi}:\boldsymbol{c}:\nabla\boldsymbol{\varphi}\,\mathrm{d}v$ satisfy $\omega^2 m^{\varphi\varphi} - k^{\varphi\varphi} = 0$. Thus, the mass spring model has the natural frequency of the structure, ω. Note that $\boldsymbol{\varphi}$ is set to satisfy $\boldsymbol{\varphi}(\boldsymbol{x}^0) = \boldsymbol{\Psi}$, where $\boldsymbol{x}^0$ is a suitable point of S and that $\boldsymbol{\Psi}$ makes $m^{\psi\psi}$ equal the total mass of S. Therefore, if $\boldsymbol{\varphi}$ satisfies

$$m^{\psi\psi} \approx m^{\psi\varphi} \approx m^{\varphi\varphi}, \tag{6}$$

then, u that is computed by solving Equation (4) is an approximate solution of the displacement at $\boldsymbol{x}^0$.

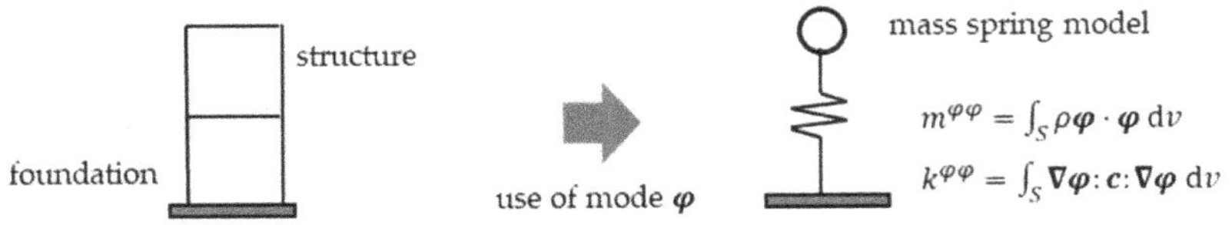

Figure 2. Mass spring model without soil spring.

3.3. Consistent Mass Spring System with Sway Spring Accounting for SSI

As the next step, we consider the following displacement function described in S and G:

$$\boldsymbol{u}(\boldsymbol{x},t) = \begin{cases} (g(t) + U(t))\boldsymbol{\Psi}(\boldsymbol{x}) + u(t)\boldsymbol{\varphi}(\boldsymbol{x}) & \text{in } S, \\ g(t)\boldsymbol{\Psi}^0(\boldsymbol{x},t) + U(t)\boldsymbol{\Psi}^1(\boldsymbol{x}) & \text{in } G, \end{cases} \tag{7}$$

where U is the translation of I, $\boldsymbol{\Psi}^0$ is the displacement field in G caused by the one-directional input ground motion in the absence of S and $\boldsymbol{\Psi}^1$ is the displacement function in G induced by U on I. Substitution of Equation (7) into $\mathcal{L}$ of Equation (1) yields

$$\begin{aligned} \mathcal{L} = {} & \tfrac{1}{2}m^{\psi\psi}\left(\dot{g} + \dot{U}\right)^2 + m^{\psi\varphi}\left(\dot{g} + \dot{U}\right)\dot{u} + \tfrac{1}{2}m^{\varphi\varphi}\dot{u}^2 - \tfrac{1}{2}k^{\varphi\varphi}u^2 \\ & + \tfrac{1}{2}M^{00}(\dot{g})^2 + M^{01}\dot{g}\dot{U} + \tfrac{1}{2}M^{11}\dot{U}^2 - \tfrac{1}{2}K^{00}g^2 - K^{01}gU - \tfrac{1}{2}K^{11}U^2 \end{aligned} \tag{8}$$

where $M^{\alpha\beta} = \int_G \rho\boldsymbol{\Psi}^\alpha\cdot\boldsymbol{\Psi}^\beta\,\mathrm{d}v$ and $K^{\alpha\beta} = \int_G \nabla\boldsymbol{\Psi}^\alpha:\boldsymbol{c}:\nabla\boldsymbol{\Psi}^\beta\,\mathrm{d}v$. For this $\mathcal{L}$ of Equation (8), $\delta\int\mathcal{L}\,\mathrm{d}t = 0$ leads to

$$
\begin{gathered}
m^{\varphi\varphi}\ddot{u} + k^{\varphi\varphi}u = -m^{\psi\varphi}\left(\ddot{g} + \ddot{U}\right), \\
K^{11}U + M^{11}\ddot{U} + m^{\psi\psi}\left(\ddot{g} + \ddot{U}\right) + m^{\psi\varphi}\ddot{u} + \tfrac{d}{dt}\left(M^{01}\dot{g}\right) + K^{01}g = 0
\end{gathered}
\tag{9}
$$

By posing suitable initial conditions for unknown u and U, we can uniquely obtain u and U.

Equation (9) coincides with the governing equation of a mass spring model with a sway spring; which has the mass, the spring constant and the sway spring constant of $m^{\varphi\varphi}$, $k^{\varphi\varphi}$ and $K^{\Phi\Phi}$, respectively, if

$$
\left|M^{11}\ddot{U} + \frac{d}{dt}\left(M^{01}\dot{g}\right) + K^{01}g\right| \ll \left|K^{11}U\right|
\tag{10}
$$

is satisfied together with Equation (6). Indeed, Equation (9) becomes

$$
\begin{gathered}
m^{\varphi\varphi}\ddot{u} + k^{\varphi\varphi}u \approx -m^{\varphi\varphi}\left(\ddot{g} + \ddot{U}\right), \\
K^{11}U \approx k^{\Phi\Phi}u,
\end{gathered}
\tag{11}
$$

which is the governing equation of the mass spring system with the sway spring with the mass and the spring constant being $m^{\varphi\varphi}$ and $k^{\varphi\varphi}$ and the sway spring constant being $K^{11} = \int_G \nabla\boldsymbol{\Psi}^1 : \boldsymbol{c} : \nabla\boldsymbol{\Psi}^1 \,dv$; see Figure 3.

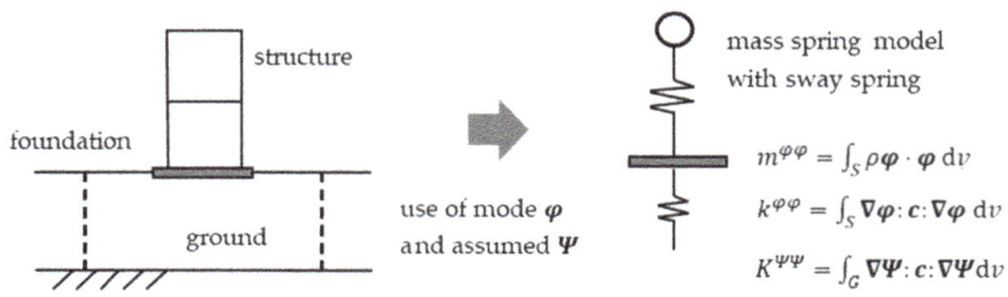

Figure 3. Mass spring model with sway spring.

There are many choices of $\boldsymbol{\Psi}^1$; a harmonic loading of I with any period is such a choice. K^{11} varies depending on the choice of $\boldsymbol{\Psi}^1$. The meta-modeling theory cannot specify a suitable choice of $\boldsymbol{\Psi}^1$ so that the solution of the resulting equation is not always a good approximate solution. However, the theory shows that Equation (10) is necessary for the sway spring, just as Equation (6) is needed for the mass spring model; see [58,59].

3.4. Extension of Consistent Multiple Mass Spring System Accounting for SSI

It is straightforward to construct a multiple mass spring system model. As an illustrative example, we consider a set of dynamic modes, $\left\{\omega^\alpha_\varphi, \boldsymbol{\varphi}^\alpha\right\}$, for S and a set of harmonic translations of I, $\{\omega^\alpha_\Psi, \boldsymbol{\Psi}^\alpha\}$, for G; $\boldsymbol{\varphi}^\alpha$ is the α-th mode with the natural frequency ω^α_φ and $\boldsymbol{\Psi}^\alpha$ is the displacement function of G when I is subjected to the harmonic translation with the frequency ω^α_Ψ. We consider the following displacement function:

$$
\boldsymbol{u}(\boldsymbol{x}, t) = \begin{cases} (g(t) + \sum_\alpha U^\alpha(t))\boldsymbol{\Psi}(\boldsymbol{x}) + \sum_\alpha u^\alpha(t)\boldsymbol{\varphi}^\alpha(\boldsymbol{x}) & in\ S, \\ g(t)\boldsymbol{\Psi}^0(\boldsymbol{x}, t) + \sum_\alpha U^\alpha(t)\boldsymbol{\Psi}^\alpha(\boldsymbol{x}) & in\ G, \end{cases}
\tag{12}
$$

Substitution of Equation (12) into $\mathcal{L}$ of Equation (1) yields

$$
\begin{aligned}
\mathcal{L} = \tfrac{1}{2}m^{\psi\psi}\left(\textstyle\sum_\alpha \dot{U}^\alpha\right)^2 + \textstyle\sum_\alpha\left(m^{\psi\alpha}\left(\textstyle\sum_\beta \dot{U}^\beta\right)\right)\dot{u}^\alpha + \tfrac{1}{2}m^{\alpha\alpha}\left(\dot{u}^\alpha\right)^2 - \tfrac{1}{2}k^{\alpha\alpha}(u^\alpha)^2 \\
+ \textstyle\sum_{\alpha,\beta}\tfrac{1}{2}M^{\alpha\beta}\dot{U}^\alpha\dot{U}^\beta - \tfrac{1}{2}K^{\alpha\beta}U^\alpha U^\beta.
\end{aligned}
\tag{13}
$$

Here, for simplicity, g is denoted by U^0 and $g + \sum_{\beta=1} U^\beta$ is replaced by $\sum_{\beta=0} U^\beta$. For $\mathcal{L}$ of Equation (13), $\delta \int \mathcal{L}\, dt = 0$ leads to

$$m^{\alpha\alpha}\ddot{u}^\alpha + k^{\alpha\alpha}u^\alpha = -m^{\psi\alpha}\left(\sum_\beta \ddot{U}^\beta\right), \tag{14a}$$

$$\sum_\beta M^{\alpha\beta}\ddot{U}^\beta + K^{\alpha\beta}U^\beta + \dot{M}^{\alpha 0}\dot{g} + m^{\psi\psi}\left(\sum_\beta \ddot{U}^\beta\right) + \sum_\beta m^{\psi\beta}\ddot{u}^\beta = 0 \tag{14b}$$

for $\alpha = 1, 2, \ldots$ and the summation of β is taken from $\beta = 0$.

Note that the last two terms in Equation (14b) are shared for all α's. Since Equation (14a) gives $\sum_\alpha \frac{(m^{\psi\alpha})^2}{m^{\alpha\alpha}}\left(\sum_\beta \ddot{U}^\beta\right) + \sum_\beta m^{\psi\beta}\ddot{u}^\beta = -\sum_\alpha \frac{m^{\psi\alpha}}{m^{\alpha\alpha}}k^{\alpha\alpha}u^\alpha$, Equation (14b) becomes

$$\sum_\beta K^{\alpha\beta}U^\beta - \sum_\alpha \frac{m^{\psi\alpha}}{m^{\alpha\alpha}}k^{\alpha\alpha}u^\alpha = -\dot{M}^{\alpha 0}\dot{g}, \tag{15}$$

if $\sum_\alpha \frac{(m^{\psi\alpha})^2}{m^{\alpha\alpha}} \approx m^{\psi\psi}$ and $\sum_\beta M^{\alpha\beta}\ddot{U}^\beta \approx 0$ are satisfied. Equation (15) is regarded as a set of the coupled equations for the sway springs.

It should be noted that sharing $\sum_\alpha \frac{m^{\psi\alpha}}{m^{\alpha\alpha}}k^{\alpha\alpha}u^\alpha$ in Equation (14b) is the consequence of Equation (13), which includes $\sum_\alpha \left(m^{\psi\alpha}\left(\sum_\beta \dot{U}^\beta\right)\right)\dot{u}^\alpha$ in $\mathcal{L}$. It should also be noted that while these displacement functions are continuous across I, they do not accompany traction functions which are continuous across I; while $\boldsymbol{\Psi}^\alpha$ produces stress in G, $\boldsymbol{\Psi}$ does not produce any stress in S; see Figure 4. Therefore, it is well expected that the displacement function of Equation (12) does not provide a good approximate solution for the original problem.

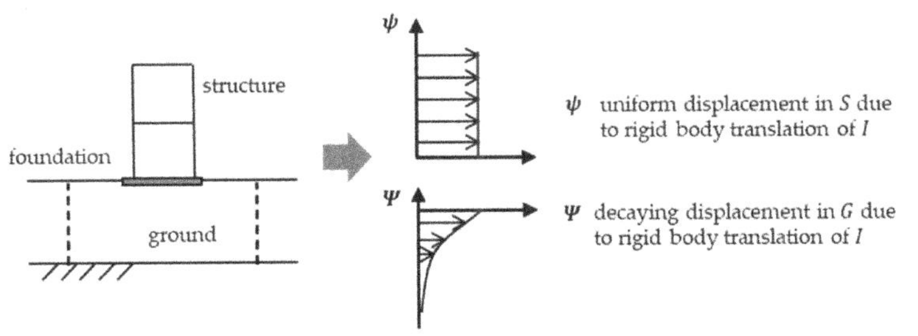

Figure 4. Displacement function induced by rigid body translation of interface.

It is also straightforward to include a rocking spring in a mass spring model. For simplicity, we consider only one mode in S and one function of translation and rotation of I for a displacement function, that is,

$$\boldsymbol{u}(\boldsymbol{x},t) = \begin{cases} (g(t) + U(t))\boldsymbol{\Psi}(\boldsymbol{x}) + \Theta(t)\boldsymbol{\xi}(\boldsymbol{x}) + u(t)\boldsymbol{\varphi}(\boldsymbol{x}) & in\ S, \\ g(t)\boldsymbol{\Psi}^0(\boldsymbol{x},t) + U(t)\boldsymbol{\Psi}^1(\boldsymbol{x}) + \Theta(t)\boldsymbol{\Xi}^1(\boldsymbol{x}) & in\ G, \end{cases} \tag{16}$$

where Θ is rotation, $\boldsymbol{\xi}$ is a linear function in S and $\boldsymbol{\Psi}^1$ is a decaying function in G; $\boldsymbol{\xi}$ and $\boldsymbol{\Psi}^1$ are the displacement functions induced by the rigid body rotation of I. Substitution of Equation (16) into Equation (1) yields

$$\mathcal{L} = \tfrac{1}{2}m^{\psi\psi}\left(\dot{g}+\dot{U}\right)^2 + \tfrac{1}{2}i^{\xi\xi}\dot{\Theta}^2 + m^{\psi\varphi}\left(\dot{g}+\dot{U}\right)\dot{u} + j^{\xi\varphi}\dot{\Theta}\dot{u} + \tfrac{1}{2}m^{\varphi\varphi}\dot{u}^2 - \tfrac{1}{2}k^{\varphi\varphi}u^2 + \tfrac{1}{2}M^{00}(\dot{g})^2 + \tfrac{1}{2}M^{11}\dot{U}^2 + \tfrac{1}{2}I^{11}\dot{\Theta}^2 + M^{01}\dot{g}\dot{U} - \tfrac{1}{2}K^{00}g^2 - \tfrac{1}{2}K^{11}U^2 - \tfrac{1}{2}R^{11}\Theta^2 - K^{01}gU \quad (17)$$

where $i^{\xi\xi}$ or $j^{\xi\varphi}$ is the integration which is taken over S in a form similar to $m^{\psi\psi}$ and I^{11} or R^{11} is the integration taken over G in a form similar to M^{00} and K^{00}; see Appendix B. For this $\mathcal{L}$ of Equation (17), $\delta\int\mathcal{L}\,\mathrm{d}t = 0$ leads to

$$\begin{aligned} m^{\varphi\varphi}\ddot{u} + k^{\varphi\varphi}u &= -m^{\psi\varphi}\left(\ddot{g}+\ddot{U}\right) - j^{\xi\varphi}\ddot{\Theta}, \\ K^{11}U + M^{11}\ddot{U} + m^{\psi\psi}\left(\ddot{g}+\ddot{U}\right) + m^{\psi\varphi}\ddot{u} + \tfrac{d}{dt}(M^{01}\dot{g}) + K^{01}g &= 0, \\ I^{11}\ddot{\Theta} + R^{11}\Theta + j^{\xi\varphi}\ddot{u} &= 0. \end{aligned} \quad (18)$$

Equation (18) coincides with the governing equation of a mass spring model with a sway spring and a rocking spring, if

$$\left| I^{11}\ddot{\Theta} - \frac{j^{\xi\varphi}}{m^{\varphi\varphi}}\left(m^{\psi\varphi}\left(\ddot{g}+\ddot{U}\right) + j^{\xi\varphi}\ddot{\Theta}\right)\right| \ll \left|R^{11}\Theta\right|. \quad (19)$$

is satisfied, together with Equations (6) and (10). Indeed, Equation (18) becomes

$$\begin{aligned} m^{\psi\psi}\ddot{u} + k^{\varphi\varphi}u &= -m^{\psi\psi}\left(\ddot{g}+\ddot{U}\right) - j^{\xi\varphi}\ddot{\Theta}, \\ K^{11}U &\approx k^{\phi\phi}u, \\ R^{11}\Theta &\approx \tfrac{j^{\xi\varphi}}{m^{\varphi\varphi}}k^{\phi\phi}u. \end{aligned} \quad (20)$$

which is the governing equation of the mass spring system with the sway and rocking springs, with the mass and the spring constant, $m^{\psi\psi}$ and $k^{\varphi\varphi}$ and the sway and rocking spring constant, K^{11} and R^{11}. Just like K^{11}, this R^{11} is defined as $R^{11} = \int_G \nabla\Xi^1 : \boldsymbol{c} : \nabla\Xi^1 \, \mathrm{d}v$ in terms of Ξ^1.

We have to emphasize that Equations (6), (10) and (19) are used in order to reduce an FE model of high fidelity to a mass spring model with sway and rocking springs. The mass spring model could be a good and simple model if these three equations are satisfied. Still, it has a limitation of applicability since the mathematical approximations for the displacement functions are required in reducing the FE model to the mass spring model.

Using a larger number of the soil-springs does not contribute to increasing the agreement by themselves. This is because the harmonic displacements, $\boldsymbol{\Psi}^\alpha$ and Ξ^α, have coupling effects with the ground motion, $\boldsymbol{\Psi}^0$ (i.e., $\int_G \rho\boldsymbol{\Psi}^0 \cdot \boldsymbol{\Psi}^\alpha \, \mathrm{d}v$, $\int_G \nabla\boldsymbol{\Psi}^0 : \boldsymbol{c} : \nabla\boldsymbol{\Psi}^\alpha \, \mathrm{d}v$, $\int_G \rho\boldsymbol{\Psi}^0 \cdot \Xi^\alpha \, \mathrm{d}v$ and $\int_G \nabla\boldsymbol{\Psi}^0 : \boldsymbol{c} : \nabla\Xi^\alpha \, \mathrm{d}v$) and we have to include terms that account for the effects. These terms change depending on $\boldsymbol{\Psi}^0$ and hence we have to tune the soil-springs for a different ground motion. In the present paper, we regard the need of tuning the model parameters as a limitation of the applicability of the mass spring model with soil-springs.

3.5. Conventional Treatment of SSI

In the previous section, the underlying approximation that is made in constructing the soil-spring is clarified by rigorously formulating the problem of SSI. It is the approximation that determines the applicability and limitations of the soil-spring. The key approximation is the introduction of a rigid body foundation. While the modeling is simple, it is not a trivial task to determine the properties of the soil-spring. Indeed, the properties change depending on the frequency of input ground motion; the dependence of the soil-spring properties on the frequency leads to the quasi-non-linear characteristics of soil, which is favored by conventional numerical analysis. However, we point out that the dependence of the soil-spring on the frequency is the consequence of the mathematical approximations that lead to the soil-spring. Based on the continuum mechanics theory, the dependence should be consistent with the mechanical properties of the soil. It is not a simple task to develop such a soil-spring.

The dynamic characteristics of a structure, such as a natural frequency or mode, is obtained by solving an eigen-value problem (or a null boundary condition problem), to which fixed displacement conditions are prescribed at the base, while this is the reason that a rigid body foundation is introduced. Assuming the rigid-body foundation is reasonable if the foundation is stiffer compared with other parts of the structure. It also has the advantage of separating a soil-structure system into a structural problem and a soil problem. Based on the continuum mechanics, a rigid-body (or sufficiently stiff) foundation leads to stress concentration at the edge of the foundation, where traction changes in a discontinuous manner. Local stress concentration should be ignored in considering overall structural responses. However, it will play a key role in inducing local damages when stronger ground motion is input.

Another advantage is that continuity of traction at every point at the interface does not have to be satisfied. The equilibrium of the reaction force and moment of the structure and the soil is sufficient, which reduces the problem size considerably. However, this advantage might hide the limitation of the applicability, the spatially non-uniform responses of the foundation ought to be considered in more accurately evaluating the structural seismic responses due to stronger ground motion. The quality of evaluating floor response becomes poorer if a soil spring that is attached to the rigid-body foundation is used.

The degree of mathematical approximations that are made for the soil-spring is relatively large and hence, on the viewpoint of the continuum mechanics theory, the applicability of the soil-spring is considerably limited. It is true that the spring model can predict overall structural responses subjected to ground motions that do not include large components in a certain domain of frequencies. However, for the evaluation of local seismic responses, we cannot deny the low applicability of the spring model when a wide range of ground input motions, such as ground motions rich in high-frequency components (which is called a killer pulse) or ground motions rich in low-frequency components are considered.

In closing this section, we emphasize that soil-springs can be constructed from a Lagrangian of a soil-structure system just by applying suitable mathematical approximations to a displacement function. A simplified model that uses the soil-spring constructed in this manner is consistent with an analysis model of high fidelity which is used to solve the wave equation. The exact solution of the simplified model is thus regarded as an approximate solution of the high-fidelity model, even though it is often misunderstood that the simplified model solves different physical problems. This characteristic of the simplified model results in the limited applicability of the simplified model in evaluating the seismic structural responses in higher resolution or subjected to stronger ground motions which are rich in particular frequency components.

4. Use of HPC for Numerical Analysis of SSI

Based on the continuum mechanics theory, solving the wave equation is the unique solution to accurately evaluate the effects of SSI on the structural seismic responses for a soil-structure system. Using computers of high performance (fast CPU's and large computing memories) is a unique solution to solve the wave equation without numerical approximation and physical assumption; as mentioned, the numerical analysis requires an analysis model of a few million DOF and several thousand time steps at least. Unlike conventional methods that use simplified models, the use of HPC for the numerical analysis of SSI does not have limited applicability.

The use of HPC for the numerical analysis of SSI has been initiated in recent years. In this section, we review key concepts and techniques which are needed to realize the use of HPC. At the end of this section, we provide an illustrative example of the seismic structural response analysis for an actual structure, using an FEM program that is enhanced with HPC.

4.1. Advantage of Using Solid Element

In solving the wave equation for an important facility such as a nuclear power plant (NPP), high reliability is needed for an analysis model in order to estimate its seismic response with higher accuracy and in higher resolution. A simple solution of making such a high fidelity model is to use a solid element in FEM; a solid element does not have the limitation in the applicability, unlike a structural element, such as a beam, plate or shell element and a spring element. It is guaranteed that in solving the wave equation, a solution of the solid element model converges to the exact solution as the mesh size and the time increment decreases; the speed of convergence can be controlled by the mesh type and the time integration scheme. This is the reason that solid element high fidelity models are used in practice as they are more reliable than structural element models. Structural element models can reduce computational costs but they have the limitation in the applicability.

In the non-linear regime of the seismic responses, the characteristic of the solid element model (i.e., the solution converges to the exact solution) becomes more important. Non-linear structural elements are often tuned for a certain class of loading but they are not versatile. Using non-linear structural elements that are tuned to reproduce certain experimental results of non-linear force-displacement relations is a smart choice when we calculate responses that are subjected to loading similar to the experiment. However, it is difficult to even verify the non-linear element behavior subjected to a different class of loadings. Non-linear solid element is fully verified if a three-dimensional tensorial constitutive relation, which is observed in the experiment of material samples, is implemented.

Numerical analysis of SSI partially has taken advantage of such characteristics of the solid element model. There are many works that employ solid elements for soil. Still applying the solid elements to the structure is possible to improve the accuracy of the analysis because the estimation of the inertial effect of SSI could be more precise in the non-linear regime. The key to higher accuracy is applying the solid elements to the structure, as well as to the soil. The advantage of the application of solid element to the structure is greater when the structure has more complicated vibration modes, which are often occurred in the large-scale structures constructed by many kinds of members such as NPP buildings. In such a case, the size of the models is usually large and the use of the HPC technique is more important.

It should be mentioned that using solid elements to the structure improves the accuracy of computing higher frequency components and local behavior. It should be also mentioned that the size of an analysis model for a soil-structure system needs to be large, in order to reduce the effects of artificial boundary conditions on the side surfaces of the model. The numerical treatment of the artificial boundary conditions is well established, even though it does not fully absorb reflection on the boundary. Making a larger soil domain is much easier to improve the numerical treatment of the artificial boundary conditions.

4.2. Solver Treatment of Tensorial Constitutive Relation

To realize the use of HPC, we have to pay attention to the mathematics of solving the wave equation. In FEM, a displacement function is discretized (or expressed in terms of a vector for the nodal displacement) and a matrix equation for the nodal vector is derived from a set of partial differential equations of the wave equation. Most of the numerical computation of FEM is used in solving this matrix equation and an algorithm that is used to solve the matrix equation is often called a solver. It is the solver that determines the performance of FEM; for ordinary solvers (such as a Gauss-Seidel method), the numerical computation required increases linearly to the square of the matrix dimension, which means that the computing time becomes 100 times longer if the DOF of the FEM model becomes 10 times larger.

To overcome the disadvantage of the ordinary solvers, a conjugate gradient method (CG method) is employed as a solver because the numerical computation increases almost linearly with respect to the matrix dimension for the CG method. Unlike the ordinary solvers, the CG method is iterative and it gives a solution to the matrix equation when the error of the matrix equation is sufficiently small. Several numerical techniques, such as pre-conditioning or a domain decomposition method, are developed for the faster convergence of the CG method. General purpose FEM programs that are implemented with the CG method are available and there are commercial FEM programs that have an option of using the CG method.

A drawback of the CG method is that the number of iterations to reach a converged solution depends on the quality of an analysis model; the number increases as the model includes more elements of bad configuration. Therefore, in constructing an analysis model of high fidelity, we must minimize the number of elements of bad configuration. Besides, for FEM programs, we need to develop a program of automated meshing to construct a high-fidelity model of higher quality. In general, however, it is not easy to make automated meshing for a thin body with complicated configuration such as ground layers.

4.3. Treatment of Tensorial Constitutive Relation

When a general purpose FEM which is enhanced with HPC is available, in order to carry out seismic response analysis for a soil-structure system, we have to implement at least the following two modules: (1) a module of computing tensorial constitutive relations for concrete and soil and (2) a module of treating artificial boundary conditions. The first module needs detailed coding because, unlike metallic materials, concrete and soil have negative stiffness (which is often called softening even though the meaning of softening in standard plasticity theory is different) when inelastic deformation becomes large. Numerous research achievements are available to develop the second module.

The tensorial constitutive relations of concrete and soil are mathematically more complicated than the directional non-linear relation between force and displacement (or moment and flexure) of structural components. They can compute complicated behaviors of the structural components which change depending on the loading history since the behaviors of the structural components are the results of accumulating local material responses and all the local material responses are controlled by the tensorial constitutive relations. It is taken for granted in the field of solid computational mechanics for automobile, ship-building or airplane industry that most complicated responses of a structural component can be reproduced/predicted by employing the tensorial constitutive relations of the material in the numerical analysis, which is regarded as their primary advantage of not having the limitation in the applicability.

Implementing the module of the tensorial constitutive relations to the FEM program enhanced with HPC causes an extra modification of the program. The FEM program employs the CG method, which is applicable to a matrix equation whose matrix is positive-definite. If the positive-definiteness of the matrix is lost due to the softening of some parts of structure or soil domains, then we have to provide an alternative matrix equation whose matrix is positive-definite to the CG method.

We explain the modification of the FEM program for the case of concrete softening; our FEM program employs the constitutive relation of Maekawa et al. [60], which is a de-facto standard in Japan. Like ordinary tensorial elasto-plastic constitutive relation model, it is formulated in the following form:

$$d\sigma = \mathbf{c}^{EP} : d\epsilon, \tag{21}$$

where $d\sigma$ and $d\epsilon$ are an increment of stress and strain tensors, denoted by σ and ϵ, respectively, $\mathbf{c}^{EP}$ is an elasto-plasticity tensor, and: stands for the second contraction. It is $\mathbf{c}^{EP}$ that has negative components in softening, which results in the loss of the positive-definiteness of the matrix equations. In non-linear FE analysis, Equation (21) gives the stiffness of each element, which is used for constructing a global stiffness matrix for non-linear iteration analysis such as Newton method. Yamashita et al. [61] derived an alternative set of equations by reformulating Equation (21), as

$$\begin{aligned} \mathrm{d}\sigma &= c^E : \mathrm{d}\epsilon + \mathrm{d}\sigma^*, \\ \mathrm{d}\sigma^* &= -c^E : \mathrm{d}\epsilon^P + (\nabla c^E : \epsilon^E) : \mathrm{d}\epsilon^E, \end{aligned} \tag{22}$$

where c^E is an elasticity tensor, $\mathrm{d}\epsilon^E$ and $\mathrm{d}\epsilon^P$ are an increment of elastic and plastic strain tensors, denoted by ϵ^E and ϵ^P, satisfying $\mathrm{d}\epsilon = \mathrm{d}\epsilon^E + \mathrm{d}\epsilon^P$ and ∇c^E is the derivative of c^E which is given as a function of ϵ^E. An advantage of Equation (22) instead of Equation (21) is that c^E (which plays a role of apparent stiffness) is always positive-definite.

The work of implementing such reformulated constitutive equation for HPC-FEM was initiated by authors; see [62]. In the work, the non-linear analysis method of solving Equation (22), assuming that $\mathrm{d}\sigma^*$ is negligibly small, is implemented. This method is similar to the modified Newton method as it uses apparent stiffness. Neglecting $\mathrm{d}\sigma^*$ results in a larger number of iterations for the non-linear analysis, compared with the Newton method. However, it is more numerically stable and reaches a converged solution compared with the Newton method. It should be mentioned that without neglecting $\mathrm{d}\sigma^*$ in Equation (22), we could improve the performance of the non-linear analysis method even though more sophisticated coding is necessary.

4.4. Hybrid Model Consisting of Solid and Structural Elements

A problem of numerically analyzing a solid element model of high fidelity for a soil-structure system is the computational cost. When computing powers were limited, we had to apply some modeling techniques to reduce the computational cost, giving up the use of the high fidelity solid element model as it is. Typical techniques are the use of structural elements to the structure and the reduction of the soil domain. As mentioned, these techniques are used in many cases if limited to the numerical analysis of NPP's seismic responses because these enable the direct method for evaluating SSI without HPC technique. We call the model which consists of solid elements and structural elements as a hybrid model. The HPC techniques and the model reduction techniques including hybrid modeling are important for higher applicability of the numerical analysis of SSI.

It is necessary to implement the module of computing the tensorial constitutive relations to a structural element, such as a beam, plate or shell element for a hybrid model. The implementation must be done carefully so that the consistency of the structural element to the solid element is maintained; a solution similar to the solid element is computed if a fully solid element model is replaced by a hybrid model. In particular, the treatment of strain requires special attention because in a structural element, the strain that is computed from a displacement function is different from the strain that corresponds to the computed stress via the tensorial constitutive relations. This difference is easily understood if we consider the simplest structural element of a linear bar (or truss) element for a linearly isotropic elastic material. This element computes uniaxial normal strain and stress components which are related via Young's modulus but the uniaxial stress component accompanies normal strain components in the transverse directions. If a solid element is used for a bar, a uniaxial stress component that accompanies a set of normal strain components is computed. Therefore, we must choose strain computed from stress via the tensorial constitutive relation in order to maintain consistency.

It is true that the difference between strain computed from displacement and strain computed from stress seems a contradiction. However, we can readily prove that the difference is due to mathematical approximations that are made to compute an approximate solution of a Lagrangian of the continuum mechanics theory. Physical relations such as a material constitutive relation should not be changed, even though an apparent relation between stress and strain, such as one-dimensional stress-strain relation via Young's modulus for a bar element, is derived as the consequence of the mathematical approximations.

To maintain the consistency of the structural element to the solid element in the viewpoint of continuum mechanics, it is difficult to apply structural elements to the members of large deformation in the non-linear regime under existing conditions. However, it seems possible to apply the structural elements to the rigid members which are rigidly connected to the members of relatively small deformation. At least, for the main seismic members, the solid elements should be employed when the validation on the non-linearity of the members against the large seismic loads is difficult.

4.5. Example of Usage of HPC for Numerical Analysis of SSI

Development of HPC-FEM which has the tensorial constitutive relation reformulated for HPC solver is initiated by the authors. In this section, an example of the seismic structural response analysis for an NPP building using the developed FEM-program is provided focusing mainly on its performance. Figure 5 illustrates the model outline. The surrounding ground size is 600 m × 800 m horizontally and 200 m vertically, which consists of the uniform rock layer. The height of the building is approximately 80 m; the precise measurements of each part are not allowed due to a safety issue. The mesh size for reinforced concrete walls against seismic load is set as less than 1.0 m and each wall is divided into four elements in the depthwise direction. Table 1 lists the example of a concrete property of seismic walls and we applied the constitutive relation of Maekawa et al. mentioned in Section 4.2 for concrete material. We applied the elastic property for the soil or rock domain. For modeling the damping, we employed element Rayleigh damping; for instance, the damping ration of the RC walls is 5% and the Rayleigh damping coefficients for mass and stiffness matrices are 2.1543 and 8.4127 × 10^{-4} respectively. The structure rests directly on the soil and no embedment is considered. Rigid contact is considered between foundation and soil. Table 2 lists the computation scale, which shows that the scale of this analysis is approximately 1.5 million degrees of freedom. This is not huge compared with the recent research of the computational science but still larger than the problems of conventional seismic response analysis of the reinforced concrete structures including SSI. The input seismic wave was the one, which could be used in standard practice in Japan. The duration time of the input seismic wave was 22.08 s. The displacement boundary conditions that were compatible with the input seismic wave were posed to the bottom, as well as on each side of the surrounding ground; the one-dimensional wave propagation in the ground was computed and the computed displacement was used for the sides. The seismic input motion is shown in Figure 6. This simple boundary condition is possible because the domain of surrounding soil is enough large. The time increment was set as $dt = 0.005$ s and the total number of time steps was 4416. For more information about the model, see [62].

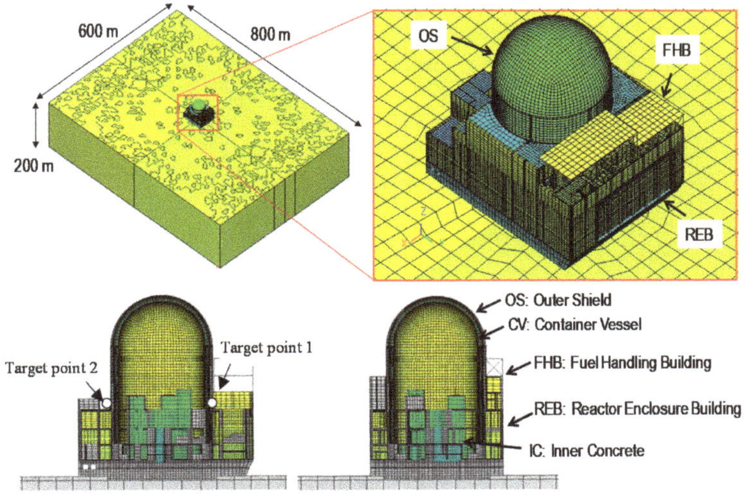

Figure 5. Outline of the model of the nuclear power plant (NPP) buildings.

Table 1. Example of the property of the concrete material on seismic walls.

Young Modulus	Poison Ratio	Compressive Strength
26,100 N/mm^2	0.2	36.8 MPa

Table 2. Computation Scale.

Node	Element	DOF
405,326	444,060	1,472,529

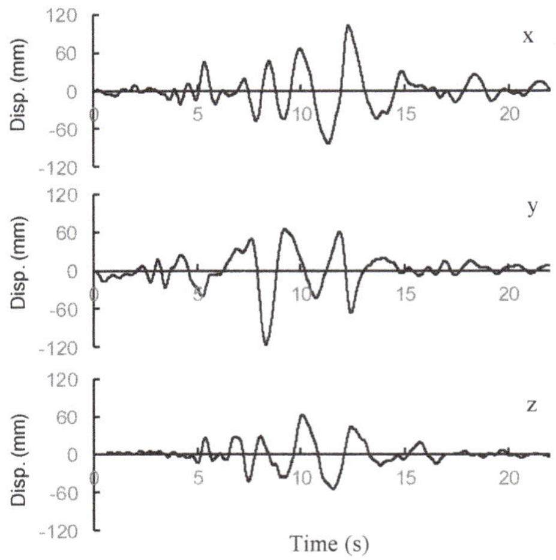

Figure 6. Input seismic motion of displacement posed to the bottom of the soil domain.

In order to save the degrees of freedom, the authors employed a hybrid model of solid elements and structural elements. Figure 7 illustrates the part to which the structural elements are applied. Shell elements and beam elements are shown as green and red, respectively. The structures with structural elements are rigid or elastic without complex seismic response. The solid elements were applied to the reinforced concrete walls which were against a seismic load which had a possibility to experience a large deformation. Moreover, the model took advantage of multi-point constraints (MPC) at the part of the connection without any damage. As a result, the degrees of freedom were suppressed to 1.5 million.

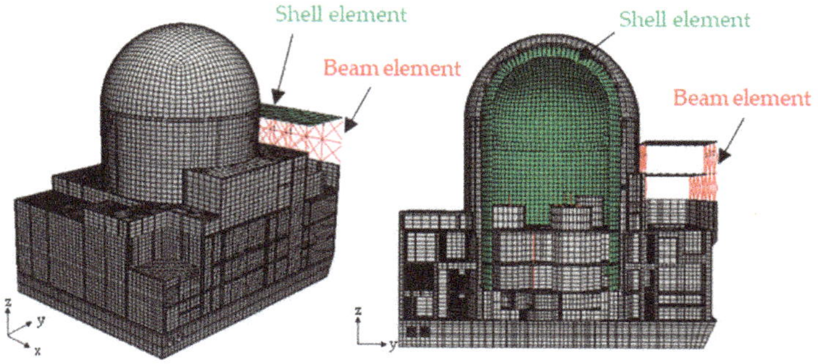

Figure 7. Application of hybrid model.

The computational performance was mainly examined. The elapsed time of this analysis was approximately 10 h, when we used a parallel computer of 64 cores. Though the computation of 64 cores is not significantly large, this elapsed time is short enough for practical use. This result shows the possibility of evaluating the seismic response of a whole soil-structure system using a high-fidelity model. Figure 8 shows the scalability of the computation, which is one of the most important performances of the HPC program. For this analysis, a shorter seismic input wave, which is extracted from the input seismic wave as 2.5 s duration including its principal shock, was applied. This result shows the scalability of elapsed time and the computation time of the solver is high. Moreover, the result also shows that the computation time of the solver occupied the majority of the elapsed time, which means that the development of the program presupposing the usage of the HPC solver works well.

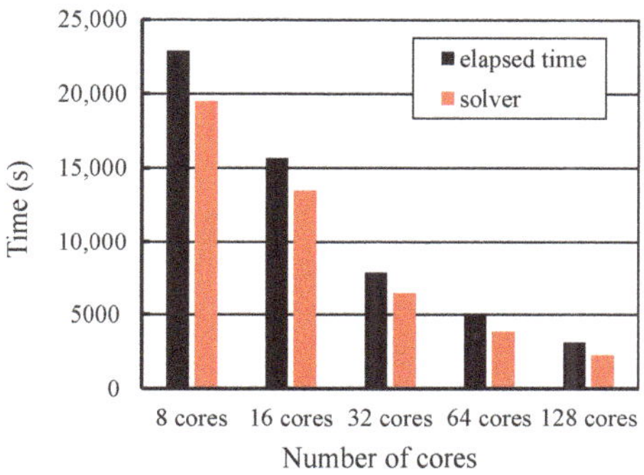

Figure 8. Scalability of the computation; relation between the elapsed time and the number of cores used in each computation.

The work briefly showed the results of the seismic response analysis. Figure 9 presents the examples of the deformation of the NPP building and the distribution of stress. Shown is at the time of the step at which the maximum shear deformation took place during the analysis. These results indicate that a numerical analysis of a high fidelity model, which is enabled by using the HPC-FEM program, visualizes the local deformation and the inertial force which are caused by the seismic load with high spatial resolution.

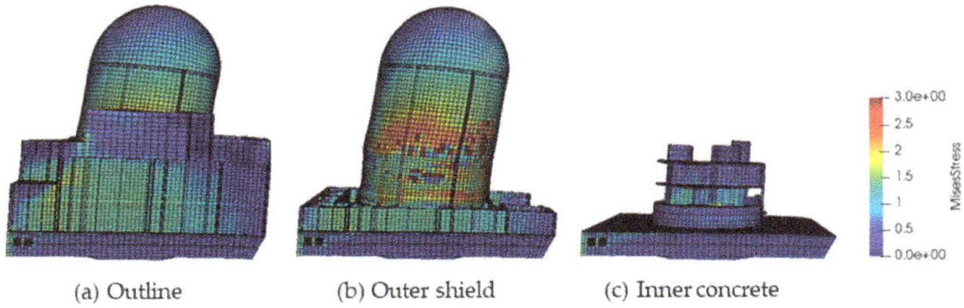

Figure 9. Deformation and the distribution of stress.

Figure 10 exhibits the acceleration response and the acceleration response spectrum. These were measured for the two target points at the top level of the reactor enclosure building; each point is at the opposite side of the outer shield shown in Figure 5. The response and spectrum are quite different even though they are on the same level and such difference cannot be computed by using a simplified model, that is, an ordinary lumped mass model, in which responses at the same level are identical. It confirms a need for the analysis of a high-fidelity model for accurate estimation of local seismic responses which are used as input for equipment and facilities located in the NPP building. The results of the linear analysis are shown in this figure for the comparison with the non-linear analysis. The acceleration response spectrum is quite different in a certain range of frequency, which confirms the need for the non-linear analysis. Precise treatment of material non-linearity is needed to make the device design more rational. It should also be noted that such high-fidelity model with the precise treatment of material non-linearity improves the inertial effect of SSI.

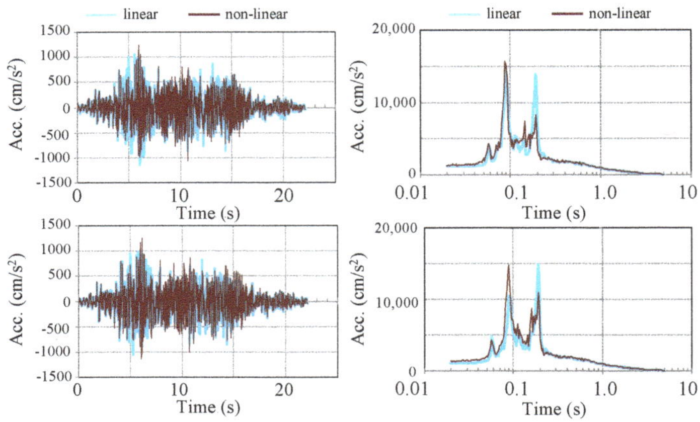

Figure 10. Acceleration time-histories and response spectra at two target points, with linear and non-linear analysis.

5. Concluding Remarks

The study of SSI is diverse and involves several research areas such as engineering seismology, soil mechanics, foundation engineering, structural engineering and so forth. The numerical analysis can integrate the achievement of all the research areas. It should be supported by the fundamental physical theory shared by the research areas and advanced computer and computational sciences. This is the reason that the present review paper employs the continuum mechanics theory and the use of HPC. It is clarified that there is a wide range of numerical analyses of SSI ranging from a conventional analysis of a simplified model to the use of HPC for a high-fidelity model. All the numerical analysis must solve a mathematical problem that is consistent with an original Lagrangian problem of a soil-structure system according to the continuum mechanics theory. Then, we can choose a suitable numerical analysis depending on the level of the accuracy and resolution of a required solution.

As is seen in the literature survey, even though the 3D solid element based FEM analysis is becoming increasingly possible for ordinary civil engineering problems, yet the conventional simplified models are still being used even for the analysis of important structures such as nuclear power plants. These models developed in the era of lesser computational resources have been helping the engineering community since long. However, there is a need for their improvement by increasing the spatial resolution and the use of high fidelity models will be an important issue of future research to meet this need.

The use of HPC for the numerical analysis of SSI started recently and we will keep using computers of increasing performances. There are many issues that could be solved with the use of HPC; for example, we have to consider various types of constitutive relations for soil and structural materials, the effects of spread or pile footing, embedded or surface foundation, slip and detaching at the interface and the effects of uncertainty in actual soil and structure properties. It is necessary to develop the automated construction of a high fidelity model from CAD data. Keeping the diversity of related fields in mind, we consider that this study will be a first step to the advancement of the SSI study using HPC and hence we focus on the clarification of the SSI problem based on the continuum mechanics theory.

Author Contributions: Conceptualization, M.R.R. and M.H.; methodology, M.R.R.; software, H.M.; validation, M.R.R., H.M. and M.H.; formal analysis, M.R.R.; resources, M.R.R.; writing—original draft preparation, M.R.R., H.M.; writing—review and editing, M.R.R., H.M.; visualization, H.M.; supervision, M.H.; project administration, M.H. All authors have read and agreed to the published version of the manuscript.

Funding: This research received no external funding.

Conflicts of Interest: The authors declare no conflict of interest.

Appendix A. Rigid Body Interface between Structure and Ground

It is of interest to derive a governing equation for the movement of a rigid-body interface between a structure and ground since elaborated mathematical treatment is needed for the rigid body movement. We use the same Lagrangian, given by Equation (1) and consider the following general displacement function:

$$u(x,t) = \begin{cases} (g(t)+U(t))\mathbf{\Psi}(x) + \mathbf{u}^S(x,t) & \text{in } S, \\ g(t)\mathbf{\Psi}^0(x,t) + \mathbf{u}^G(x,t) & \text{in } G. \end{cases} \tag{A1}$$

Here, for simplicity, we consider one-direction input and denote the displacement functions in S and G by $\mathbf{u}^S$ and $\mathbf{u}^G$, respectively; rigid-body rotation is neglected even though it is straightforward to include it in a manner similar to rigid-body translation. A key issue is the continuity of displacement across the interface, I, which is assumed to be rigid body. Setting $\mathbf{u}^S = \mathbf{0}$ on I, we have rigid body given by $g+U$ for $\mathbf{u}$ in S. Thus, we must have $\mathbf{u}^G = U\mathbf{\Psi}$ on I. This condition is implemented into $\mathcal{L}$ by addition of this condition as a constraint, that is,

$$\begin{aligned} \mathcal{L}[\mathbf{u}^S,\mathbf{u}^G,\boldsymbol{\lambda}] = \textstyle\int_S \frac{1}{2}\rho\left(\left(\dot{g}+\dot{U}\right)\boldsymbol{\psi}+\dot{\mathbf{u}}^S\right)\cdot\left(\left(\dot{g}+\dot{U}\right)\boldsymbol{\psi}+\dot{\mathbf{u}}^S\right) - \frac{1}{2}\nabla((g+U)\boldsymbol{\psi}+\mathbf{u}^S) : \mathbf{c} : \\ \nabla((g+U)\boldsymbol{\psi}+\mathbf{u}^S)\, dv + \textstyle\int_G \frac{1}{2}\rho\left(\dot{g}\mathbf{\Psi}^0+\dot{\mathbf{u}}^G\right)\cdot\left(\dot{g}\mathbf{\Psi}^0+\dot{\mathbf{u}}^G\right) - \\ \frac{1}{2}\nabla\left(g\mathbf{\Psi}^0+\mathbf{u}^G\right) : \mathbf{c} : \nabla\left(g\mathbf{\Psi}^0+\mathbf{u}^G\right) dv + \textstyle\int_I \boldsymbol{\lambda}\cdot\left(U\boldsymbol{\psi}-\mathbf{u}^G\right) ds, \end{aligned} \tag{A2}$$

where $\boldsymbol{\lambda}$ is a Lagrange multiplier. A set of coupled differential equations are derived from $\delta\int\mathcal{L}\,\mathrm{d}t = 0$ for $\mathbf{u}^S$, $\mathbf{u}^G$ and $\boldsymbol{\lambda}$. Since $\boldsymbol{\lambda}$ plays a role of the traction acting on I, both $\mathbf{u}^S$ and $\mathbf{u}^G$ are formally obtained for a given $\boldsymbol{\lambda}$ by solving the differential equation. Then, $\mathbf{u}^G = U\mathbf{\Psi}$ becomes an integral differential equation for $\boldsymbol{\lambda}$, the form of which is very complicated.

Appendix B. Computation of Equation (17)

In deriving Equation (17) from the substitution of Equation (16) into Equation (1), we drop a few terms, assuming the symmetry of S and G. For instance, $\int_S \nabla\mathbf{\Psi} : \mathbf{c} : \nabla\boldsymbol{\xi}\, \mathrm{d}v = 0$ or $\int_G \nabla\mathbf{\Psi}^\alpha : \mathbf{c} : \nabla\Xi^1\, \mathrm{d}v = 0$ is used, based on the assumption that $\mathbf{\Psi}$ or $\boldsymbol{\xi}$ ($\mathbf{\Psi}^\alpha$ or Ξ^1) is symmetric or anti-symmetric with respect to a horizontal coordinate, respectively. These integrations represent coupling effects of the displacement functions induced by rigid body translation and rigid body rotation, which vanish if certain symmetry holds for the functions; when G is symmetric in the horizontal direction, the displacement function due to the rigid body translation and rotation ought to be symmetric and anti-symmetric, respectively. Therefore, some integrations taken over S and G are dropped in Equation (17).

References

1. Kausel, E. Early history of soil-structure interaction. *Soil Dyn. Earthq. Eng.* **2010**, *30*, 822–832. [CrossRef]
2. Roesset, J. Soil structure interaction, the early stages. *J. Appl. Sci. Eng.* **2013**, *16*, 1–8.
3. Gulkan, P.; Clough, R.W. Developments in Dynamic Soil-Structure Interaction. In Proceedings of the NATO Advanced Study Institute on Developments in Dynamic Soil-Structure Interaction, Kemer, Antalya, Turkey, 8–16 July 1992.

4. Lou, M.; Wang, H.; Chen, X.; Zhai, Y. Structure–soil–structure interaction: Literature review. *Soil Dyn. Earthq. Eng.* **2011**, *31*, 1724–1731. [CrossRef]
5. Wolf, J.P. *Dynamic Soil-Structure Interaction*; Prentice-Hall: Englewood Cliffs, NJ, USA, 1985.
6. Mylonakis, G.; Gazetas, G. Seismic soil-structure interaction: Beneficial or detrimental? *J. Earthq. Eng.* **2000**, *4*, 277–301. [CrossRef]
7. Sezawa, K.; Kanai, K. Decay in the seismic vibration of a simple or tall structure by dissipation of their energy into the ground. *Bull. Earthq. Res. Inst. Jpn.* **1935**, *13*, 681–696.
8. Sezawa, K.; Kanai, K. Energy dissipation in seismic vibration of actual buildings. *Bull. Earthq. Res. Inst. Jpn.* **1935**, *13*, 925–941.
9. Reissner, E. Stationäre, axialsymmetrische, durch eine schüttelnde Masse erregte Schwingung eines homogenen elastischen Halbraum. *Ing. Arch.* **1936**, *7*, 381–396. (In German) [CrossRef]
10. Massimino, M.R.; Abate, G.; Grasso, S.; Pitilakis, D. Some aspects of DSSI in the dynamic response of fully-coupled soil-structure systems. *Riv. Ital. Geotec.* **2019**, *1*, 44–70.
11. Gazetas, G. 4th Ishihara lecture: Soil-foundation-structure systems beyond conventional seismic failure thresholds. *Soil Dyn. Earthq. Eng.* **2015**, *68*, 23–39. [CrossRef]
12. Anastasopoulos, I.; Gelagoti, F.; Spyridaki, A.; Sideri, J.; Gazetas, G. Seismic rocking isolation of an asymmetric frame on spread footings. *J. Geotech. Geoenviron. Eng.* **2014**, *140*, 133–151. [CrossRef]
13. Pecker, A.; Paolucci, R.; Chatzigogos, C.; Correia, A.A.; Figini, R. The role of non-linear dynamic soil-foundation interaction on the seismic response of structures. *Bull. Earthq. Eng.* **2014**, *12*, 1157–1176. [CrossRef]
14. Anastasopoulos, I.; Gazetas, G.; Loli, M.; Apostolou, M.; Gerolymos, N. Soil failure can be used for seismic protection of structures. *Bull. Earthq. Eng.* **2010**, *8*, 309–326. [CrossRef]
15. Gavras, A.G.; Kutter, B.L.; Hakhamaneshi, M.; Gajan, S.; Tsatsis, A.; Sharma, K.; Kohno, T.; Deng, L.; Anastasopoulos, I.; Gazetas, G. Database of rocking shallow foundation performance: Dynamic shaking. *Earthq. Spectra* **2020**, *36*, 960–982. [CrossRef]
16. Drosos, V.; Georgarakos, T.; Loli, M.; Zarzouras, O.; Gazetas, G. Soil-foundation-structure interaction with mobilization of bearing capacity: Experimental study on sand. *J. Geotech. Geoenviron. Eng.* **2012**, *138*, 1369–1386. [CrossRef]
17. Abate, G.; Gatto, M.; Massimino, M.R.; Pitilakis, D. Large scale soil-foundation-structure model in Greece: Dynamic tests vs FEM simulation. COMPDYN 2017. In Proceedings of the 6th International Conference on Computational Methods in Structural Dynamics and Earthquake Engineering, Rhodes Island, Greece, 15–17 June 2017.
18. Massimino, M.R.; Biondi, G. Some experimetnal evidences on dynamic soil-structure interaction. In Proceedings of the COMP-DYN 2015, 5th ECCOMAS Thematic Conference on Computational Methods in Structural Dynamics and Earthquake Engineering, Athens, Greece, 21–23 June 2015.
19. Lysmer, J.; Kuhlemeyer, R.L. Finite dynamic model for infinite media. *J. Eng. Mech. ASCE* **1969**, *95*, 759–877.
20. Lysmer, J.; Wass, G. Shear waves in plane infinite structures. *J. Eng. Mech. ASCE* **1972**, *98*, 85–105.
21. White, W.; Valliappan, S.; Lee, I.K. Unified boundary for finite dynamic models. *J. Eng. Mech. ASCE* **1977**, *103*, 949–964.
22. Liao, Z.P.; Wong, H.L. A transmitting boundary for the numerical simulation of elastic wave propagation. *Soil Dyn. Earthq. Eng.* **1984**, *3*, 174–183. [CrossRef]
23. Zienkiewicz, O.C.; Bielak, J.; Shen, F.Q. Earthquake input definition and the transmitting boundary condition. In *Advances in Computational Non-Linear Mechanics*; Springer: Vienna, Austria, 1988; pp. 109–138.
24. Deeks, A.J.; Randolph, M.F. Axisymmetric Time-domain Transmitting Boundaries. *J. Eng. Mech. ASCE* **1994**, *120*, 25–42. [CrossRef]
25. Lee, J.H. Non-linear soil-structure interaction analysis in poroelastic soil using mid-point integrated finite elements and perfectly matched discrete layers. *Soil Dyn. Earthq. Eng.* **2018**, *108*, 160–176. [CrossRef]
26. Coleman, J.L.; Bolisetti, C.; Whittaker, A.S. Time-domain soil-structure interaction analysis of nuclear facilities. *Nucl. Eng. Des.* **2016**, *298*, 264–270. [CrossRef]
27. Solberg, J.; Hossain, Q. Non-linear time-domain soil-structure interaction analysis of embedded reactor structures subjected to earthquake loads. *Nucl. Eng. Des.* **2016**, *304*, 100–124. [CrossRef]
28. Abell, J.A.; Orbovic, N.; McCallen, D.B.; Jeremic, B. Earthquake soil-structure interaction of nuclear power plants, differences in response to 3-D, 3 × 1-D, and 1-D excitations. *Earthq. Eng. Struct. Dyn.* **2018**, *47*, 1478–1495. [CrossRef]
29. Yoshimura, S.; Hori, M.; Ohsaki, M. *High Performance Computing for Structural Mechanics and Earthquake/Tsunamic Engineering*; Springer International Publishing: New York, NY, USA, 2016.
30. Quinay, P.E.B.; Ichimura, T.; Hori, M.; Wijerathne, M.L.L.; Nishida, A. Seismic structural response analysis considering fault-structure system: Application to nuclear power plant structures. *Prog. Nucl. Sci. Technol.* **2011**, *2*, 516–523. [CrossRef]
31. Miyamura, T.; Tanaka, S.; Hori, M. Large-Scale Seismic Response Analysis of a Super-High-Rise-Building Fully Considering the Soil–Structure Interaction Using a High-Fidelity 3D Solid Element Model. *J. Earthq. Tsunami* **2016**, *10*, 1–21. [CrossRef]
32. Veletsos, A.S.; Wei, Y.T. Lateral and rocking vibration of footings. *J. Solid Mech. Found. Div. ASCE* **1971**, *97*, 1227–1248.
33. Luco, J.E. Impedance functions for a rigid foundation on a layered medium. *Nucl. Eng. Des.* **1974**, *31*, 204–217. [CrossRef]
34. Kausel, E.; Roesset, J.M. Dynamic stiffness of circular foundations. *J. Eng. Mech. ASCE* **1975**, *101*, 771–785.
35. Luco, J.E. Vibration of a rigid disk on a layered viscoelastic medium. *Nucl. Eng. Des.* **1976**, *36*, 325–340. [CrossRef]
36. Gazetas, G.; Anastasopoulos, I.; Adamidis, O.; Kontoroupi, T. Nonlinear rocking stiffness of foundations. *Soil Dyn. Earthq. Eng.* **2013**, *47*, 83–91. [CrossRef]
37. Gazetas, G. Formulas and Charts for impedances of surface and embedded foundations. *J. Geotech. Eng. ASCE* **1991**, *117*, 1–9. [CrossRef]

38. Kubo, T.; Yamamoto, T.; Sato, K.; Jimbo, M.; Imaoka, T.; Umeki, Y. A Seismic Design of Nuclear Reactor Building Structures Applying Seismic Isolation System in a High Seismicity Region—A Feasibility Case Study in Japan. *Nucl. Eng. Technol.* **2014**, *46*, 581–594. [CrossRef]
39. Okada, Y.; Ogawa, Y.; Hiroshima, M.; Iwate, T. Application of sway-rocking model to retrofit design of bridges. In Proceedings of the 14th World Conference on Earthquake Engineering, Bejing, China, 12–17 October 2008.
40. Mylonakis, G.; Nikolaou, S.; Gazetas, G. Footings under seismic loading: Analysis and design issues with emphasis on bridge foundations. *Soil Dyn. Earthq. Eng.* **2006**, *26*, 824–853. [CrossRef]
41. Ni, P.; Petrini, L.; Paolucci, R. Direct displacement-based assessment with nonlinear soil-structure interaction for multi-span reinforced concrete bridges. *Struct. Infrastruct. Eng.* **2014**, *10*, 1211–1227. [CrossRef]
42. Yang, L.; Marshall, A.M.; Hajirasouliha, I. A simplified Nonlinear Sway-Rocking model for evaluation of seismic response of structures on shallow foundations. *Soil Dyn. Earthq. Eng.* **2016**, *81*, 14–26.
43. Wong, H.L.; Luco, J.E. *Soil-Structure Interaction: A Llinear Continuum Mechanics Approach (CLASSI)*; Department of Civil Engineering, University of Southern California: Los Angeles, CA, USA, 1980; CE 79–03.
44. Lysmer, J.; Udaka, T.; Tsai, C.F.; Seed, H.B. *FLUSH-A Computer Program for Approximate 3D Analysis of Soil-Structure Interaction Problems*; Earthquake Engineering Research Center, University of California: Berkeley, CA, USA, 1975; Report No. EERC 75–30.
45. Tabatabaie, M. SASSI FE Program for Seismic Response Analysis of Nuclear Containment Structures. In Proceedings of the International Workshop on Infrastructure Systems for Nuclear Energy, Taipei, Taiwan, 15–17 December 2008.
46. Lesgidis, N.; Kwon, O.S.; Sextos, A. A time-domain seismic SSI analysis method for inelastic bridge structures through the use of a frequency-dependent lumped parameter model. *Earthq. Eng. Struct. Dyn.* **2015**, *44*, 2137–2156. [CrossRef]
47. Carbonari, S.; Morici, M.; Dezi, F.; Leoni, G. A lumped parameter model for time-domain inertial soil-structure interaction analysis of structures on pile foundations. *Earthq. Eng. Struct. Dyn.* **2018**, *47*, 2147–2171. [CrossRef]
48. Bucinskas, P.; Andersen, L.V. Dynamic response of vehicle–bridge–soil system using lumped-parameter models for structure–soil interaction. *Comput. Struct.* **2020**, *238*, 6270. [CrossRef]
49. Maslenikov, O.R.; Chen, J.C. *Uncertainty in Soil-Structure Interaction Analysis Arising from Differences in Analytical Techniques*; U.S. Nuclear Regulatory Commission: Washington, DC, USA, 1982.
50. Nova, R.; Montrasio, L. Settlements of shallow foundations on sand. *Geotechnique* **1999**, *41*, 243–256. [CrossRef]
51. Chatzigogos, C.T.; Figini, R.; Pecker, A.; Salençon, J. A macroelement formulation for shallow foundations on cohesive and frictional soils. *Int. J. Numer. Anal. Methods Geomech.* **2011**, *35*, 902–931. [CrossRef]
52. Figini, R.; Paolucci, R.; Chatzigogos, C.T. A macro-element model for non-linear soil–shallow foundation–structure interaction under seismic loads: Theoretical development and experimental validation on large scale tests. *Earthq. Eng. Struct. Dyn.* **2012**, *41*, 475–493. [CrossRef]
53. Cavalieri, F.; Correia, A.A.; Crowley, H.; Pinho, R. Dynamic soil-structure interaction models for fragility characterisation of buildings with shallow foundations. *Soil Dyn. Earthq. Eng.* **2020**, *132*, 6004. [CrossRef]
54. Chatzigogos, C.T.; Pecker, A.; Salençon, J. Macroelement modeling of shallow foundations. *Soil Dyn. Earthq. Eng.* **2009**, *29*, 765–781. [CrossRef]
55. Cremer, C.; Pecker, A.; Davenne, L. Cyclic macro-element for soil-structure interaction: Material and geometrical non-linearities. *Int. J. Numer. Anal. Methods Geomech.* **2001**, *25*, 1257–1284. [CrossRef]
56. Nakamura, N.; Akita, S.; Suzuki, T.; Koba, M.; Nakamura, S.; Nakano, T. Study of Ultimate Seismic Response and Fragility Evaluation of Nuclear Power Building Using Nonlinear Three-dimensional Finite Element Model. *Nucl. Eng. Des.* **2010**, *240*, 166–180. [CrossRef]
57. Hori, M.; Maggededaera, L.; Ichimura, T.; Tanaka, S. Meta-Modeling for Constructing Model Consistent with Continuum Mechanics. *J. Struct. Mech. Earthq. Eng.* **2014**, *9*, 269–275. [CrossRef]
58. Roh, H.; Lee, H.; Lee, J.S. New lumped-mass-stick model based on modal characteristics of structures: Development and application to a nuclear containment building. *Earthq. Eng. Eng. Vib.* **2013**, *12*, 307–317. [CrossRef]
59. Jayasinghe, J.A.S.C.; Hori, M.; Riaz, M.R.; Wijerathne, M.L.L.; Ichimura, T. Meta-Modeling Based Consistent Mass Spring Model for Seismic Response Analysis of Bridge Structure. *J. JSCE Appl. Mech.* **2015**, *71*, 137–148. [CrossRef]
60. Maekawa, K.; Okamura, H.; Pimanmas, A. *Non-linear Mechanics of Reinforced Concrete*; Taylor & Francis: Abingdon-on-Thames, UK, 2003.
61. Yamashita, T.; Hori, M.; Oguni, K.; Okazawa, S.; Maki, T.; Takahashi, Y. Reformulation of non-linear constitutive relations of concrete for large-scale finite element method analysis. *J. JSCE* **2011**, *67*, 145–154. (In Japanese) [CrossRef]
62. Motoyama, H.; Hori, M. Development and efficiency investigation of parallel computing seismic response analysis for high-fidelity model of large-scale reinforced concrete structures. In Proceedings of the SMiRT 25—The 25th International Conference on Structural Mechanics in Reactor Technology, Charlotte, NC, USA, 4–9 August 2019.

MDPI
St. Alban-Anlage 66
4052 Basel
Switzerland
Tel. +41 61 683 77 34
Fax +41 61 302 89 18
www.mdpi.com

Geosciences Editorial Office
E-mail: geosciences@mdpi.com
www.mdpi.com/journal/geosciences

www.ingramcontent.com/pod-product-compliance
Lightning Source LLC
Chambersburg PA
CBHW040817250926
51850CB00007BA/1191
9783036568263